METHODS IN MOLECULAR BIOLOGY

Series Editor
John M. Walker
School of Life and Medical Sciences
University of Hertfordshire
Hatfield, Hertfordshire, UK

For further volumes:
http://www.springer.com/series/7651

For over 35 years, biological scientists have come to rely on the research protocols and methodologies in the critically acclaimed *Methods in Molecular Biology* series. The series was the first to introduce the step-by-step protocols approach that has become the standard in all biomedical protocol publishing. Each protocol is provided in readily-reproducible step-by step fashion, opening with an introductory overview, a list of the materials and reagents needed to complete the experiment, and followed by a detailed procedure that is supported with a helpful notes section offering tips and tricks of the trade as well as troubleshooting advice. These hallmark features were introduced by series editor Dr. John Walker and constitute the key ingredient in each and every volume of the *Methods in Molecular Biology* series. Tested and trusted, comprehensive and reliable, all protocols from the series are indexed in PubMed.

The Mitoribosome

Methods and Protocols

Edited by

Antoni Barrientos and Flavia Fontanesi

Miller School of Medicine, University of Miami, Miami, FL, USA

Humana Press

Editors
Antoni Barrientos
Miller School of Medicine
University of Miami
Miami, FL, USA

Flavia Fontanesi
Miller School of Medicine
University of Miami
Miami, FL, USA

ISSN 1064-3745 ISSN 1940-6029 (electronic)
Methods in Molecular Biology
ISBN 978-1-0716-3173-7 ISBN 978-1-0716-3171-3 (eBook)
https://doi.org/10.1007/978-1-0716-3171-3

This Humana imprint is published by the registered company Springer Science+Business Media, LLC, part of Springer Nature.
The registered company address is: 1 New York Plaza, New York, NY 10004, U.S.A.

Preface

The natural history of ribosomes spans 4 billion years of life on earth. When mitochondria arose through a fateful endosymbiosis more than 1.45 billion years ago, they had their ancestral alpha-proteobacterium's prokaryotic genome, gene expression system, and ribosomes. Although most original genes have been transferred to the nucleus during evolution, the vestigial genome present in current mitochondria encodes for a handful of proteins (e.g., 8 in *Saccharomyces cerevisiae* and 13 in mammals) that are synthesized locally by mitochondrial ribosomes (mitoribosomes). Because of their bacterial ancestry, mitoribosomes have retained many prokaryotic properties, including sensitivity to a similar set of antibiotics. However, mitoribosomes are very diverse across species. They have structurally and functionally adapted to synthesizing very hydrophobic proteins that require co-translational insertion into the membrane to prevent misfolding. These proteins are subunits of the oxidative phosphorylation (OXPHOS) system complexes that also contain numerous subunits encoded in the nuclear genome. Recent studies have uncovered regulatory systems that are in place to coordinate mitoribosomal activity with that of cytosolic ribosomes and with the import of nucleus-encoded OXPHOS subunits.

The mammalian mitoribosome was the first of its class to be purified in the 1960s followed by the purification of the fungal ribosome soon after. During the following 50 years, researchers applied biochemical methods to study yeast and mammalian mitochondrial ribosome composition, assembly, and function. Pioneering cryo-EM studies were also developed, and although at low resolution, they provided valuable information. More recently, in 2014, the development of cutting-edge cryo-electron microscopy (cryo-EM) instrumentation and software for data analyses launched the so-called cryo-EM "resolution revolution" and laid the foundations for the structure determination of the yeast mitoribosome, its porcine and human mitochondrial counterparts, and by now several other species. Over the last few years, the focus has been placed on mitoribosome assembly, a challenging process involving rRNAs encoded in the mitochondrial genome and most proteins (all in mammals) encoded in the nuclear genome. Only in 2021, eight articles were published combining biochemical and cryo-EM data to describe the maturation of the mammalian mitoribosome large subunit. Although cryo-EM studies are changing the shape of the mitoribosomal field, classical biochemistry, ribosome profiling approaches, and alternative imaging methods are essential to test translation mechanisms. The field still lacks an in vitro translation system, but progress has been made over the last couple of years, at least in studying translation initiation events in reconstituted systems. The field, therefore, is navigating exciting times.

In this volume, we aimed to bring together a selection of classical and new techniques to study the structure, assembly pathway, and protein synthesis ability of mitoribosomes across species. The book is divided into two parts. Part I is a non-protocol section that introduces general concepts about the discovery of the mitoribosomes, their evolution, assembly, and function. Part II presents a broad collection of methods to study the mitochondrial translation machinery. The protocols start by describing cryo-electron tomography and cryo-EM approaches for structural determination of mitoribosomes and their assembly intermediates.

They continue by presenting mitoribosome purification approaches to analyze mitoribosome biogenesis and interactome. Subsequently, methods are presented to study mitochondrial translation in vitro, isolated mitochondria, and whole cells. Part II is completed by methods to study mitochondrial mRNAs that are translated on mitoribosomes.

We would like to thank all the authors for their contribution to this volume and for generously sharing their protocols and expertise. We also want to express our gratitude to the series editor, John Walker, for his guidance and constant support.

Miami, FL, USA

Antoni Barrientos
Flavia Fontanesi

Contents

Contributors

RAJENDRA K. AGRAWAL • *Division of Translational Medicine, Wadsworth Center, New York State Department of Health, Empire State Plaza, Albany, NY, USA; Department of Biomedical Sciences, University at Albany, SUNY, Rensselaer, NY, USA*

CHRISTIN A. ALBUS • *Wellcome Centre for Mitochondrial Research, Newcastle University Biosciences Institute, Faculty of Medical Sciences, Newcastle upon Tyne, UK*

ANDREAS AUFSCHNAITER • *Department of Biochemistry and Biophysics, Stockholm University, Stockholm, Sweden*

ANTONI BARRIENTOS • *Department of Neurology and Department of Biochemistry and Molecular Biology, University of Miami, Miller School of Medicine, Miami, FL, USA*

BRENDAN J. BATTERSBY • *Institute of Biotechnology, Helsinki Institute of Life Science, University of Helsinki, Helsinki, Finland*

ROLANDO BERLINGUER-PALMINI • *Bioimaging Unit, Newcastle University, Faculty of Medical Sciences, Newcastle upon Tyne, UK*

LEA BERTGEN • *Cell Biology, University of Kaiserslautern, Kaiserslautern, Germany*

ARPAN BHATTACHARYA • *Department of Chemistry, University of Pennsylvania, Philadelphia, PA, USA; Department of Physiology, Perelman School of Medicine, University of Pennsylvania, Philadelphia, PA, USA*

JOAN BLANCO-FERNANDEZ • *Department of Immunobiology, University of Lausanne, Epalinges, Switzerland*

YOLANDA CAMACHO-VILLASANA • *Departamento de Genética Molecular, Instituto de Fisiología Celular, Universidad Nacional Autónoma de México, Mexico City, Mexico*

ANDREAS CARLSTRÖM • *Department of Biochemistry and Biophysics, Stockholm University, Stockholm, Sweden*

AUSTIN CHOI • *Department of Neurology, University of Miami, Miller School of Medicine, Miami, FL, USA*

ZOFIA M. CHRZANOWSKA-LIGHTOWLERS • *Wellcome Centre for Mitochondrial Research, Newcastle University Biosciences Institute, Faculty of Medical Sciences, Newcastle upon Tyne, UK*

J. CONOR MORAN • *Department of Biochemistry and Molecular Biology, University of Miami, Miller School of Medicine, Miami, FL, USA*

BARRY COOPERMAN • *Department of Chemistry, University of Pennsylvania, Philadelphia, PA, USA*

MARY COUVILLION • *Department of Genetics, Harvard Medical School, Boston, MA, USA*

SAMUEL DEL'OLIO • *Department of Molecular and Cellular Pharmacology, University of Miami, Miller School of Medicine, Miami, FL, USA*

ROBERT ENGLMEIER • *Structural Biochemistry, Bijvoet Centre for Biomolecular Research, Utrecht University, Utrecht, The Netherlands*

ALEKSANDRA FILIPOVSKA • *Harry Perkins Institute of Medical Research and ARC Centre of Excellence in Synthetic Biology, QEII Medical Centre, Nedlands, WA, Australia; Centre for Medical Research, The University of Western Australia, QEII Medical Centre, Nedlands, WA, Australia; Telethon Kids Institute, Perth Children's Hospital, Nedlands, WA, Australia*

TAMARA FLOHR • *Cell Biology, University of Kaiserslautern, Kaiserslautern, Germany*

Daniel Flores-Mireles • *Departamento de Genética Molecular, Instituto de Fisiología Celular, Universidad Nacional Autónoma de México, Mexico City, Mexico*
Flavia Fontanesi • *Department of Biochemistry and Molecular Biology, University of Miami, Miller School of Medicine, Miami, FL, USA*
Friedrich Förster • *Structural Biochemistry, Bijvoet Centre for Biomolecular Research, Utrecht University, Utrecht, The Netherlands*
Clark Fritsch • *Department of Physiology, Perelman School of Medicine, University of Pennsylvania, Philadelphia, PA, USA*
Johannes M. Herrmann • *Cell Biology, University of Kaiserslautern, Kaiserslautern, Germany*
Hauke S. Hillen • *Department of Cellular Biochemistry, University Medical Center Göttingen, Göttingen, Germany; Research Group Structure and Function of Molecular Machines, Max Planck Institute for Multidisciplinary Sciences, Göttingen, Germany; Cluster of Excellence 'Multiscale Bioimaging: from Molecular Machines to Networks of Excitable Cells' (MBExC), University of Göttingen, Göttingen, Germany*
Alexis A. Jourdain • *Department of Immunobiology, University of Lausanne, Epalinges, Switzerland*
Muhoon Lee • *Department of Computational Biology and Medical Sciences, Graduate School of Frontier Sciences, The University of Tokyo, Chiba, Japan*
Hong Li • *Department of Chemistry, University of Pennsylvania, Philadelphia, PA, USA*
Robert N. Lightowlers • *Wellcome Centre for Mitochondrial Research, Newcastle University Biosciences Institute, Faculty of Medical Sciences, Newcastle upon Tyne, UK*
Priyanka Maiti • *Department of Neurology, University of Miami, Miller School of Medicine, Miami, FL, USA*
Soneya Majumdar • *Division of Translational Medicine, Wadsworth Center, New York State Dpartment of Health, Empire State Plaza, Albany, NY, USA*
Kah Ying Ng • *Institute of Biotechnology, Helsinki Institute of Life Science, University of Helsinki, Helsinki, Finland*
Martin Ng • *Department of Chemistry, University of Pennsylvania, Philadelphia, PA, USA*
Minh Duc Nguyen • *Department of Medical Biochemistry and Biophysics, Karolinska Institutet, Solna, Sweden*
Thomas W. O'Brien • *Department of Biochemistry and Molecular Biology, University of Florida, Gainesville, FL, USA*
Martin Ott • *Department of Biochemistry and Biophysics, Stockholm University, Stockholm, Sweden; Department of Medical Biochemistry and Cell Biology, University of Gothenburg, Gothenburg, Sweden*
Hong Weng Pang • *Department of Biochemistry and Molecular Biology, University of Miami, Miller School of Medicine, Miami, FL, USA*
Xochitl Pérez-Martínez • *Departamento de Genética Molecular, Instituto de Fisiología Celular, Universidad Nacional Autónoma de México, Mexico City, Mexico*
Oliver Rackham • *Harry Perkins Institute of Medical Research and ARC Centre of Excellence in Synthetic Biology, QEII Medical Centre, Nedlands, WA, Australia; Telethon Kids Institute, Perth Children's Hospital, Nedlands, WA, Australia; Curtin Medical School and Curtin Health Innovation Research Institute, Curtin University, Bentley, WA, Australia*
Cristina Remes • *Department of Chemistry, University of Pennsylvania, Philadelphia, PA, USA*

JOANNA RORBACH • *Department of Medical Biochemistry and Biophysics, Karolinska Institutet, Solna, Sweden; Max Planck Institute Biology of Ageing, Karolinska Institutet Laboratory, Karolinska Institutet, Stockholm, Sweden*
DANIELLE L. RUDLER • *Harry Perkins Institute of Medical Research and ARC Centre of Excellence in Synthetic Biology, QEII Medical Centre, Nedlands, WA, Australia; Centre for Medical Research, The University of Western Australia, QEII Medical Centre, Nedlands, WA, Australia*
STEFAN J. SIIRA • *Harry Perkins Institute of Medical Research and ARC Centre of Excellence in Synthetic Biology, QEII Medical Centre, Nedlands, WA, Australia; Centre for Medical Research, The University of Western Australia, QEII Medical Centre, Nedlands, WA, Australia*
ILIANA SOTO • *Department of Genetics, Harvard Medical School, Boston, MA, USA*
HENRIK SPAHR • *Department of Medical Biochemistry and Biophysics, Karolinska Institutet, Solna, Sweden*
L. STIRLING CHURCHMAN • *Department of Genetics, Harvard Medical School, Boston, MA, USA*
NONO TAKEUCHI-TOMITA • *Department of Computational Biology and Medical Sciences, Graduate School of Frontier Sciences, The University of Tokyo, Chiba, Japan*
LENA THÄRICHEN • *Structural Biochemistry, Bijvoet Centre for Biomolecular Research, Utrecht University, Utrecht, The Netherlands*
HUI ZHONG • *Department of Biochemistry and Molecular Biology, University of Miami, Miller School of Medicine, Miami, FL, USA*

Part I

General Concepts

Chapter 1

Discovery of Mitochondrial Ribosomes

Thomas W. O'Brien

Abstract

In this introductory chapter, I will briefly describe how I came to discover the mammalian mitoribosome and will add a few notes on my contribution to the field.

Key words Human MRP genes, Mitochondrial ribosomal proteins, Mitochondrial disease, Human ribosome purification, Mitochondrial translation

Mammalian mitochondrial ribosomes, the first mitoribosomes to be discovered [1, 2], are 55S ribosomes. This result was not widely received because of the prevailing notion that mitochondrial ribosomes would resemble bacterial 70S ribosomes, reminiscent of their ancestral prokaryotic origins. After rigorous purification procedures to eliminate contaminating microsomal ribosomes [1] and further studies to confirm this finding [3], the 55S mitochondrial ribosome emerged as a ribosome with properties remarkably different from bacterial and eukaryotic, cytoplasmic ribosomes [4].

Despite their lower sedimentation coefficient, the 55S mitoribosomes are physically larger than bacterial 70S ribosomes. Their mass is greater and so also are their physical dimensions [5]. While they are larger than bacterial ribosomes, they actually contain smaller RNAs. Their increased size is due to the presence of protein extensions and several extra proteins, many of which are unique to mitochondrial ribosomes [4].

How I became interested in mitochondria and their ribosomes? My graduate studies on the mechanism of action of the thyroid hormone were carried out in the laboratory of Howard Klitgaard, a thyroid endocrinologist, from 1961 to 1964. Thyroxine was known to regulate basal metabolic rate. Investigators were studying the effects of chemical modifications to the thyroxin molecule on the metabolic response of animals and tissues to gain insight into the mechanism of action of the thyroid hormones. Studies from

Antoni Barrientos and Flavia Fontanesi (eds.), *The Mitoribosome: Methods and Protocols*,
Methods in Molecular Biology, vol. 2661, https://doi.org/10.1007/978-1-0716-3171-3_1,

David Green's lab in Madison implicated mitochondria in playing a role in oxidative metabolism at the time, so I decided to examine the effects of thyroxine on mitochondria. The basal metabolic rate of rats made hyperthyroid by a single injection of thyroxine (6.12 mg/kg body weight) rose by 40% within 3 days following injection [6]. I determined by electron microscopic examination of liver tissue from the hyperthyroid rats that the cellular volume fraction of mitochondria had increased by 22% relative to control euthyroid animals, indicating that the extra thyroxine had stimulated the growth of new mitochondria [6]. That observation raised my interest in mitochondrial protein synthesis and led me to apply for a postdoctoral position in the laboratory of George F. Kalf at the New Jersey College of Medicine in Jersey City, who was studying mitochondrial DNA at the time.

Ribosomes obtained from rat liver mitochondria isolated by classical methods were predominantly 80S particles resulting from the presence of contaminating microsomes and cytoplasmic ribosomes in the mitochondrial pellet [1]. With extra precautions taken to reduce this contamination, a small peak of 55S particles could be discerned in the presence of the majority 80S particles. These 55S particles were mitochondrial ribosomes, as evidenced by their radiolabeling when isolated mitochondria were pulse-labeled by incubation with radiolabeled amino acids under conditions supporting mitochondrial protein synthesis, while the contaminating 80S ribosomes remained unlabeled [1, 2].

Improvements in the isolation procedure for mitochondrial ribosomes included the addition of digitonin as a detergent, which disrupts the microsomal and outer mitochondrial membranes, reducing the amount of cosedimenting cytoplasmic ribosomes. It was noted that any bacteria introduced during the isolation of mitochondria by differential sedimentation would copurify with the mitochondria, raising the possibility of contaminating the population of ribosomes isolated from such preparations. However, the use of neutral detergents such as triton-X-100 to solubilize the mitochondria did not release any bacterial ribosomes. The use of these two detergents thus ensured a more homogeneous preparation of mitoribosomes.

Once the scientific community accepted the occurrence of mitochondrial ribosomes, many laboratories worked in the field. In my laboratory, I discovered that mitochondrial ribosomes possess a high-affinity binding site for guanine nucleotides [7], worked in multiple aspects of mitoribosome evolution [8, 9], and described some properties of human mitoribosomes [4] and their involvement in human genetic disorders [10, 11]. Over the last 50 years, the progress in the field has been nothing but spectacular.

References

1. O'Brien TW, Kalf GF (1967) Ribosomes from rat liver mitochondria. I. Isolation procedure and contamination studies. J Biol Chem 242(9):2172–2179
2. O'Brien TW, Kalf GF (1967) Ribosomes from rat liver mitochondira. II. Partial characterization. J Biol Chem 242(9):2180–2185
3. O'Brien TW (1971) The general occurrence of 55 S ribosomes in mammalian liver mitochondria. J Biol Chem 246(10):3409–3417
4. O'Brien TW (2003) Properties of human mitochondrial ribosomes. IUBMB Life 55(9): 505–513
5. Hamilton MG, O'Brien TW (1974) Ultracentrifugal characterization of the mitochondrial ribosome and subribosomal particles of bovine liver: molecular size and composition. Biochemistry 13(26):5400–5403. https://doi.org/10.1021/bi00723a024
6. O'Brien TW (1964) Electron microscopic study of the in vivo changes in rat liver mitochondria after thyroidectomy and thyroxin administration. Dissertation, Marquette University, Milwaukee, Wisconsin
7. Denslow ND, Anders JC, O'Brien TW (1991) Bovine mitochondrial ribosomes possess a high affinity binding site for guanine nucleotides. J Biol Chem 266(15):9586–9590
8. Matthews DE, Hessler RA, O'Brien TW (1978) Rapid evolutionary divergence of proteins in mammalian mitochondrial ribosomes. FEBS Lett 86(1):76–80. https://doi.org/10.1016/0014-5793(78)80102-1
9. Pietromonaco SF, Hessler RA, O'Brien TW (1986) Evolution of proteins in mammalian cytoplasmic and mitochondrial ribosomes. J Mol Evol 24(1–2):110–117. https://doi.org/10.1007/bf02099958
10. O'Brien TW, O'Brien BJ, Norman RA (2005) Nuclear MRP genes and mitochondrial disease. Gene 354:147–151
11. O'Brien TW (2002) Evolution of a protein-rich mitochondrial ribosome: implications for human genetic disease. Gene 286(1):73–79

Chapter 2

Evolution: Mitochondrial Ribosomes Across Species

Rajendra K. Agrawal and Soneya Majumdar

Abstract

The ribosome is among the most complex and ancient cellular macromolecular assemblies that plays a central role in protein biosynthesis in all living cells. Its function of translation of genetic information encoded in messenger RNA into protein molecules also extends to subcellular compartments in eukaryotic cells such as apicoplasts, chloroplasts, and mitochondria. The origin of mitochondria is primarily attributed to an early endosymbiotic event between an alpha-proteobacterium and a primitive (archaeal) eukaryotic cell. The timeline of mitochondrial acquisition, the nature of the host, and their diversification have been studied in great detail and are continually being revised as more genomic and structural data emerge. Recent advancements in high-resolution cryo-EM structure determination have provided architectural details of mitochondrial ribosomes (mitoribosomes) from various species, revealing unprecedented diversifications among them. These structures provide novel insights into the evolution of mitoribosomal structure and function. Here, we present a brief overview of the existing mitoribosomal structures in the context of the eukaryotic evolution tree showing their diversification from their last common ancestor.

Key words Mitochondrial ribosomes, Evolution, Mitochondrial rRNAs, MRPs, Cryo-EM structures

1 Introduction

The last universal common ancestor of cellular life is traced back to more than 3.5 billion years, through identification of 63 ubiquitous protein-coding genes that are present among all living organisms [1]. Most of these genes (>90%) encode either ribosomal proteins or proteins that are involved in protein synthesis, including aminoacyl-transfer RNA (tRNA)-synthases, translation factors, and ribosomal RNA (rRNA)-modifying enzymes. Phylogenetic studies of rRNAs have contributed greatly in determining molecular chronology of evolution of the three kingdoms of life [2, 3], suggesting that studies of ribosome structure and function could directly help in further elucidation of distant evolutionary relationships. A recent phylogenetic study [4] has also shown that a large majority of rRNA segments that form the monolithic core of the two unequally-sized ribosomal subunits in bacteria are conserved.

Antoni Barrientos and Flavia Fontanesi (eds.), *The Mitoribosome: Methods and Protocols*,
Methods in Molecular Biology, vol. 2661, https://doi.org/10.1007/978-1-0716-3171-3_2,

These segments include the peptidyl-transferase center (PTC) within the large ribosomal subunit (LSU) and the decoding center of the small ribosomal subunit (SSU). The PTC lies at the core of LSU and has been proposed to have first evolved from two separate strands of rRNAs in a proto-LSU leading to origin of the ribosome [5–7]. Comparison of bacterial and archaeal rRNAs reveals an increase in rRNA length by more than 200 nucleotides through multiple insertion sites of 5 to 20 nucleotides that radiate away from the two functional ribosomal cores. The size of these insertions, referred to as expansion segments, increases significantly in eukaryotes (up to ~3000 nucleotides in mammalian rRNAs), which serve as marker to study eukaryotic evolution. As the size of rRNA in eukaryotes increases, the number and size of peripheral ribosomal proteins also increase to form a second ribosomal shell [8], function of which is poorly understood [9].

Mitochondria are thought to have originated about 1.8 billion years ago via an endosymbiotic event between an α-proteobacterium and a primitive eukaryotic cell [10]. While due to rapidly evolving nature of mitochondrial genomes determination of the exact mitochondrial ancestor is still debated, deep metagenomic sequencing results [11] strongly hint at the archaean Asgard as the likely original host [12, 13]. As the endosymbiosis would require transfer of regulatory control from symbiont to host nucleus, a majority of symbiont genome were transferred to nuclear genome but the symbiont-turned mitochondria retained essential components to carry out expression of its remaining genome, including cellular nucleoprotein gene-expression machineries such as polymerases and ribosomes [14]. Most protein components of these machineries are encoded by the nuclear genome, translated in the cytoplasm, and then imported into mitochondria. Mitochondrial genomes are primarily responsible for synthesizing polypeptides that form essential components of the complexes involved in oxidative phosphorylation (or ATP generation) for the eukaryotic cell. However, there are large and somewhat dramatic variations among retained mitochondrial genomes and composition of mitochondrial ribosomes (mitoribosomes) from different eukaryotes [15].

2 Evolution of the Mitochondrial Ribosome

Mitoribosomes represent the most diverse class of ribosomes and their evolution and diversification are best revealed through their detailed structures determined by cryo-EM. The first cryo-EM structure of a mitoribosome was determined for a mammal, *Bos taurus* [16]. The ratio of rRNA to protein in the mammalian mitoribosome (1:2) is reversed as compared to that in bacterial ribosomes (2:1). The study allowed segmentation of its rRNA and mitochondrial ribosomal protein (MRP) components and provided the evidence against the 4-decade old belief that the larger and additional MRPs in the mammalian mitoribosome structurally replace the majority of missing bacterial rRNA segments. In the structure, only ~20% of missing bacterial rRNA segments were occupied by larger MRPs that are homologous to their counterparts in bacterial ribosomal proteins, while most new mito-specific MRPs acquired novel quaternary positions, giving an unusual appearance to previously familiar shapes of the ribosomes and their subunits [17]. The structure [16] revealed that the MRPs almost completely shield the inner rRNA core of the 55S mitoribosome, except for the mitoribosome's transfer RNA (tRNA)-entry and translation factor-binding regions. It provided evidence for a highly modified MRP-rich entrance of the messenger RNA (mRNA) channel [18] that could be necessary for recruitment of leaderless mitochondrial mRNAs [19], a remodeled landscape of the intersubunit space with greater surface area occupied by MRPs with an unique finger-like structure that interacts with the peptidyl (P)-site tRNA, and a dramatically altered composition of the nascent-polypeptide exit tunnel (NPET) [16, 18] with its lower two-third portion predominantly made up of MRPs that could be critical for direct insertion of nascent polypeptide chains into the inner mitochondrial membrane (IMM). A subsequent study at an improved resolution [20] allowed more precise RNA–protein separation, and provided a more precise molecular model of the 16S rRNA within the 39S LSU and placement of all bacterial homologues of MRPs within the cryo-EM density map.

A transformative technological advancement was the introduction of direct-electron detectors for cryo-EM data collection, which resulted in a major leap in resolution of structures of mammalian mitoribosome [21, 22]. The first higher resolution structures were focused on the 39S LSU and provided a molecular description to previously assigned quaternary positions for the MRP ensemble and identified unique structural features [16] [*see* ref. 23 for mammalian mitoribosome structure in preresolution revolution era]. Most importantly, these two high-resolution studies [21, 22] also revealed the unexpected presence of a tRNA molecule as a structural component within the central protuberance region of the 39S LSU. Since then several high-resolution structures of the mammalian mitoribosomes [24–26] as well as mitoribosomes from different evolutionary lineages have emerged. So far, structures of mitomonosomes or one of the mitoribosomal subunits have been published for 14 different species. The molecular composition of these structures and the number of proteins that they are responsible for synthesizing in mitochondria are listed in Table 1, encompassing a wide range of representations on the eukaryotic evolution tree (Fig. 1). The list includes mitoribosomes from the metazoan species *Homo sapiens* [24, 27], *Sus scrofa* [25], and *Bos taurus* SSU

Table 1
Comparison of molecular composition of various mitoribosomes with known high-resolution structures

Organism	PDB or EMBD ID	SSU rRNA nts (Svedberg unit)	LSU rRNA nts (Svedberg unit)	LSU tRNA/5S rRNA nts	Number of SSU proteins	Number of LSU proteins	Number of Mito-encoded proteins	References
Homo sapiens	6NU2, 6VLZ, 7L08, 7QI4, 7P2E, 7OG4, 7A5F	954 (12S)	1558 (16S)	73 (tRNA^{Val})	30	52	13	[27, 46, 48–50]
Sus scrofa	7NSI, 5AJ3	962 (12S)	1571 (16S)	73 (tRNA^{Phe})	30	52	13	[25, 47]
Bos taurus	3JD5	955 (12S)	1571 (16S)	73 (tRNA^{Val})	29	48	13	[26]
Mus musculus	7PNW	956 (12S)	1582 (16S)	67 (tRNA^{Thr})	30	48	13	[52]
Saccharomyces cerevisiae	3J6B, 5MRF, 5MRC	1649 (15S)	3296 (21S)	–	34	39	7	[29, 53]
Neurospora crassa	6YW5, 6YWS	1864 (16S)	3464 (23S)	–	35	44	28	[30]
Tetrahymena thermophila	6Z1P	196 + 1395 (14S)	278 + 2314 (21S)	–	49	45	43	[31]
Trypanosoma brucei	6HIV	620 (9S)	1176 (12S)	–	57	70	18	[32]
Trypanosoma cruzi	7AOR	613 (9S)	1150 (12S)	–	54	68	18	[33]
Leishmania tarentolae	7ANE	617 (9S)	1158 (12S)	–	54	68	18	[33]
Polytomella magna	8A22	1670 (4 fragments)	1060 (8 fragments)	5S	44	50	7	[36]
Chlamydomonas reinhardtii	7PKT, 7PKQ	1200 (4 fragments)	2035 (8 fragments)	74 (5S)	36	47	8	[37]
Brassica oleracea	6XYW	1935 (18S)	3169 (26S)	118 (5S)	37	45	34	[35]
Arabidopsis thaliana	EMDB-4408, EMDB-4409	1935 (18S)	3169 (26S)	118 (5S)	36	45	34	[34]

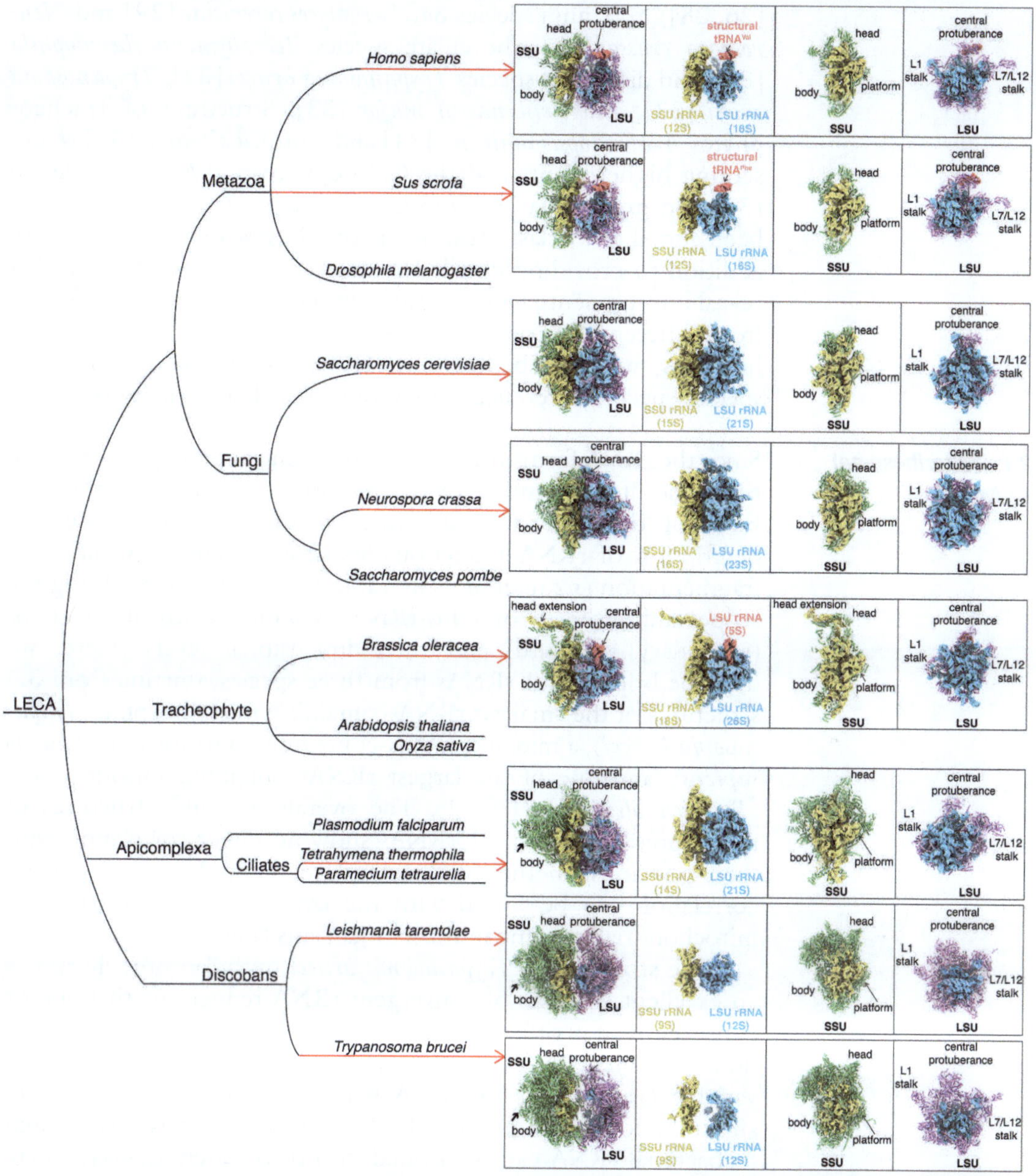

Fig. 1 The eukaryotic tree of life highlighting the branches with known mitoribosome structures. Schematic overview of the eukaryotic tree of life (left) with representative mitoribosome structures for the evolutionary branches indicated in red. For some of these branches, mitoribosome structure for more than one species is known (*see* Table 1), but only one is shown. Each structure is shown in four horizontal panels, the panel 1 starts with complete monosome, panel 2 shows the rRNA core, panels 3 and 4 show the intersubunit face of the SSU and LSU, respectively, of the same monosome. Mitoribosomes, for which structure of only one of the two subunits is known, are not included. LECA refers to Last Eukaryotic Common Ancestor. Color codes: yellow, SSU rRNA; blue and red, LSU rRNA; green, SSU MRPs; and purple, LSU MRPs. In addition to structure-specific features, head, body, and platform regions of the SSU, and L1 and L7/L12 stalk sides and central protuberance region of LSU are labeled. Black arrows point to MRP body protuberances in some of the SSUs in panel 1

[26, 28]), the fungi species *Saccharomyces cerevisiae* [29] and *Neurospora crassa* [30], the ciliate species *Tetrahymena thermophila* [31], and discobans species *Trypanosoma brucei* [32], *Trypanosoma cruzi* [33], and *Leishmania major* [33]. Structures of Tracheophytes *Arabiodopsis thaliana* [34] and *Brassica oleracea* [35] representing higher plants and the Chlorophycean *Polytomella magna* [36] and green algae *Chlamydomonas reinhardtii* [37] have also been added to the fast growing list of mitoribosome structures. In addition to providing detailed molecular structures, these studies reveal unprecedented architectural diversity among mitoribosomes from different eukaryotic species (Fig. 1). In the next three subheadings, we describe the mito-rRNA, MRPs, and point to a species-specific difference among the same class of organisms.

2.1 Mitoribosomal RNAs

Since the rRNA forms the core and most ancient component of the ribosome, its structure provides important evolutionary insights. Each of the individual structures provides detailed case-wise description of rRNA reductions, expansions, and in some cases fragmentation (*see* references in Table 1), thereby highlighting the scope and diversity of mito-rRNA evolution both at the two- (secondary structure) and three-dimensional levels. Here, we describe briefly mito-rRNAs from three species, starting from 3D structures of the smallest rRNA-containing mitoribosome (*Trypanosoma brucei*), a mid-size rRNA-containing mitoribosome (*Homo sapiens*), and one of the largest rRNA-containing mitoribosome (*Brassica oleracea*) (Fig. 2). The overall size and structures of these three representative rRNAs show no direct correlation with their positions on the evolutionary tree (Fig. 1), but a general correlation can be found with the overall size of the retained mitochondrial genome in those organisms (Fig. 2a).

The structure of *Trypanosoma brucei* mitoribosome illustrates an excellent example of convergent rRNA reductions that appear consistently in peripheral regions, involving all secondary structure domains of its rRNA in both SSU and LSU that are known to be variable (*see* ref. 38 for rRNA helix numbering and secondary structure domain assignments). The rRNA reductions as compared to bacterial rRNA can be divided into three categories: (i) where some of the bacterial helices are completely lost, (ii) where bacterial rRNA helices lost their base-pairing complementarity, and (iii) where the size of some of the rRNA helices is partially reduced in their apical regions. Despite a dramatic reduction in its rRNA size, important functional rRNA segments are retained. These include segments that would be considered peripheral in secondary and tertiary structures of rRNA, such as helices H43–44 of the LSU rRNA domain II that lie near the L7–L12 stalk region and H89–95 of the LSU rRNA domain VI that together constitute the GTPase-associated region and provide binding sites for tRNAs, translation initiation factor 2 and elongation factors, EF-G and EF-Tu in all

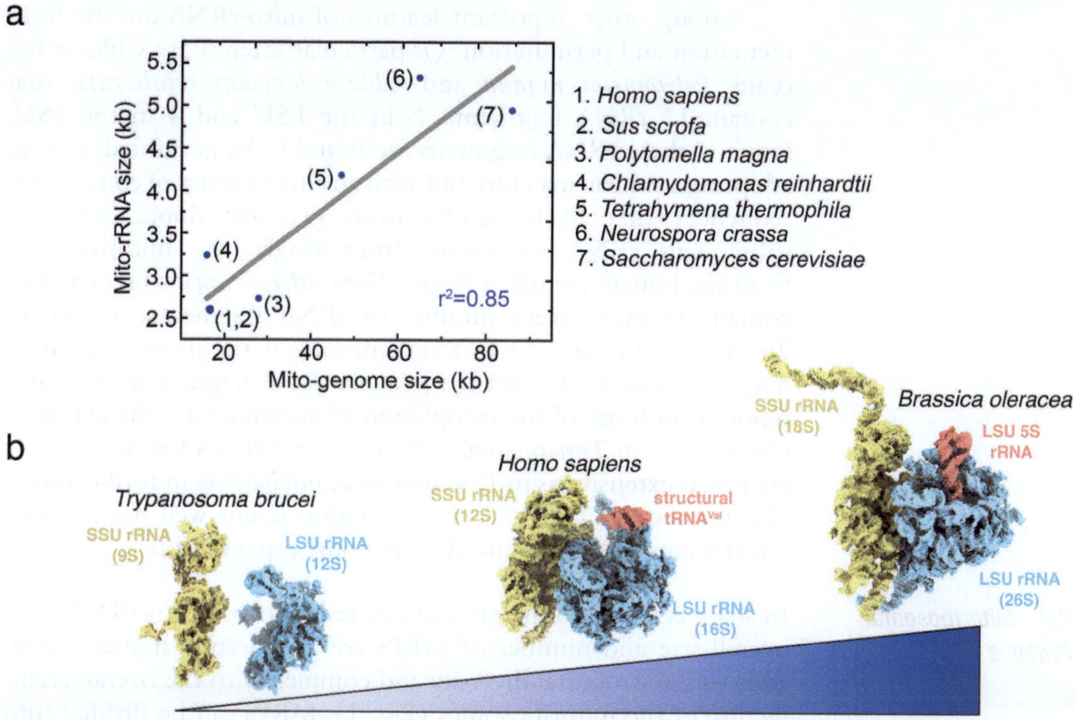

Fig. 2 Depiction of diversity among mitoribosomal RNAs. (**a**) The size of mito-rRNA plotted against the corresponding mitochondrial genome for the indicated organisms (1–7). *T. brucei*, for which the size of mito-genome is only roughly estimated to be ~23 kbp, and *B. oleracea,* which carries relatively very large mito-genome (360 kbp), are excluded from the plot but shown in panel b. (**b**) Three structures representing smallest (*T. brucei*), medium-sized (*H. sapiens*), and one of the largest (*B. oleracea*) rRNAs are shown in their increasing size order from left to right

ribosomes. To retain the helices H43–44, the linker rRNA segment that connects these two peripheral helices to the rest of the rRNA undergo significant reorganization. The functional segment of the SSU rRNA helix h44 that constitute the mRNA decoding center is also retained, despite being heavily truncated.

The structure of *Brassica oleracea* mitoribosome represents the other extreme, where the size of mito-rRNA is much larger than typical bacterial counterparts through addition of expansion segments (ES), which are also present in mito-rRNAs of fungi (*Saccharomyces cerevisiae*, *Neurospora crassa*), ciliates (*Tetrahymena thermophila*), and Tracheophyte (*Arabiodopsis thaliana*). These ESs are present in both the LSU and SSU and are exclusively peripheral, radiating away from the core as flexible double-stranded rRNA segments. One prominent and peculiar ES in the plant *Brassica oleracea* mitoribosome rRNA, whose structural and functional relevance is unknown, provides unique appearance to the head domain of its SSU (Fig. 2b).

Among other important features of mito-rRNA are the fragmentation and permutation. Of particular interest are Chlorophyceans *Polytomella magna* and *Chlamydomonas reinhardtii* that contain 12 rRNA fragments, 8 in the LSU and 4 in the SSU. Some of these rRNA fragments are found to be permuted, that is, adopting a similar structure but with an altered order of component sequences that likely results from genomic duplication and subsequent rRNA reduction. Interestingly, the mitoribosome from the human parasite (*Plasmodium falciparum*) is predicted to contain an even larger number of rRNA fragments, a total of 26 (14 in LSU and 12 in SSU), making it the most extensively fragmented rRNA known [39]. While rRNA fragmentation is also reported in some of the cytoplasmic ribosomes, e.g., the cytosolic ribosome from *Trypanosoma brucei* carries 7 rRNA fragments, they are not as extensive as in *P. magna or P. falciparum* mitoribosomes. The significance of rRNA fragmentation is not well understood, but it seems to be influenced by genome organization.

2.2 Mitoribosomal Proteins

In sharp contrast to the general size reduction of mito-rRNAs, the overall size and number of MRPs are significantly higher, which adds to the structural diversity and complexity to the overall architecture of the mitoribosomes (Fig. 1). MRPs can be divided into three categories: (i) the orthologs of bacteria ribosomal proteins, (ii) mitochondria-specific, and (iii) lineage-specific. Among bacterial orthologs, SSU protein bS20 is universally lost in all mitoribosomes. Other bacterial orthologs of MRPs are found in most mitoribosomes with varying degrees of presence. Most mitoribosomes carry more than 40 bacterial orthologs that are larger in size and carry C- and N-terminal protein extensions. In general, the loss of bacterial homologs could be correlated with loss of binding rRNA segments in the mitoribosome with reduced rRNA. Lineage-specific proteins are primarily associated with SSU and form a large protuberance on the solvent side of subunit (e.g., representative ciliates and discobans structures in Fig. 1). The lineage-specific MRPs are primarily RNA binding, but also colocalize with the regions of reduced or divergent rRNA and are often associated with enzymatic activities.

2.3 Species-Specific Structural Differences

It is generally accepted that all mammalian mitoribosomes are conserved in terms of their composition and structural organization. The high-resolution studies of human [21] and porcine [22] mitoribosomes found that one of the mitochondrial tRNAs forms a structural component of the mitoribosome LSU. Interestingly, these tRNAs were found to be different between the two species, whereas the human mitoribosome LSU carried a deacylated valyl-tRNA and the porcine mitoribosome LSU carried a deacylated

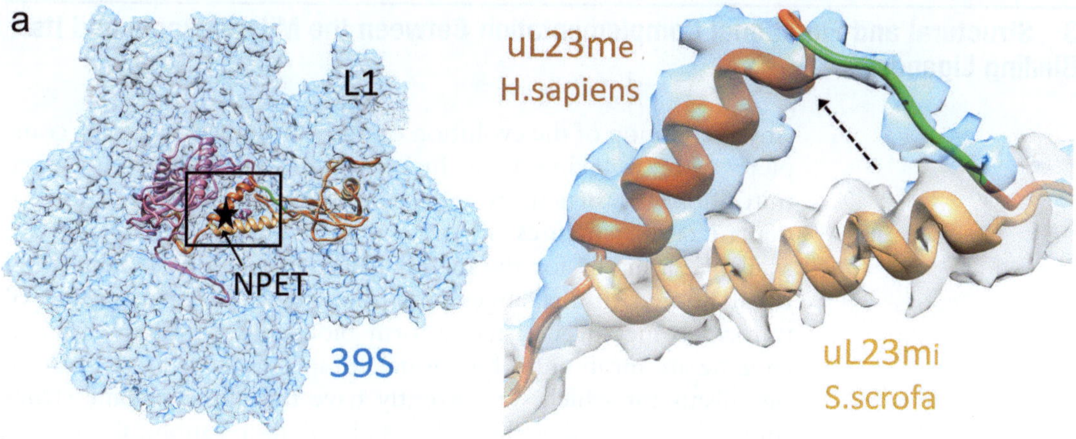

b

```
H.sapiens   MARNVVYPLYRLGGPQLRVFRTNFFIQLVRPGVAQPEDTVQFRIPMEMTRVDLRNYLEGI   60
B.taurus    MARNVLYPLYQLGNPQLRVFRTNFFIQLVRPGTAQPEDTVQFRIPMEMTRVDLRNYLERV   60
S.scrofa    MARNVLYPLYQLGNPQLRVFRTNFFIQLVRPGTAQPEDTVQFRIPMEMTRVDLRNYLERI   60

H.sapiens   YNVPVAAVRTRVQHGSNKRRDHRNVRIKKPDYKVAYVQLAHGQTFTFPDLFPEKDESPEG  120
B.taurus    YNVPVAAVRTRVQHGSNRRRDHRNVRIKQPDYKVAYVQLAHGQTFTFPDLFPKGKRPGGS  120
S.scrofa    YNVPVAAVRTRVQYGSNRRRDHRNIRIKKPDYKVAYVQLALGQTFTFPDLFPERK---GA  117

H.sapiens   SAADDLYSMLEEERQQRQSSDPRRGGVPSWFGL  153
B.taurus    SADEDLRDQVLEEQQQRQSPDPRRGGVPSWFGL  153
S.scrofa    SVDVDVRDQVLEDQRQKHSPDPRRGGVPGWFGL  150
```

Fig. 3 An example of species-specific structural change in a mammalian mitoribosomal MRP. (**a**) Near the exit point of nascent polypeptide-exit tunnel (NPET) lies the C-terminal region of MRP, uL23m. Conformational change observed between the structures of two homologous mammalian MRPs from human (*H. sapiens*) and porcine (*S. scrofa*). uL23mi and uL23me refer to structure of the protein in the initiation and elongation complexes, respectively. (**b**) Alignment of uL23m sequences from human, porcine and bovine MRPs suggests that the region of uL23m that is found to be structurally different in the initiation and elongation complexes might be also due to species related sequence differences (highlighted in green in both panels). (Adapted from ref. 27)

phenylalanyl-tRNA. A subsequent study suggested that this structural tRNA is interchangeable, depending upon the physiological state of the mitochondria [40]. More recent, higher resolution study even revealed that small differences in MRP sequences can significantly affect the local architecture of a functional site of the mitoribosome in given mammalian species. One such MRP, uL23m, that is present near the exit of the nascent polypeptide-exit tunnel, undergoes a large conformational change between the initiation and elongation phases of translation [27] (Fig. 3).

3 Structural and Functional Complementation Between the Mitoribosome and Its Binding Ligands

Any description of the evolution of mitoribosomes cannot be complete without taking their function into account. While more mitoribosomal structures are becoming available at an unprecedented rate, structures of some of the functional complexes are available primarily for the mammalian mitoribosomes. This could be attributed to challenges in purification of large quantities of mitoribosomes needed to perform biochemical experiments and to generate meaningful functional complexes from most of those organisms for which we currently have the mitoribosome structures. As reviewed earlier [23, 41], all four canonical steps of protein synthesis, namely initiation, elongation, termination, and recycling, are present in mammalian mitochondria. Currently, structures of translation initiation complexes [28, 42–44], elongation complexes [27, 45, 46], termination complexes [47], and recycling complexes [48–50] are known. These structures reveal novel features of evolution of human mitochondrial translation while hinting at complementary coevolution of various components of the mitoribosomal translational machinery. Here we outline two striking features: (i) where reduction of rRNA is complemented by unique LSU MRPs, and (ii) where mito-specific extension in translation elongation factor is introduced to maintain the functional mechanics of the mitochondrial translation (Fig. 4).

Due to reduction in rRNA size of the human mitoribosome, several of the eubacterial rRNA segments that are known to interact with tRNAs are lost in the mitoribosomal LSU [20]. The high-resolution structure of the human LSU within the 55S mitoribosome revealed that eubacterial rRNA interaction with the E-site tRNA is partially substituted by the C-terminus region of the MRP mL64 in the intersubunit space [27]. In addition, specific segments of other MRPs, such as uL11m, uL16m, L48m, and a mito-specific MRP that forms the P-site finger [16, 27], interact with A- and P-site mito-tRNAs (Fig. 4a).

As for the bacterial elongation factor G (EF-G), domain IV of mito-EF-G1 is directly involved in facilitating the translocation of anticodon arm of tRNA on the mitoribosome. However, the movement of the elbow and acceptor arm regions of the tRNAs, which is spontaneous on the bacterial ribosome, appears to be facilitated by unique mito-specific 11 amino acid C-terminal extension in the mammalian mitochondrial EF-G1 on the mammalian mitoribosome [27]. The 11 aa C-terminal extension was found to be overlapping with the position that would be occupied by acceptor arm of the A-site tRNA in the pretranslocation state (Fig. 4b),

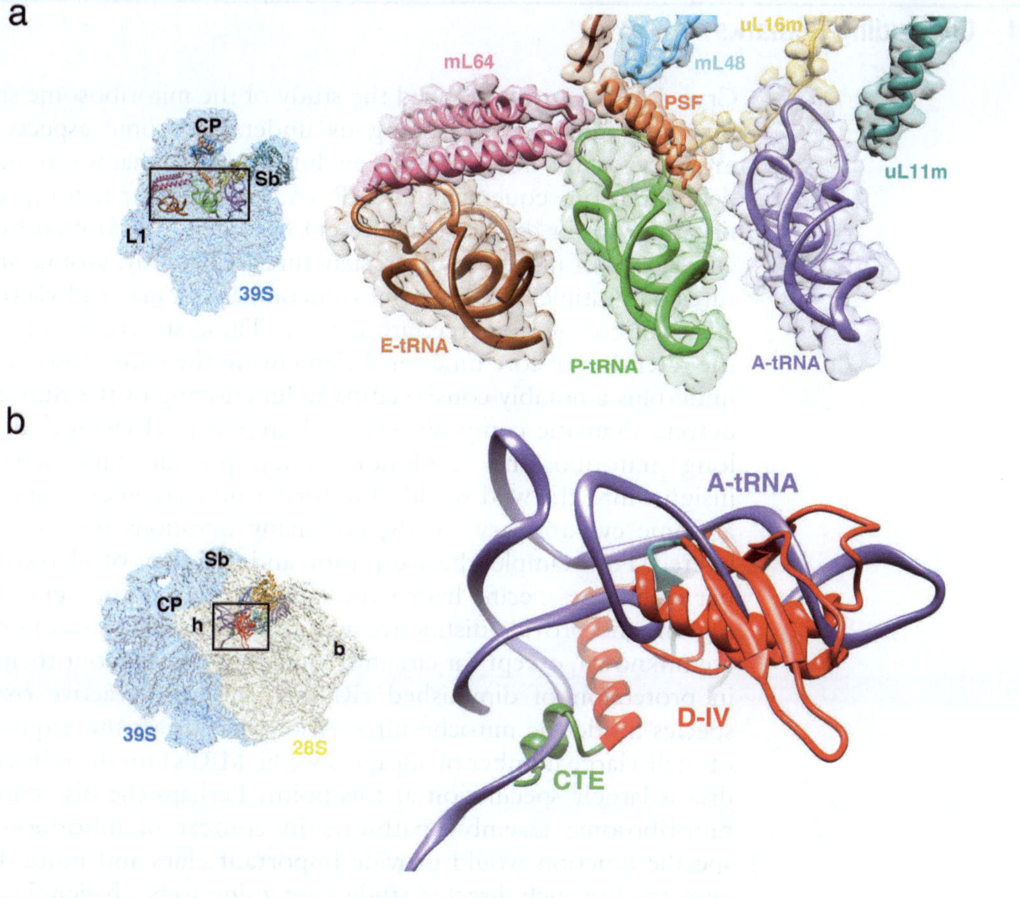

Fig. 4 Interactions among mito-specific MRPs, mito-tRNAs$_{mt}$, and mito-specific C-terminal extension of EF-G1$_{mt}$. (**a**) Interactions between three mitoribosome-bound mito-tRNAs and mito-specific MRPs or mito-specific segments in bacterial orthologs of MRPs. A (purple), P (green) and E (brown) tRNAs$_{mt}$ with mito-specific proteins, PSF (orange), mL48 (blue), mL64 (pink) and mito-specific insertion/extension to homologous MRPs uL11m (tortoise), and uL16m (yellow) are shown. (**b**) Superposition of the A-site bound tRNA$_{mt}$ (purple) onto the EF-G1$_{mt}$-bound human 55S structure revealed that while bulk of domain IV of EF-G1$_{mt}$ overlaps with the anticodon-stem-loop region, the C-terminal extension (CTE) partially overlaps with the acceptor arm of the A-site tRNA$_{mt}$. Thumbnails to the left of panels depict overall orientation of the 39S and 55S mitoribosomes. Landmarks on thumbnails: CP, central protuberance; L1, uL1m stalk; and Sb, uL11m stalk base of the 39S LSU; h, head; and b, body of the 28S SSU. (Adapted from ref. 27)

suggesting that mito-EF-G1 could play a direct role in translocation of the acceptor arm of the mito-tRNAs. These intricate interactions hint at complementary coevolution of these structural features in mitoribosome and mito-EF-G1, both presumably to compensate for generally reduced size of mito-tRNAs [51].

4 Concluding Remarks

Cryo-EM has revolutionized the study of the mitoribosome structure and function by helping us understand some aspects and intricacies of the mitoribosome evolution. A field that was primarily dependent on sequencing and rRNA secondary structure prediction analyses in the prestructure era has been transformed by the emergence of high-resolution structures. Currently, atomic structures of multiple mitoribosomes, encompassing major phyla of the eukaryotic evolution tree, are known. These structures highlight the retention of core functional elements in the mitoribosome that underpins a notably conserved basic functioning of the ribosomes despite dramatic compositional and architectural changes during long mitoribosomal evolution. They provide unprecedented insights into the wild world of mitoribosome complexity and hint at some evolutionary trends, but many questions remain unanswered. For example, the acquisition and functions of a large number of lineage-specific hydrophobic MRPs that occupy peripheral regions and provide distinctive architecture to each mitoribosome are unknown, except for circumstantial arguments about their role in protection of diminished rRNA cores from reactive oxygen species inside the mitochondria. Also, what drives the acquisition of such a large number of lineage-specific MRPs into the mitochondria is largely speculation at this point. Perhaps the dissection of mitoribosome assembly pathways in context of mitoribosome-specific function would provide important clues and more direct answers. But such directed studies are going to be challenging and are currently in their infancy at best. What we present here is an overview of the fast-emerging mitoribosome structural studies in the context of structure-function evolution. There are several non-structural aspects to mitochondria and mitoribosome evolution that go beyond the scope of this perspective article and are not covered here.

Acknowledgments

This work was supported in part by the National Institutes of Health grant R01 GM61576 (to R.K.A.). R.K.A. also acknowledges the financial support to his lab through following NIH grants: R01 GM139277, R01 AI132422, and R01 AI155473. The authors thank Dr. Nilesh Banavali for critical reading of the manuscript.

References

1. Koonin EV (2003) Comparative genomics, minimal gene-sets and the last universal common ancestor. Nat Rev Microbiol, Nature Publishing Group 1:127–136. https://doi.org/10.1038/nrmicro751
2. Woese CR (1987) Bacterial evolution. Microbiol Rev 51:221–271
3. Woese CR, Kandler O, Wheelis ML (1990) Towards a natural system of organisms: proposal for the domains Archaea, Bacteria, and Eucarya. Proc Natl Acad Sci U S A 87:4576–4579. https://doi.org/10.1073/pnas.87.12.4576
4. Bernier CR, Petrov AS, Kovacs NA, Penev PI, Williams LD (2018) Translation: the universal structural core of life. Mol Biol Evol 35:2065–2076. https://doi.org/10.1093/molbev/msy101
5. Ban N, Nissen P, Hansen J, Moore PB, Steitz TA (2000) The complete atomic structure of the large ribosomal subunit at 2.4 Å resolution. Science, American Association for the Advancement of Science 289:905–920. https://doi.org/10.1126/science.289.5481.905
6. Agmon I, Bashan A, Zarivach R, Yonath A (2005) Symmetry at the active site of the ribosome: structural and functional implications. Biol Chem 386:833–844. https://doi.org/10.1515/BC.2005.098
7. Noller HF (2012) Evolution of protein synthesis from an RNA world. Cold Spring Harb Perspect Biol 4:a003681. https://doi.org/10.1101/cshperspect.a003681
8. Melnikov S, Ben-Shem A, Garreau de Loubresse N, Jenner L, Yusupova G, Yusupov M (2012) One core, two shells: bacterial and eukaryotic ribosomes. Nat Struct Mol Biol 19: 560–567. https://doi.org/10.1038/nsmb.2313
9. Bowman JC, Petrov AS, Frenkel-Pinter M, Penev PI, Williams LD (2020) Root of the tree: the significance, evolution, and origins of the ribosome. Chem Rev, American Chemical Society 120:4848–4878. https://doi.org/10.1021/acs.chemrev.9b00742
10. Gray MW, Burger G, Lang BF (2001) The origin and early evolution of mitochondria. Genome Biol 2:REVIEWS1018. https://doi.org/10.1186/gb-2001-2-6-reviews1018
11. Martijn J, Vosseberg J, Guy L, Offre P, Ettema TJG (2018) Deep mitochondrial origin outside the sampled alphaproteobacteria. Nature, Nature Publishing Group 557:101–105. https://doi.org/10.1038/s41586-018-0059-5
12. Spang A, Saw JH, Jørgensen SL, Zaremba-Niedzwiedzka K, Martijn J, Lind AE, van Eijk R, Schleper C, Guy L, Ettema TJG (2015) Complex archaea that bridge the gap between prokaryotes and eukaryotes. Nature, Nature Publishing Group 521:173–179. https://doi.org/10.1038/nature14447
13. Zaremba-Niedzwiedzka K, Caceres EF, Saw JH, Bäckström D, Juzokaite L, Vancaester E, Seitz KW, Anantharaman K, Starnawski P, Kjeldsen KU, Stott MB, Nunoura T, Banfield JF, Schramm A, Baker BJ, Spang A, Ettema TJG (2017) Asgard archaea illuminate the origin of eukaryotic cellular complexity. Nature, Nature Publishing Group 541:353–358. https://doi.org/10.1038/nature21031
14. O'Brien TW (2002) Evolution of a protein-rich mitochondrial ribosome: implications for human genetic disease. Gene 286:73–79. https://doi.org/10.1016/s0378-1119(01)00808-3
15. Waltz F, Giegé P (2020) Striking diversity of mitochondria-specific translation processes across eukaryotes. Trends Biochem Sci, Elsevier 45:149–162. https://doi.org/10.1016/j.tibs.2019.10.004
16. Sharma MR, Koc EC, Datta PP, Booth TM, Spremulli LL, Agrawal RK (2003) Structure of the mammalian mitochondrial ribosome reveals an expanded functional role for its component proteins. Cell 115:97–108. https://doi.org/10.1016/s0092-8674(03)00762-1
17. Frank J, Agrawal RK, Verschoor A (2001) Ribosome structure and shape. In: Encyclopedia of life sciences. John Wiley & Sons, Ltd., London, pp 1–6
18. Frank J, Zhu J, Penczek P, Li Y, Srivastava S, Verschoor A, Radermacher M, Grassucci R, Lata RK, Agrawal RK (1995) A model of protein synthesis based on cryo-electron microscopy of the E. coli ribosome. Nature, Nature Publishing Group 376:441–444. https://doi.org/10.1038/376441a0
19. Temperley RJ, Wydro M, Lightowlers RN, Chrzanowska-Lightowlers ZM (2010) Human mitochondrial MRNAs--like members of all families, similar but different. Biochim Biophys Acta 1797:1081–1085. https://doi.org/10.1016/j.bbabio.2010.02.036
20. Mears JA, Sharma MR, Gutell RR, McCook AS, Richardson PE, Caulfield TR, Agrawal RK, Harvey SC (2006) A structural model for the large subunit of the mammalian mitochondrial ribosome. J Mol Biol 358:193–212. https://doi.org/10.1016/j.jmb.2006.01.094

21. Brown A, Amunts A, Bai X, Sugimoto Y, Edwards PC, Murshudov G, Scheres SHW, Ramakrishnan V (2014) Structure of the large ribosomal subunit from human mitochondria. Science, American Association for the Advancement of Science 346:718–722. https://doi.org/10.1126/science.1258026
22. Greber BJ, Boehringer D, Leitner A, Bieri P, Voigts-Hoffmann F, Erzberger JP, Leibundgut M, Aebersold R, Ban N (2014) Architecture of the large subunit of the mammalian mitochondrial ribosome. Nature, Nature Publishing Group 505:515–519. https://doi.org/10.1038/nature12890
23. Sharma MR, Kaushal PS, Gupta M, Banavali NK, Agrawal RK (2013) Insights into structural basis of mammalian mitochondrial translation. In: Duchêne A-M (ed) Translation in mitochondria and other organelles. Springer, Berlin, Heidelberg, pp 1–28. https://doi.org/10.1007/978-3-642-39426-3_1
24. Amunts A, Brown A, Toots J, Scheres SHW, Ramakrishnan V (2015) The structure of the human mitochondrial ribosome. Science, American Association for the Advancement of Science 348:95–98. https://doi.org/10.1126/science.aaa1193
25. Greber BJ, Bieri P, Leibundgut M, Leitner A, Aebersold R, Boehringer D, Ban N (2015) The complete structure of the 55S mammalian mitochondrial ribosome. Science, American Association for the Advancement of Science 348:303–308. https://doi.org/10.1126/science.aaa3872
26. Kaushal PS, Sharma MR, Booth TM, Haque EM, Tung C-S, Sanbonmatsu KY, Spremulli LL, Agrawal RK (2014) Cryo-EM structure of the small subunit of the mammalian mitochondrial ribosome. Proc Natl Acad Sci U S A, Proceedings of the National Academy of Sciences 111:7284–7289. https://doi.org/10.1073/pnas.1401657111
27. Koripella RK, Sharma MR, Bhargava K, Datta PP, Kaushal PS, Keshavan P, Spremulli LL, Banavali NK, Agrawal RK (2020) Structures of the human mitochondrial ribosome bound to EF-G1 reveal distinct features of mitochondrial translation elongation. Nat Commun, Nature Publishing Group 11:3830. https://doi.org/10.1038/s41467-020-17715-2
28. Koripella RK, Sharma MR, Haque ME, Risteff P, Spremulli LL, Agrawal RK (2019) Structure of human mitochondrial translation initiation factor 3 bound to the small ribosomal subunit. iScience 12:76–86. https://doi.org/10.1016/j.isci.2018.12.030
29. Desai N, Brown A, Amunts A, Ramakrishnan V (2017) The structure of the yeast mitochondrial ribosome. Science, American Association for the Advancement of Science 355:528–531. https://doi.org/10.1126/science.aal2415
30. Itoh Y, Naschberger A, Mortezaei N, Herrmann JM, Amunts A (2020) Analysis of translating mitoribosome reveals functional characteristics of translation in mitochondria of fungi. Nat Commun, Nature Publishing Group 11:5187. https://doi.org/10.1038/s41467-020-18830-w
31. Tobiasson V, Amunts A (2020) Ciliate mitoribosome illuminates evolutionary steps of mitochondrial translation. Wolberger, C., Ed. eLife, eLife Sciences Publications, Ltd 9: e59264. https://doi.org/10.7554/eLife.59264
32. Ramrath DJF, Niemann M, Leibundgut M, Bieri P, Prange C, Horn EK, Leitner A, Boehringer D, Schneider A, Ban N (2018) Evolutionary shift toward protein-based architecture in trypanosomal mitochondrial ribosomes. Science 362:eaau7735. https://doi.org/10.1126/science.aau7735
33. Soufari H, Waltz F, Parrot C, Durrieu-Gaillard S, Bochler A, Kuhn L, Sissler M, Hashem Y (2020) Structure of the mature kinetoplastids mitoribosome and insights into its large subunit biogenesis. Proc Natl Acad Sci U S A, Proceedings of the National Academy of Sciences 117:29851–29861. https://doi.org/10.1073/pnas.2011301117
34. Waltz F, Nguyen T-T, Arrivé M, Bochler A, Chicher J, Hammann P, Kuhn L, Quadrado M, Mireau H, Hashem Y, Giegé P (2019) Small is big in Arabidopsis mitochondrial ribosome. Nat Plants, Nature Publishing Group 5:106–117. https://doi.org/10.1038/s41477-018-0339-y
35. Waltz F, Soufari H, Bochler A, Giegé P, Hashem Y (2020) Cryo-EM structure of the RNA-rich plant mitochondrial ribosome. Nat Plants, Nature Publishing Group 6:377–383. https://doi.org/10.1038/s41477-020-0631-5
36. Tobiasson V, Berzina I, Amunts A (2022) Structure of a mitochondrial ribosome with fragmented RRNA in complex with membrane-targeting elements. Nat Commun, Nature Publishing Group 13:6132. https://doi.org/10.1038/s41467-022-33582-5
37. Waltz F, Salinas-Giegé T, Englmeier R, Meichel H, Soufari H, Kuhn L, Pfeffer S, Förster F, Engel BD, Giegé P, Drouard L, Hashem Y (2021) How to build a ribosome from RNA fragments in Chlamydomonas mitochondria. Nat Commun, Nature Publishing Group 12:7176. https://doi.org/10.1038/s41467-021-27200-z

38. Maly P, Brimacombe R (1983) Refined secondary structure models for the 16S and 23S ribosomal RNA of Escherichia coli. Nucleic Acids Res 11:7263–7286
39. Feagin JE, Harrell MI, Lee JC, Coe KJ, Sands BH, Cannone JJ, Tami G, Schnare MN, Gutell RR (2012) The fragmented mitochondrial ribosomal RNAs of Plasmodium falciparum. PLoS One, Public Library of Science 7: e38320. https://doi.org/10.1371/journal.pone.0038320
40. Rorbach J, Gao F, Powell CA, D'Souza A, Lightowlers RN, Minczuk M, Chrzanowska-Lightowlers ZM (2016) Human mitochondrial ribosomes can switch their structural RNA composition. Proc Natl Acad Sci U S A 113:12198–12201. https://doi.org/10.1073/pnas.1609338113
41. Christian BE, Spremulli LL (2012) Mechanism of protein biosynthesis in mammalian mitochondria. Biochim Biophys Acta 1819:1035–1054. https://doi.org/10.1016/j.bbagrm.2011.11.009
42. Yassin AS, Haque ME, Datta PP, Elmore K, Banavali NK, Spremulli LL, Agrawal RK (2011) Insertion domain within mammalian mitochondrial translation initiation factor 2 serves the role of eubacterial initiation factor 1. Proc Natl Acad Sci U S A 108:3918–3923. https://doi.org/10.1073/pnas.1017425108
43. Kummer E, Leibundgut M, Rackham O, Lee RG, Boehringer D, Filipovska A, Ban N (2018) Unique features of mammalian mitochondrial translation initiation revealed by cryo-EM. Nature 560:263–267. https://doi.org/10.1038/s41586-018-0373-y
44. Khawaja A, Itoh Y, Remes C, Spåhr H, Yukhnovets O, Höfig H, Amunts A, Rorbach J (2020) Distinct pre-initiation steps in human mitochondrial translation. Nat Commun 11: 2932. https://doi.org/10.1038/s41467-020-16503-2
45. Kummer E, Ban N (2020) Structural insights into mammalian mitochondrial translation elongation catalyzed by MtEFG1. EMBO J 39:e104820. https://doi.org/10.15252/embj.2020104820
46. Desai N, Yang H, Chandrasekaran V, Kazi R, Minczuk M, Ramakrishnan V (2020) Elongational stalling activates mitoribosome-associated quality control. Science 370:1105–1110. https://doi.org/10.1126/science.abc7782
47. Kummer E, Schubert KN, Schoenhut T, Scaiola A, Ban N (2021) Structural basis of translation termination, rescue, and recycling in mammalian mitochondria. Mol Cell 81: 2566–2582.e6. https://doi.org/10.1016/j.molcel.2021.03.042
48. Koripella RK, Sharma MR, Risteff P, Keshavan P, Agrawal RK (2019) Structural insights into unique features of the human mitochondrial ribosome recycling. Proc Natl Acad Sci U S A 116:8283–8288. https://doi.org/10.1073/pnas.1815675116
49. Koripella RK, Deep A, Agrawal EK, Keshavan P, Banavali NK, Agrawal RK (2021) Distinct mechanisms of the human mitoribosome recycling and antibiotic resistance. Nat Commun 12:3607. https://doi.org/10.1038/s41467-021-23726-4
50. Aibara S, Singh V, Modelska A, Amunts A (2020) Structural basis of mitochondrial translation. eLife 9:e58362. https://doi.org/10.7554/eLife.58362
51. Watanabe K (2010) Unique features of animal mitochondrial translation systems. The non-universal genetic code, unusual features of the translational apparatus and their relevance to human mitochondrial diseases. Proc Jpn Acad Ser B Phys Biol Sci 86:11–39. https://doi.org/10.2183/pjab.86.11
52. Itoh Y, Khawaja A, Laptev I, Cipullo M, Atanassov I, Sergiev P, Rorbach J, Amunts A (2022) Mechanism of mitoribosomal small subunit biogenesis and preinitiation. Nature, Nature Publishing Group 606:603–608. https://doi.org/10.1038/s41586-022-04795-x
53. Amunts A, Brown A, Bai X, Llácer JL, Hussain T, Emsley P, Long F, Murshudov G, Scheres SHW, Ramakrishnan V (2014) Structure of the yeast mitochondrial large ribosomal subunit. Science, American Association for the Advancement of Science 343:1485–1489. https://doi.org/10.1126/science.1249410

Chapter 3

Mitoribosome Biogenesis

J. Conor Moran, Samuel Del'Olio, Austin Choi, Hui Zhong, and Antoni Barrientos

Abstract

Mitoribosome biogenesis is a complex and energetically costly process that involves RNA elements encoded in the mitochondrial genome and mitoribosomal proteins most frequently encoded in the nuclear genome. The process is catalyzed by extra-ribosomal proteins, nucleus-encoded assembly factors that act in all stages of the assembly process to coordinate the processing and maturation of ribosomal RNAs with the hierarchical association of ribosomal proteins. Biochemical studies and recent cryo-EM structures of mammalian mitoribosomes have provided hints regarding their assembly. In this general concept chapter, we will briefly describe the current knowledge, mainly regarding the mammalian mitoribosome biogenesis pathway and factors involved, and will emphasize the biological sources and approaches that have been applied to advance the field.

Key words Mitochondrial ribosome, Mitochondrial translation, Mitoribosome assembly, OXPHOS deficiency, Mitochondrial disease

1 Introduction

Fifty years after the isolation and characterization of the fungal [1] and mammalian mitoribosomes [2, 3], investigations into mitoribosome composition, structure, biogenesis, and function have advanced our knowledge of the field. Recently, researchers have taken advantage of innovative biochemical approaches to decipher details of the mitoribosome assembly pathway in several organisms [4, 5]. They have also used cutting-edge cryogenic electron microscopy (cryo-EM) to unravel the structure of the yeast mitoribosome [6, 7] and its trypanosomal [8–11], porcine, and human mitochondrial counterparts in their mature state [12–16], and in several native states of assembly [8–11, 17–23].

J. Conor Moran and Samuel Del'Olio contributed equally with all other contributors.

Antoni Barrientos and Flavia Fontanesi (eds.), *The Mitoribosome: Methods and Protocols*, Methods in Molecular Biology, vol. 2661, https://doi.org/10.1007/978-1-0716-3171-3_3,

These studies have shown that proteins and RNA domains of mitochondrial and bacterial ribosomes that contribute to decoding, and peptide bond formation share a high degree of similarity, supporting the conservation of their catalytic properties. However, evolution led to the formation of mitoribosomes that differ significantly in structure and composition, compared not only to their bacterial relatives but also among different species [24–26], which has been attributed to divergent structural patching [25, 26].

The mammalian mitoribosome, the focus of this chapter, is a 55S ribonucleoprotein complex formed by a 39S large subunit (mtLSU) with 52 mitoribosomal proteins (MRPs) (Table 1), a *16S rRNA,* and a structural tRNA ($tRNA^{Val}$ in human cells), and a 28S small subunit (mtSSU) with 30 MRPs (Table 2) and a *12S rRNA*. The 55S ribosomes are protein-rich, with only 25–30% RNA compared to bacterial and eukaryotic cytoplasmic ribosomes, which are ~60% RNA [12, 14]. Whereas the ribosomal RNAs are encoded in the mitochondrial DNA (mtDNA), all MRPs are encoded in the nuclear genome, synthesized on cytoplasmic ribosomes, and imported into the mitochondrial matrix to be assembled with the subunit-specific RNAs. Mitoribosome assembly involves a growing number of ancillary factors, including RNA processing and modification enzymes, guanosine triphosphatases (GTPases), DEAD-box RNA helicases, kinases, translation factors, and other proteins acting as chaperones to fuel the process [27, 28]. They act as assembly factors guiding the maturation of mitoribosomal components and their temporal association, forming preribosomal particles during the assembly of individual subunits and facilitating subunit association during monosome formation [4, 5].

Excellent reviews on mitoribosome assembly have been reported elsewhere [29–33]. Here, we will summarize the current knowledge of the process as it occurs for the mammalian mitoribosome, and the factors involved. We will include brief comparative notes on the bacterial ribosome assembly process and the biogenesis of mitoribosomes in nonmammalian species. We will also provide a general description of the methods that have been used to advance fundamental aspects of mitoribosome biogenesis.

2 Mitoribosome Structural Features

Cryo-EM analyses of human and porcine mitoribosomes allowed for a precise map of all mitoribosomal mtSSU and mtLSU components, thus establishing their composition. They disclosed an array of mitoribosome-specific features described below [12, 14, 18, 34].

The cryo-EM studies showed that their catalytic region at the subunit interface is essentially conserved from the bacterial ribosome [12, 14]. However, significant amounts of rRNA and several bacterial proteins (uS4, uS8, uS13, uS19, and bS20) were lost in the mammalian mitoribosome, whereas conserved homologs of

Table 1
Mitoribosome small subunit proteins

Mitoribosome small subunit							
	Old nomenclature				Old nomenclature		
New name	Yeast	Human	Bacteria	New name	Human	Yeast	Bacteria
bS1m	Mrp51	MRPS28	S1	mS23	MRPS23	Rsm25	–
uS2m	Mrp4	MRPS2	S2	mS25	MRPS25	–	–
uS3m	Var1	MRPS24	S3/S24	mS26	MRPS26	Pet123	–
bS4m	Nam9	–	S4	mS27	MRPS27	–	–
uS5m	Mrps5	MRPS5	S5	mS29	MRPS29	Rsm23	–
bS6m	Mrp17	MRPS6	S6	mS31	MRPS31	–	–
uS7m	Rsm7	MRPS7	S7	mS33	MRPS33	Rsm27	–
uS8m	Mrps8	–	S8	mS34	MRPS34	–	–
uS9m	Mrps9	MRPS9	S9	mS35	MRPS35	Rsm24	–
uS10m	Rsm10	MRPS10	S10	mS37	MRPS37	Mrp10	–
uS11m	Mrps18	MRPS11	S11	mS38	MRPS38	Cox24	–
uS12m	Mrps12	MRPS12	S12	mS39	MRPS39	–	–
uS13m	Sws2	–	S13	mS40	MRSP18B	–	–
uS14m	Mrp2	MRPS14	S14	mS41	–	Fyv4	–
uS15m	Mrps28	MRPS15	S15	mS42	–	Rsm26	–
bS16m	Mrps16	MRPS16	S16	mS43	–	Mrp1	–
uS17m	Mrps17	MRPS17	S17	mS44	–	Mrp13	–
bS18m	Rsm18	MRPS18C	S18	mS45	–	Mrps35	–
uS19m	Rsm19	–	S19	mS46	–	Rsm28	–
bS21m	Mrp21	MRPS21	S21	mS47	–	Ehd3	–
mS22	–	MRPS22					

The table shows the unified nomenclature for mitochondrial ribosome proteins [104], where proteins with a prefix "u" (for universal) are observed in all kingdoms of life, proteins with a prefix "b" are bacterial in origin and do not have a eukaryotic (or archaeal) homolog, and proteins with a prefix "m" are mitochondrion-specific. The old nomenclature is presented as a reference

bacterial proteins frequently acquired N- or C- terminal extensions [24] to accommodate novel positions rather than compensate for the missing rRNA [15]. Furthermore, out of 82 MRPs, 36 mitochondrion-specific proteins were recruited to occupy mainly peripheral locations distributed over the solvent-accessible surface [24] (Fig. 1). In the mtLSU, they form clusters at the central protuberance (CP), the L7/L12 stalk, and adjacent to the polypeptide exit site. In the mtSSU, two mitochondrion-specific proteins form the

Table 2
Mitoribosome large subunit proteins

Mitoribosome large subunit							
	Old nomenclature				Old nomenclature		
New name	Human	Yeast	Bacteria	New name	Human	Yeast	Bacteria
uL1m	MRPL1	Mrpl1	L1	bL35m	MRPL35	Ynl122C	L35
uL2m	MRPL2	Rml2	L2	bL36m	MRPL36	RTC6	L36
uL3m	MRPL3	Mrpl9	L3	mL37	MRPL37	–	–
uL4m	MRPL4	YML6	L4	mL38	MRPL38	Mrpl35	–
uL5m	–	Mrpl7	L5	mL39	MRPL39	–	–
uL6m	–	Mrpl6	L6	mL40	MRPL40	Mrpl28	–
bL9m	MRPL9	Mrpl50	L9	mL41	MRPL41	Mrpl27	–
uL10m	MRPL10	Mrpl11	L10	mL42	MRPL42	–	–
uL11m	MRPL11	Mrpl19	L11	mL43	MRPL43	Mrpl51	–
bL12m	MRPL12	Mnp1	L7/L12	mL44	MRPL44	Mrpl3	–
uL13m	MRPL13	Mrpl23	L13	mL45	MRPL45	–	–
uL14m	MRPL14	Mrpl38	L14	mL46	MRPL46	Mrpl17	–
uL15m	MRPL15	Mrpl10	L15	mL48	MRPL48	–	–
uL16m	MRPL16	Mrpl16	L16	mL49	MRPL49	Img2	–
bL17m	MRPL17	Mrpl8	L17	mL50	MRPL50	Mrpl13	–
uL18m	MRPL18	–	L18	mL51	MRPL51	–	–
bL19m	MRPL19	Img1	L19	mL52	MRPL52	–	–
bL20m	MRPL20	–	L20	mL53	MRPL53	Mrpl44	–
bL21m	MRPL21	Mrpl49	L21	mL54	MRPL54	Mrpl37	–
uL22m	MRPL22	Mrpl22	L22	mL57	–	Mrpl15	–
uL23m	MRPL23	Mrp20	L23	mL58	–	Mrpl20	–
uL24m	MRPL24	Mrpl40	L24	mL59	–	Mrpl25	–
bL27m	MRPL27	Mrp7	L27	mL60	–	Mrpl31	–
bL28m	MRPL28	Mrpl24	L28	mL61	–	Mrp49	–
uL29m	MRPL47	Mrpl4	L29	mL62	MRPL58	–	–
uL30m	MRPL30	Mrpl33	L30	mL63	MRPL57	–	–
bL31m	–	Mrpl36	L31	mL64	MRPL59	–	–
bL32m	MRPL32	Mrpl32	L32	mL65	MRPS30	–	–
bL33m	MRPL33	Mrpl39	L33	mL66	MRPS18A	–	–
bL34m	MRPL34	Mprl34	L34	mL67	–	Mhr1	–

The table shows the unified nomenclature for mitochondrial ribosome proteins [104]. The old nomenclature is presented as a reference

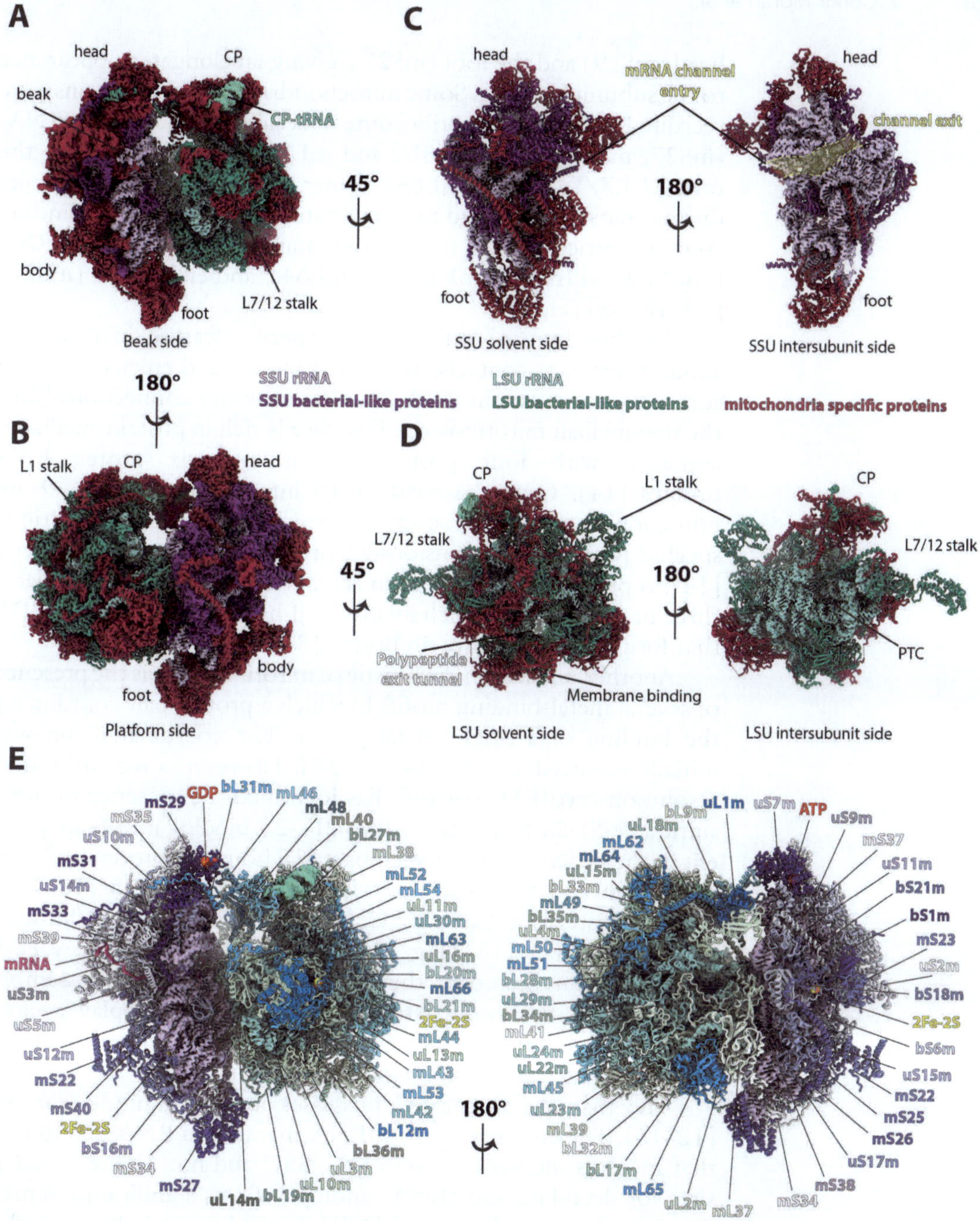

Fig. 1 The human mitochondrial ribosome. High-resolution 2.2 Å consensus map of the human mitoribosome viewed from the beak side (**a**) and platform side (**b**) highlighting mitochondrial-specific proteins in red. The mtSSU 12S rRNA is displayed in light purple and proteins conserved in bacteria are shaded dark purple. In the mtLSU, the 16S rRNA is colored light blue, the CP-tRNA in cyan, and bacterial-like proteins in sea green. Solvent and intersubunit views of the mtSSU (**c**) and mtLSU (**d**) with proteins shown as cartoons using the atomic model and RNA components as density maps. The path of mRNA during translation is indicated in yellow and the polypeptide exit tunnel in white. Atomic model of mitoribosome proteins individually annotated and colored (**e**). RNA components are shown as density maps and colored as in panels A–D. Bound nucleotides ATP and GDP are labeled in red, iron-sulfur clusters in yellow, and the mRNA in pink. EM map used is EMD-13980 and atomic coordinates are PDB-7QI4

head (mS29) and the foot (mS27), giving an elongated appearance to the subunit [12, 14]. Some mitochondrion-specific proteins were recruited to stabilize mitoribosome structures, including the rRNAs (mS27, mS38 in the mtSSU, and mL37 in the mtLSU), or the mtLSU L7/L12 stalk (mL63). Other proteins play relevant roles during translation, including preinitiation (mS37), mRNA binding (the pentatricopeptide repeat-containing protein mS39), tRNA translocation (mL40, mL48, and mL64), and elongation (mL53) [16, 18, 35] (Fig. 2).

Another human mitoribosome-specific feature involves the subunit interface contacts. Whereas bacterial and eukaryotic ribosomes typically contain RNA-RNA intersubunit connections [36], the mammalian mitoribosome interface is rich in protein-mediated contacts, with four protein–protein and six protein–RNA bridges [14]. Out of a total of 16 intersubunit bridges, 9 are mitochondrion-specific, several of which involve mitochondrion-specific proteins or extensions of proteins conserved in bacteria [14]. A prominent contribution is made by mS29, which mediates three bridges, one of which involves a β-hairpin supported by GDP that forms a protein: protein bridge [37] (Fig. 1e).

Another unusual characteristic of mitoribosomes is the presence of several metal-binding motifs in which a protein pair coordinates the binding of a single metal group. The coordinated ion was initially reported as zinc [14, 16, 38]. However, a recent higher-resolution cryo-EM structure has identified the presence of iron-sulfur (Fe-S) clusters instead. Two 2Fe-2S-binding motifs are present in the mtSSU: one between bacterial homolog proteins bS18m and bS6m and the other between bS16m and mS25 [18, 39]. A third cluster is in the mtLSU, between mL66 and uL10m near the L7/L12 stalk [18] (Fig. 1e). Although the role of these Fe-S clusters remains unknown, they likely serve to stabilize the assembly and overall mitoribosome structure. They could also play a regulatory role by sensing changes in the redox environment to adapt mitochondrial translation accordingly.

The mammalian mtSSU possesses multiple unique features [12–16]. The entrance of the mRNA channel, an RNA-rich tunnel that involves the neck region of the SSU and houses the A and P sites for decoding and tRNA binding [40], is significantly remodeled compared to the bacterial SSU. In the bacterial ribosome, the mRNA channel entrance is formed by proteins uS3, uS4, and uS5, in which basic residues from uS3 and uS4 confer a helicase-like activity responsible for unwinding secondary structures of the translating mRNA [40, 41]. However, the human mitoribosome does not contain a homolog to uS4, and that of uS3, although present as uS3m, lacks the C-terminal domain [16]. Instead, the opening of the channel is widened from 9 Å to 15 Å by mitochondria-specific extensions of uS5m, and the mitoribosome has acquired mS39, which resides near the mRNA entrance and

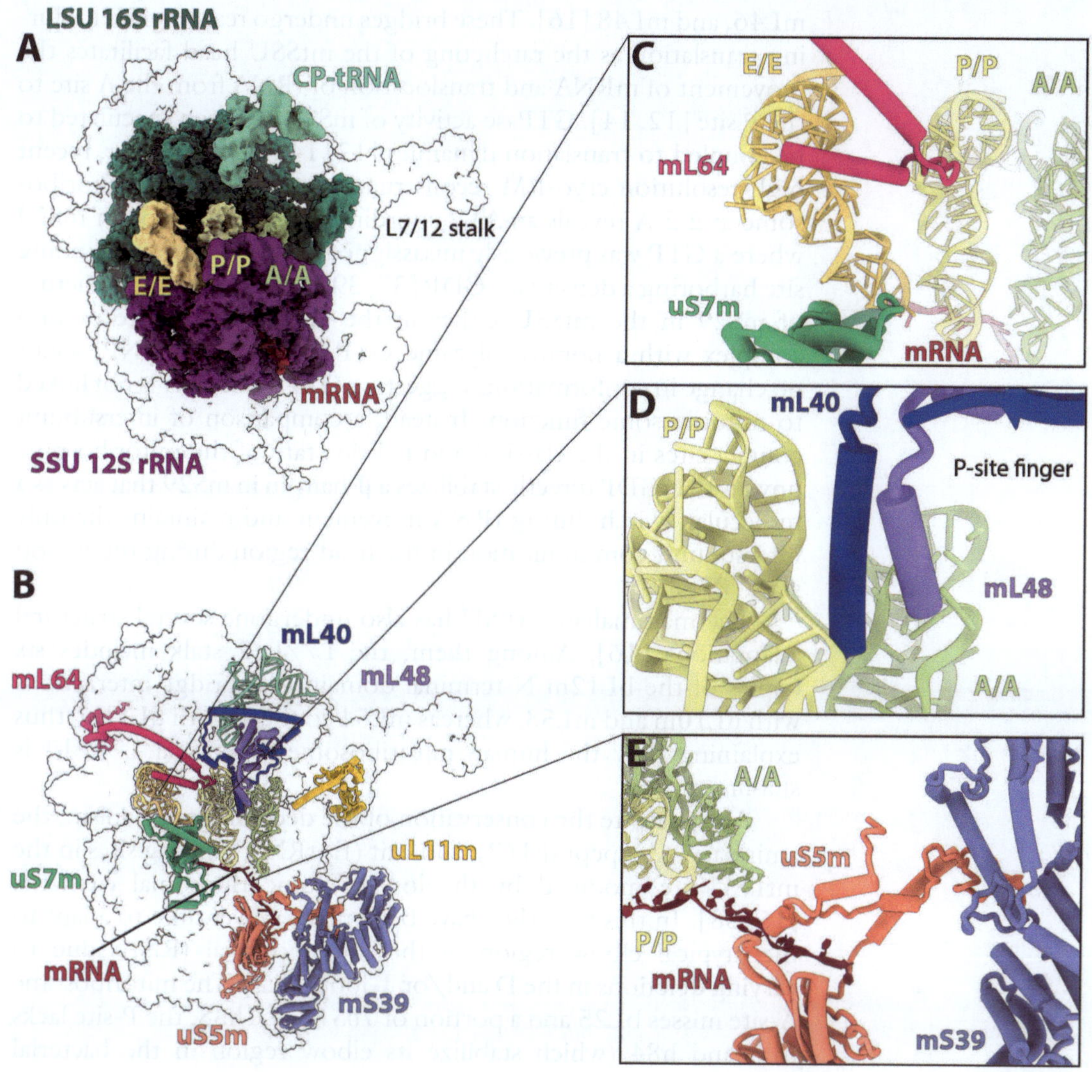

Fig. 2 Features of tRNA binding in the human mitoribosome. (**a**) Cryo-EM map of the human mitoribosome highlighting RNA elements. The mtSSU and mtLSU rRNAs are shown in dark purple and sea green, respectively, while the tRNAs are indicated in cyan (CP-tRNA), light green (A-site tRNA), yellow (P-site tRNA), and orange (E-site tRNA). Mitoribosomal proteins are transparent and shown as a surface outline. (**b**) Depiction of proteins involved in tRNA binding, stabilization, and movement during translation. Individual proteins are colored and labeled. Proteins are shown using atomic coordinates with helices as tubes and RNA elements as cartoons with bases as ladders. (**c**, **d**, **e**) Zoomed-in view of protein–RNA interactions. Nearby structural elements were removed for clarity. EM map is EMD-11397 and atomic coordinates are PDB-6ZSG

serves as a binding platform for mRNA bound by the LRPPRC/SLIRP complex [16, 35] (Fig. 2e). The mRNA channel exit (formed by bS1m, bS21m, and mS37) has also experienced significant structural rearrangements, presumably to adapt to mammalian mitochondrial transcripts, which lack Shine Dalgarno sequences to assist with start codon selection [16].

The signature protein of the mtSSU head region is the intrinsic GTPase mS29, which establishes intersubunit communication via mitochondria-specific bridges with the mtLSU CP proteins mL40, mL46, and mL48 [16]. These bridges undergo rearrangement during translation as the ratcheting of the mtSSU head facilitates the movement of mRNA and translocation of tRNA from the A site to the P site [12, 14]. GTPase activity of mS29 has been speculated to be coupled to translation dynamics [12, 14, 29]. However, recent high-resolution cryo-EM reconstruction of the human mitoribosome at 2.2 Å reveals an ATP coordinated by the P loop of mS29 where a GTP was previously misassigned and an additional binding site harboring a density for GDP [37, 39]. Comparing the structure of mS29 in the mtSSU either in the GDP-bound state or in a complex with a nonhydrolyzable GTP analog (GMPPNP) yields no change in conformation, suggesting GTPase activity is not linked to mitoribosome function. Instead, a comparison of intersubunit contact sites in the classical and hybrid state of the mitoribosome unveils that GDP directly stabilizes a β-hairpin in mS29 that acts as a molecular switch during tRNA movement and maintains the only intersubunit communication in the head region during the hybrid state [37, 39].

The mammalian mtLSU has also undergone several structural adaptations [16]. Among them, the L7/L12 stalk includes six copies of the bL12m N-terminal domain that bridge interactions with uL10m and mL53, whereas mL54 connects with uL11m, thus explaining how the human mitoribosome functional L7/L12 is stabilized [16].

Also, despite the conservation of the decoding mechanism, the aminoacyl (A), peptidyl (P), and exit (E) tRNA-binding sites in the mtLSU are modified by the loss of some ribosomal elements [15, 38]. In this way, they have become more versatile to adapt to the atypical elbow region of the mitochondrial tRNAs due to varying deletions in the D and/or T-loops [42]. The mitoribosome A-site misses bL25 and a portion of *16S rRNA* h38, the P-site lacks uL5 and h84 (which stabilize its elbow region in the bacterial ribosome), and the E-site lacks h76 and h77 of the L1 stalk (which stabilize the tRNAs elbows) [15, 38]. The mammalian ribosome has adopted a unique structure, the so-called P-site finger, formed by mL40 and mL48 [16] to compensate for the missing mt-tRNA-binding sites and stabilize the A-site and P-site tRNA elbows (Fig. 2d).

Furthermore, the mammalian mtLSU CP lacks a 5S rRNA but is characterized by the presence of a structural $tRNA^{Val}$ in the human mitoribosome [16] and $tRNA^{Phe}$ in the porcine mitoribosome [12]. Analyses of additional species have shown that each mammal favors one of these mt-tRNA species in all tissue types [43]. However, the human mitoribosome shows a high degree of plasticity, being able to incorporate mt-$tRNA^{Phe}$ when mt-$tRNA^{Val}$ is not available [43].

Finally, the polypeptide exit tunnel is adapted to the transit and delivery of hydrophobic nascent polypeptides [13, 15]. The tunnel exit site is formed by uL23m, uL29m, uL22m, uL24m, and bL17m, proteins conserved in bacteria that form a ring around the exit site. Notably, the mitochondrion-specific mL45 surrounds the exit tunnel and tethers the mitoribosome to the inner mitochondrial membrane [13, 15]. Membrane anchoring aligns the polypeptide delivery site with the OXA1L translocon to enable cotranslational membrane insertion of newly-synthesized proteins [6, 44].

3 Mitoribosome Assembly Pathway and Factors Involved

3.1 General Concepts

Mitoribosome biogenesis follows a maturation pathway involving, for each subunit, the processing, maturation, and folding of rRNAs [45–48] coupled with the cooperative incorporation to the rRNA of mitoribosome protein sets and preassembled modules, forming structural clusters [4, 5, 9, 11, 49]. Biochemical and genetic studies have provided clues about the mitoribosome assembly process as it occurs in mammalian and yeast cells [4, 5]. Structural insights into mitoribosomal assembly have been provided by cryo-EM analyses of mitoribosomes from human HEK293T cells [17] and the human parasite *Trypanosoma brucei* [11, 49] in native states of assembly. A comparison of the current mammalian mitoribosome assembly line with the assembly pathways described for ribosomes from bacteria [50, 51] and mitochondria from *S. cerevisiae* [4] and *T. brucei* [9, 11, 49] can be found elsewhere [29, 33].

The biogenetic process requires numerous trans-acting factors (Tables 3, 4, and 5), including RNA-modifying enzymes, GTPases, RNA helicases, or phosphatases, some of which are conserved in bacterial systems, acting in each step of the process [29]. In humans, several of these factors were recognized as disease-causative in patients suffering from mitochondrial disorders [29]. This section will review our current knowledge of mitoribosome assembly as it occurs in mammals.

Table 3
Mammalian mitoribosome assembly factors: rRNA processing

Assembly factors					
Mammals	**Yeast**	**Bacteria**	**Protein class**	**Role in mitoribosome assembly**	**Ref.**
12S and 16S rRNA processing					
RNase P	RNase P	RNase P	Endoribonuclease	Catalyzes tRNA-5′ cleavage, required to liberate 12S and 16S rRNAs and $tRNA^{Val}$ from RNA precursor	[46]
ELAC2	Trz1	RNase Z	Endoribonuclease	Catalyzes tRNA-3′ cleavage, required to liberate 12S and 16S rRNAs and $tRNA^{Val}$ from RNA precursor	[84]

Table 4
Mammalian mitoribosome assembly factors: mtSSU biogenesis

Assembly factors					
Mammals	**Yeast**	**Bacteria**	**Protein Class**	**Role in Mitoribosome assembly**	**Ref.**
12S rRNA modification					
TFB1M	–	KsgA	Methyltransferase	Catalyzes $m^6{}_2A$ 936/937 in the 12S rRNA h45	[105, 106]
TRMT2B	Trm2	TrmA	Methyltransferase	Catalyzes m^5U 429 in the 12S rRNA h	[107, 108]
NSUN4	–	RsmB	Methyltransferase	Catalyzes m^5C 842 in the 12S rRNA h44	[64]
METTL15	–	RsmH	Methyltransferase	Catalyzes m^4C 839 in the 12S rRNA h44	[93]
METTL17	–	–	Methyltransferase	Catalyzes m^4C840 and regulates m^5C842 in 12S rRNA h44	[109]
mtSSU assembly					
ERAL1	–	Era	GTPase	It binds the hairpin in 12S rRNA h45, where two conserved adenines undergo dimethylation by TFB1M Its absence of ERAL1 or its accumulation result in decreased 12S rRNA levels and mtSSU assembly defects	[63, 86, 88]
MTG3 (NOA1)	Mtg3	YqeH	GTPase	Involved in undefined step of 12S rRNA maturation It may act concurrently with ERAL1	[90, 91, 110]
RCC1L-V3 (WBSCR16-V3)			GEF (Guanine exchange factor)	It could target ERAL1 or MTG3	[92]
YBEY	–	YbeY	Metallo-endoribonuclease	Supports the incorporation of uS11m into the head region of the maturating mtSSU	[89]
RBFA	–	RbfA	Ribosome binding and rRNA processing	Maintains the 12S rRNA h45 partially unfolded to allow binding of methyltransferase TFB1M while the mRNA channel is blocked Remains bound to the maturing mtSSU until it is displaced by mS37	[18]

(continued)

Table 4 (continued)

Assembly factors					
Mammals	**Yeast**	**Bacteria**	**Protein Class**	**Role in Mitoribosome assembly**	**Ref.**
mtIF3	Aim23	IF3	Mitochondrial translation initiation factor	Occupies the subunit interface during the late stages of mtSSU assembly Interacts with mS37 on the last step of mtSSU assembly	[18]
GRSF1	–	–	RNA binding	Interacts with the mtSSU and a 22S subassembly accumulates when GRSF1 is absent	[55]

3.2 Methods to Study Mitoribosome Biogenesis

Several approaches have been used to advance our understanding of the mitoribosome assembly pathway in mammalian cells.

3.2.1 Pulse-Chase SILAC

The global measurement of assembly and turnover of protein-containing complexes within cells has advanced with the development of stable isotope labeling with amino acids in cell culture (SILAC). SILAC allows the incorporation of "light" (^{12}C or ^{14}N) amino acids or "heavy" (^{13}C or ^{15}N) amino acids into cells or organisms and the quantitation of proteins and peptides containing these amino acid tags using mass spectrometry. Use of these labels in pulse or pulse-chase scenarios has enabled measurements of dynamic processes such as human mitoribosome assembly [52].

A set of SILAC pulse-labeling experiments in human HeLa cells determined the rates of mitochondrial import of MRPs and their assembly into intact 55S mitoribosomes. These experiments provide a basis for distinguishing MRPs that bind at early versus late stages in the assembly pathway, generating a useful—albeit low-resolution—working model for mitoribosome assembly [5].

The study showed that, for each mitoribosomal subunit, the protein components are synthesized in excess and imported into mitochondria, where their stoichiometric accumulation is regulated by degradation of the nonassembled free protein fractions [5]. Furthermore, the data suggested that complete human mitoribosome assembly requires ~2 to 3 h, indicating that mitoribosome biogenesis is a slow process. This contrasts with the bacterial ribosome assembly process, which is fast, requiring ~2 min for production of a single ribosome [36, 53], and efficient, with the vast majority of assembly events resulting in mature, translationally active complexes [54]. The nuclear genetic origin of the human MRP components and their required synthesis in cytosolic ribosomes and import into mitochondria in excess may explain their slow assembly.

Table 5
Mammalian mitoribosome assembly factors: mtLSU biogenesis

Assembly factors					
Mammals	**Yeast**	**Bacteria**	**Protein class**	**Role in mitoribosome assembly**	**Ref.**
16S rRNA modification					
MRM1	Mrm1	RmlB	Methyltransferase	Methylates Gm 1145 in the 16S rRNA h81	[47]
MRM2	Mrm2	RrmJ	Methyltransferase	Methylates Um 1369 in the 16S rRNA h92 (A loop)	[47, 48, 78]
MRM3	–	–	Methyltransferase	Methylates Gm 1370 in the 16S rRNA h92 (A loop)	[47, 48]
TRMT61B	Trm61	TrmI	Methyltransferase	Methylates m^1A 947 in the 16S rRNA h71	[111]
RPUSD4	Pus5	RluA	Pseudouridine synthase	Pseudouridylates Ψ 1397 in the 16S rRNA h90 Part of pseudouridine synthase module that binds the 16S rRNA	[45, 101, 112]
mtLSU assembly					
DDX28	Mrh4	–	DEAD-box RNA helicase	Interacts with the 16S during the mid-late stages of mtLSU assembly Binds to the mtLSU CP and stabilizes it with 16S rRNA h88 in an immature conformation	[19, 57, 58]
DHX30	–	–	DEAH-box RNA helicase	Required for mtLSU maturation	[57]
MRM3	–	–	Methyltransferase	Interacts with DDX28 on an mtLSU intermediate in which an MRM3 homodimer stabilizes a conformation of 16S rRNA h90–93 in domain V	[19]
GTPBP5 (MTG2)	Mtg2	Obg (CgtA)	GTPase	Binds to the 16S rRNA and is involved in the PTC maturation process together with NSUN4 Required for efficient MRM2-dependent methylation of the 16S rRNA Subunit antiassociation role	[20–22, 98]
GTPBP6	–	HflX	GTPase	Associates with an mtLSU intermediate and facilitates P-loop and PTC folding Triggers the release of GTPBP5 and MTERF4-NSUN4 from late mtLSU assembly intermediate Subunit antiassociation role Also acts as a ribosome recycling factor	[21]

(continued)

Table 5 (continued)

Assembly factors					
Mammals	**Yeast**	**Bacteria**	**Protein class**	**Role in mitoribosome assembly**	**Ref.**
GTPBP7 (MTG1)	Mtg1	RbgA	GTPase	Binds to the 16S rRNA and is involved in the maturation process of the PTC by promoting 16S rRNA h89-h93 folding Verifies the MRM2-catalyzed 2′-O-methylation of U1369 (in 16S rRNA h92 (A loop) Interacts with mS27, a putative GEF, and might be involved in intersubunit bridge formation	[20, 23, 97]
GTPBP8	Mrx8	RbgB	GTPase	Associates with mitochondrial ribosomes	[28, 58]
GTPBP10	–	Obg	GTPase	Interacts with 16S rRNA between states containing DDX28 or NSUN4/mTERF4 Stabilizes helix 92 in the mtLSU domain V Subunit antiassociation role	[19, 28]
MALSU1 (C7orf30)	–	–	DUF143 domain-containing protein	Binds to uL14m, fuels late stages of mtLSU assembly, and prevents premature subunit association Forms a module with L0R0F8 and mtACP	[17, 94]
LOR8F8 (AltMid51, MIEF1) mtACP	– Acp	– Acp	LYR-motif-containing protein Acyl carrier protein	A MALSU1–mtACP–LOR8F8 complex associate to late mtLSU assembly intermediates and prevent premature subunit association	[17, 113, 114]
mTERF3	–	–	Mitochondrial transcription termination factor family	Required for 16S rRNA stability and mtLSU assembly	[115]
RCC1L-V1 (WBSCR16-V1)	–	–	GEF (guanine exchange factor)	It could target GTPBP10 Part of pseudouridine synthase module that binds the 16S rRNA	[45, 92, 112]
FASTKD2	–	–	FAST Kinase-domain-containing protein	Part of pseudouridine synthase module that binds the 16S rRNA Controls 16S mt-rRNA abundance	[57, 96]

(continued)

Table 5 (continued)

Assembly factors					
Mammals	**Yeast**	**Bacteria**	**Protein class**	**Role in mitoribosome assembly**	**Ref.**
NGRN	–	–	Lacks common motifs		[45, 112]
RPUSD3		RluA	Pseudouridine synthase		
TRUB2	Pus4?	TruB			
mTERF4	–	–	Mitochondrial transcription termination factor family	Forms a complex with NSUN4 that associates with an mtLSU intermediate and facilitates PTC folding Stabilizes the maturation state of the domain V of the 16S rRNA while keeping domain IV helices 68–71 in an immature conformation, thus acting as an antiassociation factor	[17, 19, 21, 22, 116]
NSUN4	–	–	Methyltransferase	Forms a complex with mTERF4 that associates with an mtLSU intermediate and facilitates PTC folding N-terminal tail facilitates PTC loop folding together with GTPBP5	[17, 21, 22]
mtEFTu	EF-Tu	EF-Tu	Mitochondrial translation elongation factor	Interacts with and may accommodate GTPBP5 transiently, in a late mtLSU intermediate	[20]
p32	p32	–	Other	Affects 55S accumulation	[117, 118]
MPV17L2	Sym1	MPV17 Family	Other	Unclear role. In its absence, the mtLSU is unstable and the mtSSU is trapped in the nucleoid	[66]

3.2.2 Screens for Mitochondrial RNA-Binding and Mitoribosome-Interacting Proteins

To identify potential mitoribosome assembly factors, some researchers have screened for mitochondrial RNA-binding proteins [55, 56], and others performed proteomics analyses of the mitoribosome interactome [28, 45, 57, 58]. Other research groups have taken advantage of the better characterized bacterial ribosome assembly pathway [59, 60] and screens in the amenable *Saccharomyces cerevisiae* yeast model [27, 61, 62] to identify factors conserved in mammals. These assembly factors have been later characterized in human cultured cells or mouse models [46, 63, 64].

As an example, the DEAD-box RNA helicase Mrh4 was first identified in yeast as a 21S rRNA-binding mtLSU late-stage assembly factor required for the incorporation of subunits bL33m and uL16m [65]. The mammalian homolog of Mrh4 is DDX28, which was subsequently characterized in human cultured cells, as a 16S rRNA-interacting protein acting at the mid-late stages of mtLSU assembly [57, 58]. Further proteomics studies of the DDX28 protein interactome identified many ribosome assembly factors, including helicases, multiple GTPases, and RNA modifying enzymes (Table 5) [28].

3.2.3 Screens for Mutations in Ribosomal Components and Assembly Factors in Patients Suffering from Mitochondrial Disorders Associated with Multienzymatic OXPHOS Deficiencies

Studies in patients with mitochondrial translation efficiency disorders have identified new assembly factors and the relevance of MRPs for mitoribosome assembly across tissues [66–69]. These are multisystemic mitochondrial diseases such as Leigh syndrome, cardio- and encephalo-myopathies, liver disease, and Perrault syndrome [29].

Mutations in mitoribosome RNAs or proteins are a frequent cause of disorders owing to mitochondrial protein synthesis deficiencies ([70–74] and reviewed in [29, 67, 75]). Presently, disease-causing mutations have been found in 13 mtSSU proteins (bS1m, uS2m, bS6m, uS7m, uS9m, uS11m, uS14m, bS16m, mS22, mS23, mS25, mS34, and mS39) and 4 mtLSU proteins (uL3m, bL12m, uL24m, and mL44). Most of these proteins incorporate at the early stages of mitoribosome assembly, which may disrupt the mitoribosome structure and subsequent assembly steps [74]. Furthermore, disease-causing mutations have so far been found in four rRNA processing, stabilization, and maturation factors (ELAC2, MRPP3, FASTKD2, MRM2, and TFB1M), five mitoribosome assembly chaperones (ERAL1, p32, GTPBP5, GTPBP10, and DHX30), and one mitochondrial protease (CLPP) [29].

An example of the value of these discoveries to the understanding of the human mitoribosome assembly process is the case of mutations in mS22. They result in cardiomyopathy, renal tubulopathy, antenatal skin edema, and muscle hypotonia [76], associated with multiple OXPHOS enzymatic deficiencies and decreased *12S rRNA* levels in muscle mitochondria [76, 77]. The mitochondrion-specific mtSSU protein mS22 is a unique early-mid stage assembly protein without contact with the 12S rRNA but interactions with bS16m and mS40 [16]. In patient fibroblasts, the steady-state levels of uS11m and bS16m were drastically decreased, whereas uS2m was found stable, as for fibroblasts depleted of bS16m [77], therefore supporting the interactions between these proteins and their hierarchical order of assembly.

3.2.4 Determination of Cryo-EM Structures of Stable Assembly Intermediates

Cryo-EM studies have provided insights into the timing of rRNA folding and protein incorporation during the final steps of ribosomal maturation when stable assembly intermediates accumulate in sufficient amounts to allow their characterization.

So far, only one report has documented structures of late-stage assembly intermediates of the human mtLSU isolated from a native pool within a nonmodified wild-type human cell line (HEK293T) [17]. This study described two intermediates in which subunit uL14m had an adjacent density that also formed electrostatic interactions with the sarcin-ricin stem-loop (SRL; h95) of the 16S rRNA. The interacting portion of the density was identified as the mitochondrial assembly of ribosomal large subunit 1 (MALSU1), which belongs to the ribosome silencing factor (RsfS) family (Table 5). Notably, the density also contained two previously uncharacterized MALSU1 interactors, the LYR-motif-containing protein L0R0F8 (also known as AltMid51 or MIEF1) and the acyl carrier protein mtACP [17]. The MALSU1–mtACP–LOR8F8 module was proposed to prevent premature subunit association.

Other studies applied a genetic perturbation to wild-type human cells to enrich mitoribosome subassemblies. For example, depletion of TRMT2B, a methyltransferase that catalyzes the formation of 5-methyluridine m^5U429 in the 12S rRNA, produced a viable cell line with stable rRNA that allowed the purification of intermediate pre-SSU particles in sequential states of assembly and translation initiation [18]. The mtSSU stable assembly intermediates involve five factors: the ribosome binding and rRNA processing factor A (RBFA), the methyltransferases TFB1M and METTL15, the initiation factor mtIF3, and the mtSSU protein mS37 [18]. The study portrays the role of these proteins in rRNA folding, the recruitment of a translation factor to support ribosome assembly, and the ranking of mS37 as the last MRP incorporated to yield a mature mtSSU [18].

Compared to mtSSU assembly, the process of mtLSU maturation has been more extensively characterized. Between 2021 and 2022, as many as six different research groups published cryo-EM studies of the mtLSU mid and late assembly intermediates [19–23, 78]. In three studies, mtLSU subassemblies were characterized from HEK293T cells knockout (KO) for the methyltransferase MRM2 [78], or the GTPases GTPBP5 [20] or GTPBP6 [21]. Other studies used HEK293T cells expressing FLAG-tagged versions of GTPBP7 [22], MALSU1 [19], or GTPBP10 [19] to perform affinity purification of these mitoribosome biogenesis factors followed by cryo-EM. In another study, researchers included the nonhydrolyzable GTP analog β,γ-methyleneguanosine 5′--triphosphate (GMPPCP) during mitoribosome purification to trap GTPBP7 and other GTPases that participate in mitoribosome assembly and function [23]. Notably, these studies revealed the molecular function and hierarchical order of action for all the targeted assembly factors and others, such as the helicase DDX28 or the mitochondrial transcription termination factor family protein MTERF4 in complex with the dual function 12S rRNA methyltransferase NSUN4. They also disclosed a role for the mitochondrial translation elongation factor mtEFTu in mtLSU assembly, which, together with the role of mtIF3 in mtSSU assembly, consolidates a theme on translation factors being recruited to facilitate or probe late stages of mitoribosome maturation. The results of all these studies have been integrated into a stepwise model of mtLSU biogenesis [33] that we will summarize in the next section.

3.3 Cotranscriptional Mitoribosome Assembly

In bacterial and eukaryotic systems, the processing and modification of rRNAs and the association of ribosomal proteins and assembly factors with nascent pre-rRNA occurs cotranscriptionally, which facilitates guiding of ribosome assembly by RNA folding (reviewed in [54, 79]).

In mammalian mitochondria, the process also starts cotranscriptionally which serves to coordinate the biogenesis of the two mitoribosome subunits. Notably, in mammalian mtDNAs, the genes for rRNAs locate in the so-called heavy (H) strand and are flanked by tRNAs in a sequence mt-tRNA^{Phe}/mtSSU rRNA/mt-tRNA^{Val}/mtLSU rRNA. Either tRNA^{Val} or tRNA^{Phe} has been found to be structural components of the mtLSU in mammals [43] and, therefore, their genomic location reminds that of a bacterial operon [80]. Mitochondrial transcription is polycistronic. Transcription from a single H-strand promoter generates a transcript that spans almost the entire genome [81–83] and is subsequently processed. The mitochondrial ribonuclease P (RNase P) and the mitochondrial RNase Z, known as ELAC2, catalyze tRNA-5′ cleavage and tRNA-3′ cleavage, respectively [46, 84], thereby liberating the rRNAs and most mRNAs [85].

Mitoribosome assembly has been proposed to start with a subset of 27 mtLSU proteins, forming a subcomplex on an unprocessed RNA containing the *16S rRNA*. This complex formation is required for precursor RNA processing by RNase P and ELAC2, resulting in liberation of the *12S rRNA* and further mtSSU assembly [46]. The identity of these mtLSU proteins awaits confirmation since they have been only reported in a mouse KO for the RNase P MRPP3 component [46]. In support of this model, silencing of the mtLSU assembly factor MPV17L2 causes not only a decrease in mtLSU without accumulation of subassemblies, but also a severe mtSSU depletion and accumulation of mtSSU proteins in aggregated nucleoids [66]. MPV17L2 could be required for early mtLSU assembly steps needed to facilitate rRNA precursor processing and release from the mtDNA nucleoids, where transcription occurs, before proceeding with its maturation within the mitochondrial RNA granule compartment [5, 66].

3.4 mtSSU Assembly

The current knowledge regarding mtSSU protein incorporation during subunit biogenesis and assembly factors involved is schematically summarized in Fig. 3.

3.4.1 Early and Intermediate Stages of mtSSU Assembly

According to the SILAC proteomics study, mtSSU protein assembly proceeds by the early incorporation of two large protein modules [5]. The module formed by uS5m, bS16m, mS22, mS27, mS34, and mS40 binds to the mtSSU lower body/foot, contacting the 5′ and 3′ rRNA domains. The module formed by uS7m, uS9m, mS29, mS31, mS35, and mS39 localizes to the head, extending through the major 3′ domain in the 12S rRNA. A smaller set of proteins (mS23, uS2m, and bS1m) interacts with both modules. To complete the early-mid stages, another three proteins (uS11m, uS17m, and uS12m) bind independently [5]. These early proteins bind to the mtSSU outer surface and are essential for the recruitment of the late assembly proteins. A discrepancy with the bacterial assembly pathway is the incorporation of proteins binding to the major 3′ domain of 12S rRNA in the head region (uS7m and uS9m), which in bacteria occurs at the mid-late stages [50].

Several assembly factors sustain early mtSSU biogenesis (Fig. 3). The human GTPase ERAL1 (Era G-protein-like 1) acts as an RNA chaperone to stabilize the 12S rRNA before maturation and assembly. It coimmunoprecipitates with early assembly MRPs mS22 and mS31 [86] and binds to h45 at the 3′ terminus of the 12S mt-rRNA [63], which contains two highly conserved adenines that undergo subsequent methylation catalyzed by TFB1M (mitochondrial transcription factor B) [87]. TFB1M remains bound to the maturing mtSSU until the late stages of assembly [18]. The ATP-dependent protease CLPP tightly controls ERAL1 levels to avoid excess accumulation that blocks mitoribosome formation [88]. ERAL1 also interacts with the endoribonuclease YBEY, although probably due to the GTP-binding dependence of this interaction, their association is labile [89]. In contrast, YBEY stably binds the multifunctional protein p32 and is required for uS11m incorporation into the assembly pathway [89].

Several other factors are required for fueling the early-mid stages of mtSSU biogenesis, although their functions remain ill-defined. The GTPase MTG3 also participates in mtSSU assembly [90, 91], perhaps at the early stages as its bacterial counterpart [91]. The putative GDP/GTP exchange factor RCC1L (regulator of chromatin condensation 1-like, also known as WBSCR16) interacts with the mitoribosome. Two alternative splicing isoforms, $RCC1L^{V1}$ and $RCC1L^{V3}$ are associated with the mtLSU and mtSSU, respectively. ERAL1, MTG3, and mS29 could be the targets of $RCC1L^{V3}$ [92]. The G-rich sequence-binding factor 1 (GRSF1) stabilizes the *12S rRNA* [55, 56] and a small fraction cosediments with the mtSSU [58]. *GRSF1* silencing causes the formation of a mtSSU subassembly that sediments as a 22S subunit

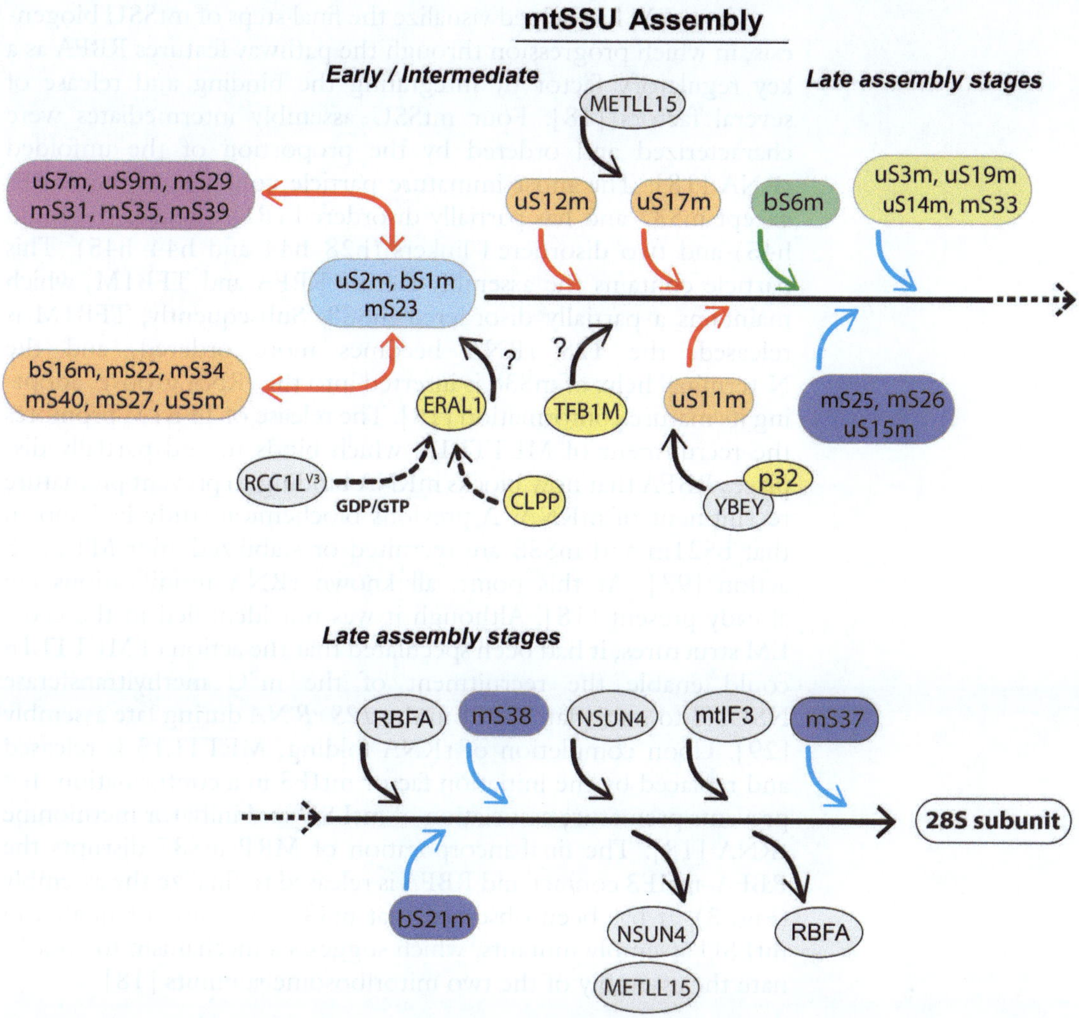

Fig. 3 Mitoribosome SSU assembly. Model of human 28S mtLSU biogenesis depicting a hierarchical and module-based protein assembly pathway, including known assembly factors at their approximate stage of incorporation. All modules are color-coded. The assembly process is divided in three stages: early (red arrows), intermediate (green arrows) late (blue arrows) (see explanation in the text) Boxes, highlighted with the same color used in the spiral assembly, represent different protein clusters at different assembly stages

but contains all the canonical 28S mtSSU proteins [55], which suggests a direct role for GRSF1 in mtSSU assembly.

3.4.2 Late Stages of mtSSU Assembly

Most late-binding proteins localize to the interface with the mtLSU [5]. Late-binding proteins include some that incorporate as single units (e.g., mS38) and two clusters. One binds in the head region (uS14m, uS10m, uS3m, and mS33) in association with the early uS7m-mS29 group, and the second (uS15m, mS25, and mS26) near the early bS16m-mS22 cluster. The late assembly of uS15m is intriguing since uS15 assembles early with uS15 in bacteria [50].

Cryo-EM has helped visualize the final steps of mtSSU biogenesis, in which progression through the pathway features RBFA as a key regulatory factor by integrating the binding and release of several factors [18]. Four mtSSU assembly intermediates were characterized and ordered by the proportion of the unfolded rRNA [18]. The most immature particle contains all the MRPs except mS37 and has partially disordered rRNA helices (h44 and h45) and two disordered linkers (h28–h44 and h44–h45). This particle contains the assembly factors RBFA and TFB1M, which maintains a partially disordered mS38. Subsequently, TFB1M is released, the 12S rRNA becomes more ordered, and the N-terminal helix of mS38 is inserted into the rRNA groove, adopting its mature conformation [18]. The release of TFB1M promotes the recruitment of METTL15, which binds to and partially displaces RBFA that now blocks mRNA binding to prevent premature recruitment of mRNA. A previous biochemical study had shown that bS21m and mS38 are recruited or stabilized after MTTL15 action [93]. At this point, all known rRNA modifications are already present [18]. Although it was not identified in the cryo-EM structures, it had been speculated that the action of METTL15 could enable the recruitment of the m^5C methyltransferase NSUN4 to methylate C911 in the *12S rRNA* during late assembly [29]. Upon completion of rRNA folding, METTL15 is released and replaced by the initiation factor mtIF3 in a conformation that prevents premature association of mtLSU and initiator methionine tRNA [18]. The final incorporation of MRP mS37 disrupts the RBFA–mtIF3 contact and RBFA is released to finalize the assembly (Fig. 3). It has been observed that mS37 levels are attenuated in mtLSU assembly mutants, which suggests a mechanism to coordinate the assembly of the two mitoribosome subunits [18].

3.5 mtLSU Assembly

The current knowledge regarding mtSSU protein incorporation during subunit biogenesis and assembly factors involved is schematically summarized in Fig. 4.

3.5.1 Early and Intermediate Stages of mtLSU Assembly

Based on a SILAC proteomics study, mtLSU assembly (Fig. 4) involves the incorporation of individual proteins and modules during three defined stages, early, intermediate, and late [5].

During the early phase, three large protein clusters, including 24 proteins, were identified to assemble with similar kinetics, suggesting coordinated binding. These proteins mainly localize in a region encompassing the 5′ rRNA domain. A first cluster, formed by rRNA-binding MRPs uL3m and bL19m, and other proteins (uL14m, bL17m, uL22m, and bL32m) anchors mL39, and then mL45. The protein mL45 could tether the mtLSU to the inner membrane during subsequent assembly steps. The assembly of this module involves the action of the assembly factor MALSU1 assisting the insertion of uL14m [94, 95]. The DEAD-box RNA

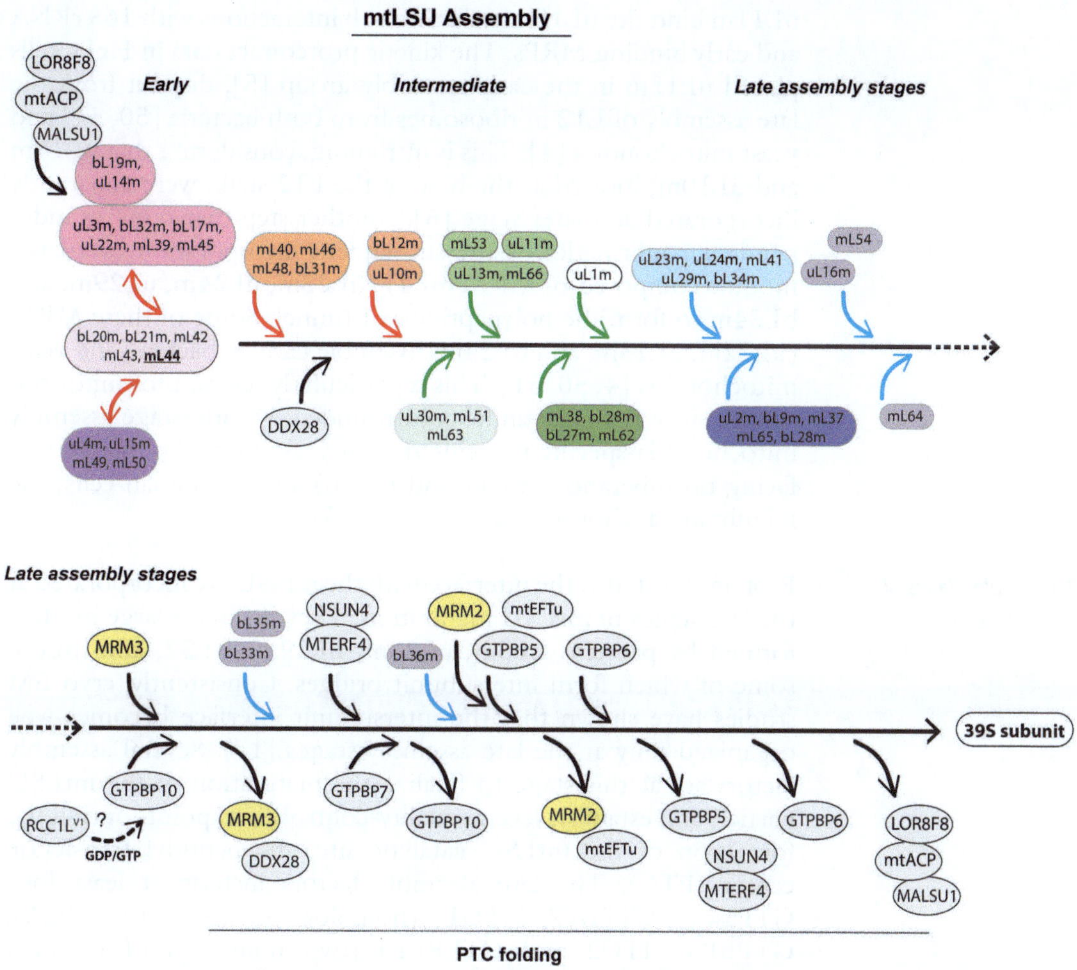

Fig. 4 Mitoribosome LSU assembly. Model of human 39S mtLSU assembly pathway, including known assembly factors at their approximate stage of incorporation. All modules are color-coded. Boxes represent different protein clusters at different assembly stages: early (red arrows), intermediate (green arrows), and late (blue arrows) (see explanation in the text)

helicase DDX28 and the Fas-activated serine/threonine (FAST) kinase family protein FASTKD2 [57, 58, 96] bind to and stabilize the 16S rRNA, suggesting an early action, although their precise role is yet to be disclosed. A second early MRP module includes the mRNA-binding protein bL20m and bL21m, mL42, mL43, and mL44. A third module is formed by the mRNA-binding heterodimer uL4m–uL15m, which recruits mL49 and mL50. The fourth early-assembly module includes proteins mL40, mL46, and mL48, which associate with the tRNA^{Val} in the mtLSU CP [5].

At the intermediate stage, the fourth early module facilitates the incorporation of mL38, uL18m, and bL27m, the second group of tRNA^{Val} surrounding proteins. It is currently unknown how the tRNA^{Val} is recruited. At this stage, the dimers uL13m-mL66 and uL11m bind the uL10m stalk through interactions with 16S rRNA and early binding MRPs. The kinetic proteomics data in HeLa cells placed uL12m in the early-assembly group [5], distinct from the late assembly of L12 in ribosomes from both bacteria [50, 51] and yeast mitochondria [4]. This is intriguing, considering that uL11m and uL10m, located at the base of the L12 stalk, were apparently incorporated at a later stage [5]. Another step that requires independent confirmation is the subsequent incorporation of a large module composed of MRPs mL41, uL23m, uL24m, uL29m, and bL34m to form the polypeptide exit tunnel. Some of these MRPs (uL22m, uL23m, and uL24m) assemble early in bacteria and yeast mitochondria [4, 50, 51]. This is particularly relevant for mitoribosomes, since the exit tunnel is surrounded by early-stage assembly mitochondria-specific proteins that in yeast form the membrane-facing protuberance (mL44 and mL50) and, in human cells, the membrane-anchoring site (mL45) [4, 5].

3.5.2 Late Stages of mtLSU Assembly

Proteins located at the interface with the mtSSU are incorporated at the late stages of mtLSU maturation. They include a large module formed by proteins uL2m, uL28m, uL29m, mL37, and mL65, some of which form intersubunit bridges. Consistently, cryo-EM studies have shown that the intersubunit interface becomes well organized only at the late assembly stages [17]. Several assembly factors act at this stage to finalize the maturation of the mtLSU particle and establish several quality-control checkpoints during the formation of the mtLSU catalytic site, the peptidyl transferase center (PTC). The late assembly factors include at least four GTPases: GTPBP7/MTG1 (homolog of bacterial RbgA), GTPBP5/MTG2 and GTPBP10 (two homologs of bacterial Obg), and GTPBP6 (homolog of bacterial HflX) (reviewed in [30]). Proteomics, biochemical, and structural studies have suggested sequential recruitment of assembly factors (Fig. 4).

The earliest human mtLSU intermediate analyzed by cryo-EM lacks bL33m, bL35m, and bL36m and displays an immature CP and delocalized 16S rRNA domains IV and V. It is bound by the assembly factors DDX28, GTPBP10, the 2′-O-methyltransferase MRM3, and the MALSU1–mtACP–LOR8F8 module [17, 19, 94]. The MALSU1 module remains bound to the early-assembled uL14m, bL19m, and the 16S rRNA sarcin-ricin loop (SRL, h95) until mtLSU assembly is finalized [17]. In this way, it probably blocks the formation of intersubunit bridge B8 and impedes premature subunit joining [17]. The DEAD-box RNA helicase DDX28 binds early to the 16S rRNA, but it also remains bound to the growing mtLSU particle until the late maturation stages [58]. Cryo-EM showed that DDX28 binds to the 16S rRNA h88 and stabilizes the CP in an immature conformation. GTPBP10 interacts with the SRL and the L12 stalk base and shifts the 16S rRNA h89 to facilitate MRM3 reaching h92 of domain V (A loop) to methylate G1370 [19]. The absence of GTPBP10 in a KO cell line prevents the incorporation of bL33m, bL34m, bL35m, and bL36m [28] in agreement with the cryo-EM studies. Notably, GTPBP10 may be the target of the GTP/GDP exchange factor RCC1L^{V1} [92]. Although it is believed that premature binding of translation components is prevented during ribosome assembly, this pre-mtLSU maturation intermediate featured a deacylated tRNA in the ribosomal E-site [19]. Because the tRNA is released before mtLSU maturation is completed, it could either stabilize assembly intermediates or monitor the CP folding state even before its full maturation [19].

Biochemical studies suggest that the next assembly factor being recruited should be GTPBP7. GTPBP7 interacts with domain VI helices in the *16S rRNA* and with bL19m, which induces a conformational change and remodeling of the bL19m-containing domain, thereby facilitating the incorporation of the late assembly proteins bL36m and bL35m to complete the formation of mature mtLSU [97]. The interaction of GTPBP7 with the mitoribosome must be labile, because it could be captured in an assembly intermediate by cryo-EM only in the presence of GMPPCP (nonhydrolyzable GTP analog) [23]. This intermediate contains GTPBP7 locked in a prehydrolysis conformation, bound in a heterodimer to the MTERF4-NSUN4 complex. In this structure, GTPBP7 interacts with h92 and directly contacts U3039 in the 16S rRNA A loop to monitor its MRM2-dependent methylation status. The recruitment of MRM2 forces GTPBP7 to change its conformation and indicates that GTPBP7 may bind to different locations on the mtLSU in different states [20, 23]. Although the functional relevance of this observation is not fully clear, biochemical studies have shown that GTPBP7 remains bound to the mtLSU until maturation is completed [97]. Only when subunit joining is about to occur does GTPBP7 interact with the mtSSU protein mS27, a putative guanosine triphosphate exchange factor (GEF) that catalyzes fast GDP-GTP exchange to enable the release of GTPBP7/MTG1 from the ribosome and facilitate the formation of the mB6 intersubunit bridge between bL19m and mS27 [97].

The subsequent recruitment of GTPBP5 could be accommodated by mtEFTu, similarly, as it delivers aminoacyl tRNAs to the A site [20]. The presence of GTPBP5 leads to an assembly state in which the 16S rRNA A loop is displaced from the active site of MRM2 [20, 98], the methyltransferase that catalyzes the 2′-O-methyl modification at position U1369, an essential PTC component [48, 99]. Cryo-EM captured a particle containing GTPBP7, GTPBP5, and MRM2 [23], suggesting that they can act simultaneously, contrary to their bacterial homologs [100].

Following *16S rRNA* maturation by MRM2 and MRM3-catalyzed methylation, and probably also its pseudouridylation by the pseudouridine synthase RPUSD4 [45, 101], the modifying enzymes dissociate, and bL36m is the last MRP recruited to the maturing mtLSU [98]. GTPBP5 and MTERF4-NSUN4 remain bound to this intermediate, in which NSUN4 and GTPBP5 cooperate to facilitate the folding of the PTC [21, 22]. Although it was not detected in the cryo-EM structures, GTPBP7 is also expected to be present in this intermediate.

Next, the MTERF4-NSUN4 and GTPBP5 assembly factors are released, and GTPBP6, homologous to the bacterial ribosome splitting factor HflX, is recruited to the same positions where GTPBP5 binds in the mtLSU to further PTC folding [21, 102]. The release of GTPBP6 leads to a final intermediate in which all elements are present but retains the MALSU1–mtACP–LOR8F8 module bound [17], acting as the last antiassociation complex that prevents premature subunit joining.

4 Conclusions and Perspectives

The field of mammalian mitoribosome biogenesis has rapidly expanded over the last decade due to technical advances in mass spectrometry, cryoelectron microscopy, and gene editing approaches to modify human cell lines. However, the substrates and precise mechanisms of action for the many auxiliary assembly factors remain, in most cases, poorly understood. This is particularly true for factors that act at the early or intermediate stages of assembly, at which preribosomal particles are unstable. New proteomics studies are identifying potentially novel assembly factors or proteins that interact with the maturing or fully assembled ribosome. For example, the mechanism of $tRNA^{Val\ or\ Phe}$ incorporation to the mtLSU CP or the iron-sulfur cluster chaperones that deliver the metal cofactors to the coordinating pairs of proteins in the two subunits remain to be identified. Also, little is known regarding how misassembled ribosomes are degraded. Furthermore, the cofactor insertion machinery and other components of the mitoribosome extended proteome may play essential roles in the integration of mitoribosome assembly with other mitochondrial and cellular processes such as mtDNA replication and transcription, Fe-S cluster biogenesis, and adaptation to changes in the energy status of the cell, nutrient availability, or stress. It will also assist with identifying targets to produce safer antibiotics that do not interfere with the mitoribosome and develop therapeutics to treat cancer [29, 103].

Acknowledgments

This research was supported by an NIGMS-MIRA award [R35GM118141 to A.B.] and an NICHD F30 predoctoral M. D./Ph.D. fellowship [F30HD107939 to J.C.M.].

References

1. Küntzel H, Noll H (1967) Mitochondrial and cytoplasmic polysomes from Neurospora crassa. Nature 215(5108):1340–1345. https://doi.org/10.1038/2151340a0
2. O'Brien TW, Kalf GF (1967) Ribosomes from rat liver mitochondria. I. Isolation procedure and contamination studies. J Biol Chem 242(9):2172–2179
3. O'Brien TW, Kalf GF (1967) Ribosomes from rat liver mitochondira. II. Partial characterization. J Biol Chem 242(9):2180–2185
4. Zeng R, Smith E, Barrientos A (2018) Yeast mitoribosome large subunit assembly proceeds by hierarchical incorporation of protein clusters and modules on the inner membrane. Cell Metab 27(3):645–656
5. Bogenhagen DF, Ostermeyer-Fay AG, Haley JD et al (2018) Kinetics and mechanism of mammalian mitochondrial ribosome assembly. Cell Rep 22(7):1935–1944
6. Amunts A, Brown A, Bai X et al (2014) Structure of the yeast mitochondrial large ribosomal subunit. Science 343:1485–1489
7. Desai N, Brown A, Amunts A et al (2017) The structure of the yeast mitochondrial ribosome. Science 355(6324):528–531
8. Lenarčič T, Niemann M, Ramrath DJF et al (2022) Mitoribosomal small subunit maturation involves formation of initiation-like complexes. Proc Natl Acad Sci U S A 119(3). https://doi.org/10.1073/pnas.2114710118
9. Tobiasson V, Gahura O, Aibara S et al (2021) Interconnected assembly factors regulate the biogenesis of mitoribosomal large subunit. EMBO J 40(6):e106292. https://doi.org/10.15252/embj.2020106292
10. Soufari H, Waltz F, Parrot C et al (2020) Structure of the mature kinetoplastids mitoribosome and insights into its large subunit biogenesis. Proc Natl Acad Sci U S A 117(47):29851–29861. https://doi.org/10.1073/pnas.2011301117
11. Saurer M, Ramrath DJF, Niemann M et al (2019) Mitoribosomal small subunit biogenesis in trypanosomes involves an extensive assembly machinery. Science 365(6458):1144–1149
12. Greber BJ, Bieri P, Leibundgut M et al (2015) Ribosome. The complete structure of the 55S mammalian mitochondrial ribosome. Science 348(6232):303–308
13. Greber BJ, Boehringer D, Leibundgut M et al (2014) The complete structure of the large subunit of the mammalian mitochondrial ribosome. Nature 515(7526):283–286
14. Amunts A, Brown A, Toots J et al (2015) The structure of the human mitochondrial ribosome. Science 348(6230):95–98
15. Brown A, Amunts A, Bai XC et al (2014) Structure of the large ribosomal subunit from human mitochondria. Science 346(6210):718–722
16. Aibara S, Singh V, Modelska A et al (2020) Structural basis of mitochondrial translation. Elife 9. https://doi.org/10.7554/eLife.58362
17. Brown A, Rathore S, Kimanius D et al (2017) Structures of the human mitochondrial ribosome in native states of assembly. Nat Struct Mol Biol 24(10):866–869
18. Itoh Y, Khawaja A, Laptev I et al (2022) Mechanism of mitoribosomal small subunit biogenesis and preinitiation. Nature 606(7914):603–608. https://doi.org/10.1038/s41586-022-04795-x
19. Cheng J, Berninghausen O, Beckmann R (2021) A distinct assembly pathway of the human 39S late pre-mitoribosome. Nat Commun 12(1):4544. https://doi.org/10.1038/s41467-021-24818-x
20. Cipullo M, Gesé GV, Khawaja A et al (2021) Structural basis for late maturation steps of the human mitoribosomal large subunit. Nat Commun 12(1):3673. https://doi.org/10.1038/s41467-021-23617-8
21. Hillen HS, Lavdovskaia E, Nadler F et al (2021) Structural basis of GTPase-mediated mitochondrial ribosome biogenesis and recycling. Nat Commun 12(1):3672. https://doi.org/10.1038/s41467-021-23702-y
22. Lenarčič T, Jaskolowski M, Leibundgut M et al (2021) Stepwise maturation of the peptidyl transferase region of human mitoribosomes. Nat Commun 12(1):3671. https://doi.org/10.1038/s41467-021-23811-8

23. Chandrasekaran V, Desai N, Burton NO et al (2021) Visualizing formation of the active site in the mitochondrial ribosome. Elife 10. https://doi.org/10.7554/eLife.68806
24. Smits P, Smeitink JA, van den Heuvel LP et al (2007) Reconstructing the evolution of the mitochondrial ribosomal proteome. Nucleic Acids Res 35(14):4686–4703. https://doi.org/10.1093/nar/gkm441
25. Petrov AS, Wood EC, Bernier CR et al (2019) Structural patching fosters divergence of mitochondrial ribosomes. Mol Biol Evol 36(2):207–219
26. van der Sluis EO, Bauerschmitt H, Becker T et al (2015) Parallel structural evolution of mitochondrial ribosomes and OXPHOS complexes. Genome Biol Evol 7(5):1235–1251. https://doi.org/10.1093/gbe/evv061
27. De Silva D, Fontanesi F, Barrientos A (2013) The DEAD-Box protein Mrh4 functions in the assembly of the mitochondrial large ribosomal subunit. Cell Metab 18:712–725
28. Maiti P, Kim HJ, Tu YT et al (2018) Human GTPBP10 is required for mitoribosome maturation. Nucleic Acids Res 13. https://doi.org/10.1093/nar/gky938
29. Ferrari A, Del'Olio S, Barrientos A (2021) The diseased mitoribosome. FEBS Lett 595(8):1025–1061. https://doi.org/10.1002/1873-3468.14024
30. Maiti P, Lavdovskaia E, Barrientos A et al (2020) Role of GTPases in driving mitoribosome assembly. Trends Cell Biol 3(4):284–297
31. Lopez Sanchez MIG, Krüger A, Shiriaev DI et al (2021) Human mitoribosome biogenesis and its emerging links to disease. Int J Mol Sci 22(8). https://doi.org/10.3390/ijms22083827
32. Kummer E, Ban N (2021) Mechanisms and regulation of protein synthesis in mitochondria. Nat Rev Mol Cell Biol 22(5):307–325. https://doi.org/10.1038/s41580-021-00332-2
33. Lavdovskaia E, Hillen HS, Richter-Dennerlein R (2022) Hierarchical folding of the catalytic core during mitochondrial ribosome biogenesis. Trends Cell Biol 32(3):182–185. https://doi.org/10.1016/j.tcb.2021.09.004
34. Kummer E, Leibundgut M, Rackham O et al (2018) Unique features of mammalian mitochondrial translation initiation revealed by cryo-EM. Nature 560(7717):263–267
35. Singh V, Itoh Y, Huynen MA et al (2022) Activation mechanism of mitochondrial translation by LRPPRC-SLIRP. bioRxiv:2022.2006.2020.496763. https://doi.org/10.1101/2022.06.20.496763
36. Chen Y, Hagopian K, Bibus D et al (2013) The influence of dietary lipid composition on liver mitochondria from mice following 1 month of calorie restriction. Biosci Rep 33(1):83–95. https://doi.org/10.1042/BSR20120060
37. Singh V, Itoh I, Del'Olio S et al (2023) The complete structure of human mitoribosome, roles of mito-specific protein elements, cofactors and rRNA modifications. Nat Commun In press
38. Greber BJ, Ban N (2016) Structure and function of the mitochondrial ribosome. Annu Rev Biochem 85:103–132
39. Itoh Y, Singh V, Khawaja A, Naschberger A, Nguyen MD, Rorbach J, Amunts A (2022) Structure of the mitoribosomal small subunit with streptomycin reveals Fe-S clusters and physiological molecules. eLife 11:e77460
40. Yusupova GZ, Yusupov MM, Cate JH et al (2001) The path of messenger RNA through the ribosome. Cell 106(2):233–241. https://doi.org/10.1016/s0092-8674(01)00435-4
41. Takyar S, Hickerson RP, Noller HF (2005) mRNA helicase activity of the ribosome. Cell 120:49–58
42. Helm M, Brulé H, Friede D et al (2000) Search for characteristic structural features of mammalian mitochondrial tRNAs. RNA 6(10):1356–1379. https://doi.org/10.1017/s1355838200001047
43. Rorbach J, Gao F, Powell CA et al (2016) Human mitochondrial ribosomes can switch their structural RNA composition. Proc Natl Acad Sci U S A 113(43):12198–12201. https://doi.org/10.1073/pnas.1609338113
44. Greber BJ, Boehringer D, Leitner A et al (2014) Architecture of the large subunit of the mammalian mitochondrial ribosome. Nature 505(7484):515–519. https://doi.org/10.1038/nature12890
45. Antonicka H, Choquet K, Lin ZY et al (2017) A pseudouridine synthase module is essential for mitochondrial protein synthesis and cell viability. EMBO Rep 18(1):28–38
46. Rackham O, Busch JD, Matic S et al (2016) Hierarchical RNA processing is required for mitochondrial ribosome assembly. Cell Rep 16(7):1874–1890
47. Lee KW, Bogenhagen DF (2014) Assignment of 2′-O-methyltransferases to modification sites on the mammalian mitochondrial large subunit 16 S ribosomal RNA (rRNA). J Biol Chem 289(36):24936–24942. https://doi.org/10.21074/jbc.C24114.581868. Epub 582014 Jul 581829
48. Rorbach J, Boesch P, Gammage PA et al (2014) MRM2 and MRM3 are involved in

biogenesis of the large subunit of the mitochondrial ribosome. Mol Biol Cell 9:01–0014
49. Jaskolowski M, Ramrath DJF, Bieri P et al (2020) Structural insights into the mechanism of mitoribosomal large subunit biogenesis. Mol Cell 79(4):629–644
50. Chen SS, Williamson JR (2013) Characterization of the ribosome biogenesis landscape in *E. coli* using quantitative mass spectrometry. J Mol Biol 425(4):767–779
51. Davis JH, Tan YZ, Carragher B et al (2016) Modular assembly of the bacterial large ribosomal subunit. Cell 167(6):1610–1622
52. Snider J, Wang D, Bogenhagen DF et al (2019) Pulse SILAC approaches to the measurement of cellular dynamics. Adv Exp Med Biol 1140:575–583. https://doi.org/10.1007/978-3-030-15950-4_34
53. Lindahl L (1975) Intermediates and time kinetics of the *in vivo* assembly of *Escherichia coli* ribosomes. J Mol Biol 92(1):15–37
54. Davis JH, Williamson JR (2017) Structure and dynamics of bacterial ribosome biogenesis. Philos Trans R Soc Lond Ser B Biol Sci 372(1716). https://doi.org/10.1098/rstb.2016.0181
55. Antonicka H, Sasarman F, Nishimura T et al (2013) The mitochondrial RNA-binding protein GRSF1 localizes to RNA granules and is required for posttranscriptional mitochondrial gene expression. Cell Metab 17(3):386–398
56. Jourdain AA, Koppen M, Wydro M et al (2013) GRSF1 regulates RNA processing in mitochondrial RNA granules. Cell Metab 17(3):399–410
57. Antonicka H, Shoubridge EA (2015) Mitochondrial RNA granules are centers for posttranscriptional RNA processing and ribosome biogenesis. Cell Rep 10(6):920–932
58. Tu YT, Barrientos A (2015) The human mitochondrial DEAD-box protein DDX28 resides in RNA granules and functions in mitoribosome assembly. Cell Rep 10(6):854–864
59. Shajani Z, Sykes MT, Williamson JR (2011) Assembly of bacterial ribosomes. Annu Rev Biochem 80:501–526
60. Britton RA (2009) Role of GTPases in bacterial ribosome assembly. Annu Rev Microbiol 63:155–176
61. Barrientos A, Korr D, Barwell KJ et al (2003) *MTG1* codes for a conserved protein required for mitochondrial translation. Mol Biol Cell 14(6):2292–2302
62. Paul MF, Alushin GM, Barros MH et al (2012) The putative GTPase encoded by *MTG3* functions in a novel pathway for regulating assembly of the small subunit of yeast mitochondrial ribosomes. J Biol Chem 287(29):24346–24355
63. Dennerlein S, Rozanska A, Wydro M et al (2010) Human ERAL1 is a mitochondrial RNA chaperone involved in the assembly of the 28S small mitochondrial ribosomal subunit. Biochem J 430(3):551–558
64. Metodiev MD, Spahr H, Loguercio Polosa P et al (2014) NSUN4 is a dual function mitochondrial protein required for both methylation of 12S rRNA and coordination of mitoribosomal assembly. PLoS Genet 10(2):e1004110
65. De Silva D, Poliquin S, Zeng R et al (2017) The DEAD-box helicase Mss116 plays distinct roles in mitochondrial ribogenesis and mRNA-specific translation. Nucleic Acids Res 45(11):6628–6643
66. Dalla Rosa I, Durigon R, Pearce SF et al (2014) MPV17L2 is required for ribosome assembly in mitochondria. Nucleic Acids Res 42(13):8500–8515
67. De Silva D, Tu YT, Amunts A et al (2015) Mitochondrial ribosome assembly in health and disease. Cell Cycle 14(14):2226–2250
68. Boczonadi V, Horvath R (2014) Mitochondria: impaired mitochondrial translation in human disease. Int J Biochem Cell Biol 48(100):77–84
69. Gopisetty G, Thangarajan R (2016) Mammalian mitochondrial ribosomal small subunit (MRPS) genes: a putative role in human disease. Gene 589(1):27–35
70. Pulman J, Ruzzenente B, Bianchi L et al (2019) Mutations in the MRPS28 gene encoding the small mitoribosomal subunit protein bS1m in a patient with intrauterine growth retardation, craniofacial dysmorphism and multisystemic involvement. Hum Mol Genet 28(9):1445–1462
71. Bugiardini E, Mitchell AL, Rosa ID et al (2019) MRPS25 mutations impair mitochondrial translation and cause encephalomyopathy. Hum Mol Genet 28(16):2711–2719
72. Borna NN, Kishita Y, Kohda M et al (2019) Mitochondrial ribosomal protein PTCD3 mutations cause oxidative phosphorylation defects with Leigh syndrome. Neurogenetics 20(1):9–25
73. Di Nottia M, Marchese M, Verrigni D et al (2020) A homozygous MRPL24 mutation causes a complex movement disorder and affects the mitoribosome assembly. Neurobiol Dis 141:104880
74. Gardeitchik T, Mohamed M, Ruzzenente B et al (2018) Bi-allelic mutations in the mitochondrial ribosomal protein MRPS2 cause sensorineural hearing loss, hypoglycemia,

and multiple OXPHOS complex deficiencies. Am J Hum Genet 102(4):685–695. https://doi.org/10.1016/j.ajhg.2018.1002.1012. Epub 2018 Mar 1022

75. O'Brien TW, O'Brien BJ, Norman RA (2005) Nuclear MRP genes and mitochondrial disease. Gene 354:147–151
76. Saada A, Shaag A, Arnon S et al (2007) Antenatal mitochondrial disease caused by mitochondrial ribosomal protein (MRPS22) mutation. J Med Genet 44(12):784–786
77. Emdadul Haque M, Grasso D, Miller C et al (2008) The effect of mutated mitochondrial ribosomal proteins S16 and S22 on the assembly of the small and large ribosomal subunits in human mitochondria. Mitochondrion 8(3):254–261
78. Rebelo-Guiomar P, Pellegrino S, Dent KC et al (2022) A late-stage assembly checkpoint of the human mitochondrial ribosome large subunit. Nat Commun 13(1):929. https://doi.org/10.1038/s41467-022-28503-5
79. Klinge S, Woolford JL Jr (2019) Ribosome assembly coming into focus. Nat Rev Mol Cell Biol 20(2):116–131. https://doi.org/10.1038/s41580-018-0078-y
80. Srivastava AK, Schlessinger D (1990) Mechanism and regulation of bacterial ribosomal RNA processing. Annu Rev Microbiol 44:105–129. https://doi.org/10.1146/annurev.mi.44.100190.000541
81. Montoya J, Christianson T, Levens D et al (1982) Identification of initiation sites for heavy-strand and light-strand transcription in human mitochondrial DNA. Proc Natl Acad Sci U S A 79(23):7195–7199
82. Terzioglu M, Ruzzenente B, Harmel J et al (2013) MTERF1 binds mtDNA to prevent transcriptional interference at the light-strand promoter but is dispensable for rRNA gene transcription regulation. Cell Metab 17(4):618–626. https://doi.org/10.1016/j.cmet.2013.03.006
83. Litonin D, Sologub M, Shi Y et al (2010) Human mitochondrial transcription revisited: only TFAM and TFB2M are required for transcription of the mitochondrial genes in vitro. J Biol Chem 285(24):18129–18133. https://doi.org/10.1074/jbc.C110.128918
84. Siira SJ, Rossetti G, Richman TR et al (2018) Concerted regulation of mitochondrial and nuclear non-coding RNAs by a dual-targeted RNase Z. EMBO Rep 19(10). https://doi.org/10.15252/embr.201846198
85. Jedynak-Slyvka M, Jabczynska A, Szczesny RJ (2021) Human mitochondrial RNA processing and modifications: overview. Int J Mol Sci 22(15). https://doi.org/10.3390/ijms22157999
86. Uchiumi T, Ohgaki K, Yagi M et al (2010) ERAL1 is associated with mitochondrial ribosome and elimination of ERAL1 leads to mitochondrial dysfunction and growth retardation. Nucleic Acids Res 38(16):5554–5568
87. Seidel-Rogol BL, McCulloch V, Shadel GS (2003) Human mitochondrial transcription factor B1 methylates ribosomal RNA at a conserved stem-loop. Nat Genet 33(1):23–24
88. Szczepanowska K, Maiti P, Kukat A et al (2016) CLPP coordinates mitoribosomal assembly through the regulation of ERAL1 levels. EMBO J 35(23):2566–2583
89. Summer S, Smirnova A, Gabriele A et al (2020) YBEY is an essential biogenesis factor for mitochondrial ribosomes. Nucleic Acids Res 48(17):9762–9786. https://doi.org/10.1093/nar/gkaa148
90. Kolanczyk M, Pech M, Zemojtel T et al (2011) NOA1 is an essential GTPase required for mitochondrial protein synthesis. Mol Biol Cell 22(1):1–11
91. He J, Cooper HM, Reyes A et al (2012) Human C4orf14 interacts with the mitochondrial nucleoid and is involved in the biogenesis of the small mitochondrial ribosomal subunit. Nucleic Acids Res 40:6097–6108
92. Reyes A, Favia P, Vidoni S et al (2020) RCC1L (WBSCR16) isoforms coordinate mitochondrial ribosome assembly through their interaction with GTPases. PLoS Genet 16(7):e1008923
93. Van Haute L, Hendrick AG, D'Souza AR et al (2019) METTL15 introduces N4-methylcytidine into human mitochondrial 12S rRNA and is required for mitoribosome biogenesis. Nucleic Acids Res 47(19):10267–10281. https://doi.org/10.1093/nar/gkz735
94. Fung S, Nishimura T, Sasarman F et al (2013) The conserved interaction of C7orf30 with MRPL14 promotes biogenesis of the mitochondrial large ribosomal subunit and mitochondrial translation. Mol Biol Cell 24(3):184–193
95. Rorbach J, Gammage PA, Minczuk M (2012) C7orf30 is necessary for biogenesis of the large subunit of the mitochondrial ribosome. Nucleic Acids Res 40:4097–4109
96. Popow J, Alleaume AM, Curk T et al (2015) FASTKD2 is an RNA-binding protein required for mitochondrial RNA processing and translation. RNA 21(11):1873–1884. https://doi.org/10.1261/rna.052365.115
97. Kim H-J, Barrientos A (2018) MTG1 couples mitoribosome large subunit assembly and

intersubunit bridge formation. Nucleic Acid Res 46:8435
98. Maiti P, Antonicka H, Gingras A-C et al (2020) Human GTPBP5 (MTG2) fuels mitoribosome large subunit maturation by facilitating 16S rRNA methylation. Nucleic Acids Res 48(14):7924–7943
99. Lee KW, Okot-Kotber C, Lacomb JF et al (2013) Mitochondrial rRNA methyltransferase family members are positioned to modify nascent rRNA in foci near the mtDNA nucleoid. J Biol Chem 288(43):31386–31399
100. Nikolay R, Hilal T, Schmidt S et al (2021) Snapshots of native pre-50S ribosomes reveal a biogenesis factor network and evolutionary specialization. Mol Cell 81(6):1200–1215. e1209. https://doi.org/10.1016/j.molcel.2021.02.006
101. Zaganelli S, Rebelo-Guiomar P, Maundrell K et al (2017) The pseudouridine synthase RPUSD4 is an essential component of mitochondrial RNA granules. J Biol Chem 292(11):4519–4532
102. Lavdovskaia E, Denks K, Nadler F et al (2020) Dual function of GTPBP6 in biogenesis and recycling of human mitochondrial ribosomes. Nucleic Acids Res. https://doi.org/10.1093/nar/gkaa1132
103. Kim HJ, Maiti P, Barrientos A (2017) Mitochondrial ribosomes in cancer. Semin Cancer Biol 47:67–81
104. Ban N, Beckmann R, Cate JH et al (2014) A new system for naming ribosomal proteins. Curr Opin Struct Biol 24:165–169. https://doi.org/10.1016/j.sbi.2014.01.002
105. Metodiev MD, Lesko N, Park CB et al (2009) Methylation of 12S rRNA is necessary for in vivo stability of the small subunit of the mammalian mitochondrial ribosome. Cell Metab 9(4):386–397
106. Liu X, Shen S, Wu P et al (2019) Structural insights into dimethylation of 12S rRNA by TFB1M: indispensable role in translation of mitochondrial genes and mitochondrial function. Nucleic Acids Res 47(14):7648–7665. https://doi.org/10.1093/nar/gkz505
107. Powell CA, Minczuk M (2020) TRMT2B is responsible for both tRNA and rRNA m(5)U-methylation in human mitochondria. RNA Biol 17(4):451–462. https://doi.org/10.1080/15476286.2020.1712544
108. Laptev I, Shvetsova E, Levitskii S et al (2020) Mouse Trmt2B protein is a dual specific mitochondrial metyltransferase responsible for m (5)U formation in both tRNA and rRNA. RNA Biol 17(4):441–450. https://doi.org/10.1080/15476286.2019.1694733
109. Shi Z, Xu S, Xing S et al (2019) Mettl17, a regulator of mitochondrial ribosomal RNA modifications, is required for the translation of mitochondrial coding genes. FASEB J 33(11):13040–13050. https://doi.org/10.1096/fj.201901331R
110. Tang T, Zheng B, Chen SH et al (2009) hNOA1 interacts with complex I and DAP3 and regulates mitochondrial respiration and apoptosis. J Biol Chem 284(8):5414–5424. https://doi.org/10.1074/jbc.M807797200
111. Bar-Yaacov D, Frumkin I, Yashiro Y et al (2016) Mitochondrial 16S rRNA is methylated by tRNA methyltransferase TRMT61B in all vertebrates. PLoS Biol 14(9):e1002557. https://doi.org/10.1371/journal.pbio.1002557
112. Arroyo JD, Jourdain AA, Calvo SE et al (2016) A genome-wide CRISPR death screen identifies genes essential for oxidative phosphorylation. Cell Metab 24(6):875–885. https://doi.org/10.1016/j.cmet.2016.08.017
113. Rathore A, Chu Q, Tan D et al (2018) MIEF1 microprotein regulates mitochondrial translation. Biochemistry 57(38):5564–5575
114. Dibley MG, Formosa LE, Lyu B et al (2020) The mitochondrial acyl-carrier protein interaction network highlights important roles for LYRM family members in complex i and mitoribosome assembly. Mol Cell Proteomics 19(1):65–77. https://doi.org/10.1074/mcp.RA119.001784
115. Wredenberg A, Lagouge M, Bratic A et al (2013) MTERF3 regulates mitochondrial ribosome biogenesis in invertebrates and mammals. PLoS Genet 9(1):e1003178
116. Camara Y, Asin-Cayuela J, Park CB et al (2011) MTERF4 regulates translation by targeting the methyltransferase NSUN4 to the mammalian mitochondrial ribosome. Cell Metab 13(5):527–539
117. Fogal V, Richardson AD, Karmali PP et al (2010) Mitochondrial p32 protein is a critical regulator of tumor metabolism via maintenance of oxidative phosphorylation. Mol Cell Biol 30(6):1303–1318. https://doi.org/10.1128/mcb.01101-09
118. Yagi M, Uchiumi T, Takazaki S et al (2012) p32/gC1qR is indispensable for fetal development and mitochondrial translation: importance of its RNA-binding ability. Nucleic Acids Res 40(19):9717–9737. https://doi.org/10.1093/nar/gks774

Chapter 4

Translation in Mitochondrial Ribosomes

Zofia M. Chrzanowska-Lightowlers and Robert N. Lightowlers

Abstract

Mitochondrial protein synthesis is essential for the life of aerobic eukaryotes. Without it, oxidative phosphorylation cannot be coupled. Evolution has shaped a battery of factors and machinery that are key to production of just a handful of critical proteins. In this general concept chapter, we attempt to briefly summarize our current knowledge of the overall process in mitochondria from a variety of species, breaking this down to the four parts of translation: initiation, elongation, termination, and recycling. Where appropriate, we highlight differences between species and emphasize gaps in our understanding. Excitingly, with the current revolution in cryoelectron microscopy and mitochondrial genome editing, it is highly likely that many of these gaps will be resolved in the near future. However, the absence of a faithful in vitro reconstituted system to study mitochondrial translation is still problematic.

Key words Mitochondria, Protein synthesis, Translation, mt-mRNAs, Mitoribosomes, Initiation, Elongation, Termination, Recycling

1 Introduction

Since the initial report of the isolation of a contamination-free 55S mitoribosomal preparation from rat liver mitochondria, the field of mitochondrial translation has expanded enormously [1, 2]. The progress in our understanding has been substantial. Translation factors have been identified, as have rRNA chaperones and modifying enzymes, mitoribosomal assembly pathways, and insertases to facilitate cotranslational integration of proteins into the inner/cristae membranes and factors that can help resolve stalled mitoribosomes. A wealth of intense research effort has yielded these discoveries. Some of the most striking advances have recently been made possible by the evolution of cryo-EM techniques and interpretive software. In a single step, this took us from mitoribosomal structures, that at the time of their production were impressive but lacked more than a low definition shape of the large and small subunits, to near atomic resolution!

Antoni Barrientos and Flavia Fontanesi (eds.), *The Mitoribosome: Methods and Protocols*,
Methods in Molecular Biology, vol. 2661, https://doi.org/10.1007/978-1-0716-3171-3_4,

Early pioneers exploring how mitochondria could make their own proteins were limited in what tools were available. Without the power of transfection, a technique open to other avenues of molecular genetics, there was no opportunity to use reporter constructs to analyze, for example, what features were critical to the recognition of the start codon to initiate translation, how internal start sites were distinguished within the sequences of the bicistronic transcripts, what determined whether or not the 3′ terminus of a transcription unit was modified by the addition of a poly/oligoadenylate tail, or a CCA trinucleotide, or left naked. Isolation of mitoribosomes was the first critical step, with seminal work carried out in the late 1960s/1970s by Tom O'Brien and colleagues [2–7] working with rat liver mitochondria but soon extended by others to *Neurospora* [8, 9] and other sources (reviewed in [10]). Exploring how mitochondria could synthesize proteins [11], what made up a mitoribosome [4, 12, 13], characterizing the protein components [14–18] and how these assembled, all represented the slightly more accessible targets. To investigate these areas, more fundamental chemistry and biochemistry techniques were employed. The subsequent increase in available techniques and their sophistication has allowed us to develop a much more detailed understanding of many of the stages of intramitochondrial translation.

2 How Are Mitochondrial Transcripts Prepared for Translation?

In mammals, the multicopy, circular mitochondrial DNA (mtDNA) is transcribed into large polycistronic RNA units that are processed and matured in membrane-less condensates termed mitochondrial RNA granules, or MRGs [19, 20]. The majority of the 13 open reading frames are interspersed by mt-tRNA sequences that spontaneously fold and act to nucleate cleavage by mitochondrial RNase P and Elac2, releasing immature mRNAs that are matured by the addition of a short, approximately 40 nucleotide polyadenylated 3′ tail by a mitochondrial poly(A) polymerase [21–25]. These mt-mRNAs have no 5′ cap, only an unmodified phosphate group and generally have an exceedingly short (1–3 nucleotides) or no 5′ untranslated region (UTR), with the exception of the downstream elements of the bicistronic units (*RNA7* and *RNA14*) [21]. Similarly, most of these species also carry no 3' UTR and in numerous cases require polyadenylation to complete a UAA termination codon [21, 26].

In contrast to the mt-mRNAs in mammals, plant mt-mRNAs contain both 5′ and 3′ UTRs that can be of considerable length. These species are processed to form variants with multiple 5′ termini [27] and can be subject to remarkable levels of RNA editing (C to U with a few exceptions) [28, 29], mediated by a large number of sequence-specific RNA-binding proteins, termed pentatricopeptide repeat containing proteins, or PPRs [30]. Further, unlike in the mammal, these transcripts contain introns in protein-coding genes which require both *cis*- and *trans*-splicing for their removal [31–33]. The 3′ termini of plant mt-mRNA are often polyadenylated or even in some cases polycytidylated as seen in a number of green algae [34, 35]. Critically, unlike the situation for a subset of mammalian mt-mRNAs, polyadenylation appears to serve as a signal for mRNA degradation rather than acting to stabilize them [24, 36, 37].

In the budding yeast *Saccharomyces cerevisiae*, all processed mt-mRNAs are monocistronic and have 5′ UTRs that play crucial roles in translational initiation, but there is no evidence of mRNA polyadenylation [24, 38]. Instead, there is a dodecamer sequence (5′-AAUAA(U/C)AUUCUU-3′) that is present toward the end of the 3′ untranslated region of mt-mRNAs [39] and appears to play a role in the stability of the transcripts [40].

The mt-mRNAs of the trypanosomatid protozoa specifically *Leishmania major* and *Trypanosoma brucei* differ again. They are derived from the transcription of the mtDNA that is found in a combination of numerous mini and maxicircles. Similar to the mammalian system, the mt-mRNAs in these organisms, are transcribed as polycistronic RNA precursors from which the RNA transcription units are excised through an endonucleolytic process [41]. These pre-mRNAs are then extensively edited through a mechanism that inserts and deletes uridines, orchestrated by a band of small guide RNAs [42]. The length of the poly(A) tail is dependent on the editing status but fully mature mt-mRNAs in the kinetoplast have long poly(A/U) tails of 200–300 residues that stabilize them and signal their competence for use in translation (reviewed in [43]).

3 Critical Factors and Mechanisms of Translation Initiation

For protein synthesis to commence, the start codon of the mt-mRNA must be selected and the initiator tRNA placed at the peptidyl-site (P-site) of the mitoribosome, thus establishing the correct reading frame of the transcript. How this occurs is a question that remains incompletely resolved despite being a topic of interest for over 20 years [44]. Pioneering work by the Spremulli group investigated how the mammalian mt-mRNAs are loaded onto the mitoribosomal subunits. Since no Shine-Dalgarno mechanism exists between the mitochondrial rRNA/mRNA, the group used RNA SHAPE chemistry to better understand this process, and to determine what role is played by the secondary structures near the start codons of all 13 open reading frames. Their studies revealed that the first 35 nucleotides of the mt-mRNAs form structures with free energies equal to, or less than a single typical base pair (approx. −3 kcal/mol). This implies that the start codons of these mRNAs, which lie at the very 5′ termini, are accessible within single stranded motifs, making them potentially poised for mitoribosome binding [45]. Subsequent work by the same group tested the interactions of *in vitro* transcribed mt-mRNA sequences with small 28S subunits (mt-SSU) isolated from bovine mitoribosomes. They observed the preferential selection of a start codon that was located at the very 5′-terminus of the transcript and that the efficiency of the interaction was seen to be significantly reduced by the presence of any nucleotides preceding the start codon [46]. Although mt-mRNA loading onto a preinitiation complex that includes the mt-SSU and various release factors is the more conserved mechanism across species, recent *in vitro* evidence has suggested that loading of mt-mRNAs onto mammalian mitoribosomes may also occur onto preformed monosomes [47]. This is an intriguing possibility, but further experimentation is required to challenge this model.

Across different organisms, the most commonly used initiation codon is AUG but this is not the only codon used within the various mitochondrial systems. Alternative start codons that are read as methionines include AUA, AUU, AUC, GUG, UUA, UUG, and CUG (*see* Table 1) [48]. Another anomaly is that although standard translation systems have two tRNAs for methionine, one used for initiation and the second for elongation, mammalian mitochondria are unusual as they encode only a single mt-tRNA^{Met}. A fraction of the latter becomes formylated, and it is this fMet-tRNA^{Met} that is used to initiate extension of the polypeptide chain. It has been shown, however, that in the absence of formyl methionine, initiation can occur but with lower efficiency, compromising protein synthesis and therefore the assembly of oxidative phosphorylation (OXPHOS) complexes [49]. The formylation is nevertheless important as illustrated by the serious clinical consequences arising from mutations in the mitochondrial methionyl-tRNA transformylase [50].

How are these initiation codons correctly loaded onto mitoribosomes? Since those early experiments there has been much progress. The mt-mRNAs that have been processed are chaperoned by an RNA-binding complex made up of two proteins, LRPPRC (leucine-rich pentatricopeptide repeat-containing protein) and SLIRP (SRA stem-loop-interacting RNA-binding protein) [51]. This interaction reduces any secondary structure, stimulates the poly(A) polymerase activity, and stabilizes this now fully matured transcript [25, 51, 52]. Advances in cryo-EM in recent years, which has facilitated structural imaging of large complexes, have certainly contributed to our more detailed understanding. Very recent cryo-EM data have managed to generate structures of the LRPPRC-SLIRP bound to the mRNA-containing mitoribosome through pincer-like interactions with mS31 and mS39 [53]. This indicates how the LRPPRC-SLIRP mediates the transfer of the mt-mRNA to the mitoribosome via four of the LRPPRC helical repeats [53].

Table 1
Alternative initiation codons

Organism	AUA	AUU	AUC	GUG	UUA	UUG	CUG
Vertebrate	Bovine Human Mouse	Human Mouse	Mouse	Quail Chicken Zebrafish			
Invertebrates	START	START	Apis (bee)	Polyplacophora		Ascaris, Caenorhabditis	
Yeast	*Saccharomyces, Candida, Hansenula, Kluyveromyces*						
Starfishes sea urchins	Isoleucine						
Mold, Protozoan, Coelenterate	Leishmania Tetrahymena Paramecium	Leishmania Tetrahymena Paramecium	Paramecium	Paramecium	Trypanosoma	Trypanosoma	Trypanosoma
Ascidian	START						
Flatworms, roundworms							
Chlorophycean							
Trematode	START						
Rhabdopleuridae	Isoleucine			START			
Cephalodiscidae							
Green alga Scenedesmus obliquus							
Thraustochytrium aureum		START		START	STOP		

As the translational mechanisms in eubacteria and organelles were assumed to be similar, the prediction was that three nuclear-encoded initiation factors (IFs) would be required. Initial bioinformatic studies by Spremulli and colleagues revealed the presence of an initiation factor that bore similarity in the central portion to the eubacterial IF3 but had notable extensions at both the N and C termini [54]. Subsequently it was identified that mammalian mitochondria use only two initiation factors, initiation factor 2 (mtIF2) and initiation factor 3 (mtIF3) with no evidence or requirement for an equivalent of IF1, which is also the case in *Drosophila* mitochondria (reviewed in [55]). Biochemical and molecular modeling followed more recently by cryo-EM analysis confirms that the lack of an IF1 is functionally compensated by a 37 amino acid insertion into mtIF2 [56, 57]. Early on it was perceived that mtIF3 enabled the start codon to be positioned in the P-site of the mitoribosome, and that the contact sites between mtIF3 and various mt-SSU proteins assisted binding of the mt-mRNA to the mt-SSU [56, 58]. A recent paper reported that two preinitiation steps occurred, termed mtPIC-1 and mtPIC-2 (Fig. 1a, b). Investigations into the mechanism of mtPIC-1 showed the mtIF3 interaction to be specifically with mitoribosomal protein mS37. This maintains the mt-SSU in a favorable conformation for the interaction with mtIF2, and the intermediate conformational change of mtPIC-2 where mtIF3 is displaced in favor of the initiator mt-tRNA, such that association of the mitoribosomal large subunit (mt-LSU) with the leaderless mt-mRNA can occur (Fig. 1c)

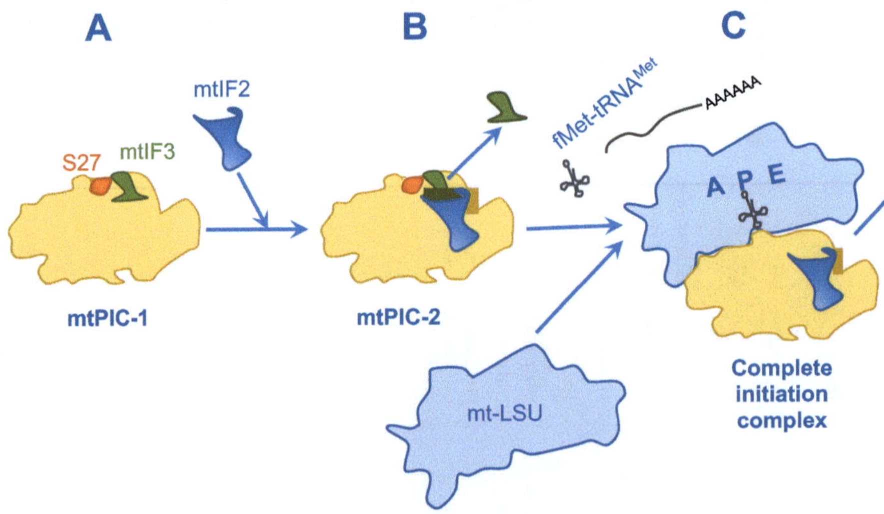

Fig. 1 Schematic of translation initiation in mammalian mitochondria. The mitochondrial initiation factor 3 (mtIF3) remains present in the small subunit of the mitoribosome following disassembly of the monosome, thus preventing premature reassociation with the large submit. This constitutes the preinitiation complex 1 (mtPIC-1) (panel **A**). Association with mtIF2 forms mtPIC-1 and triggers displacement of mtIF3 (panel **B**). Positioning of the formylated mt-tRNAMet, mt-mRNA, within the assembled monosome generates the complete initiation complex (panel **C**)

[47]. Mammalian mtIF3 has mitochondrial specific extensions at the N- and C- termini, which have been shown to modulate affinity of the mtIF3 to several mitoribosomal subunits and affect the fidelity of mt-LSU binding to the initiation complex [59–61]. Studies of a conditional *Mtif3* knockout in mice demonstrated an additional role for mtIF3 in initiation, as a decreased association between the mt-mRNAs and the mitoribosomes was observed, along with decreased removal of mt-tRNA^{Met} from mt-SSUs that lacked an associated mt-mRNA [62]. Together, the work confirms that mtIF3 plays a central role in the process of translation, but whether it has further functions remains to be established.

A strategy used extensively by yeast is the implementation of protein factors (translational activators) that interact with the 5' UTRs to guide the mt-mRNAs to the mitoribosome [63]. Consistent with the lack of 5' UTRs in mammalian mt-mRNAs, these translational activators are absent with the exception of TACO1, a protein that is important for promoting the translation of *MTCO1* [64, 65]. A thorough understanding of how TACO1 promotes translation is still lacking.

Interestingly, contrary to chloroplasts where some genes contain Shine-Dalgarno-like sequences, plant mitochondria more closely resemble the mammalian condition where all from the green lineage have entirely lost Shine-Dalgarno sequences from their mRNA sequences and also the anti-Shine-Dalgarno sequences from the 18S rRNA in the mt-SSU [66]. Although many plants contain 5' UTRs that may help direct assembly onto initiation complexes, in members of chlorophyta such as *Chlamydomonas* this feature is absent and the mt-mRNAs resemble the leaderless mt-mRNAs seen in mammals [34]. From studies thus far, it seems that the mitochondrial translation processes among plants, even within those of the green lineage, are very divergent in terms of initiation processes, ribosome composition, and structures.

Trypanosomes have short 5'UTRs preceding initiation codons that can be encoded or generated upon correct editing [67]. As with green plants and mammals, these do not possess a recognizable Shine-Dalgarno sequence. Further, there may be upstream redundant initiation codons. These features combined with the lack of a 5′ cap structure, and the lack of a start codon at the very 5′-terminus as found in mammalian mt-mRNAs and the eukaryotic cytosol respectively, all make recognition of the correct start site a challenge for the trypanosomal mitoribosome. Codon recognition looks to face a situation more similar to that in yeast mitochondria where there are long 5' UTRs containing potentially redundant out of frame AUG triplets [68]. An additional step that must precede positioning the start codon with the mitoribosome, is the completion of all U insertion/deletion editing, which in certain cases generates the correct start codon. In yeast, translational activators assist in the recognition step, but there are none identified as yet in trypanosomes. The closest resemblance to factors essential for translational mRNA activation are the 2 mt-SSU proteins associated with the polyadenylation complex, KRIPP1 and KRIPP8. Depletion of each of these proteins caused a reduction of the translationally competent A/U tailed transcripts encoding cytochrome *c* oxidase subunit 1 concomitantly with synthesis specifically of this gene product. Loss of KRIPP8 additionally elicited the same effect on transcripts encoding cytochrome *b* as well as on *MT-CO1* [69]. Curiously, these are the two genes that are universally conserved in the mitochondrial genome (reviewed in [70]). The trypanosome life cycle includes parasitic phases in mammalian and insect cells that differ in metabolic requirement. Consistent with this is the need for KRIPP1/8 during the actively respiring insect stage and not in the mammalian phase [69], thereby demonstrating a developmental regulation of mitochondrial translation. A newly identified mechanism for regulating initiation of translation involves the shortening of the 3′-terminus of the required tRNAs by LCCR4, which removes partially or completely the CCA-tail. This happens in the cytosol of trypanosomes under conditions of nutritional stress and once the stress is relieved, rapid readdition of the CCA facilitates recovery of protein synthesis [71]. There is still relatively little detailed information on the translation processes including initiation in the cytosol of trypanosomes and rather less still of what goes on in the mitochondrial compartment [72]. However, since intramitochondrial protein synthesis in trypanosomes depends entirely on tRNAs imported from the cytosol, this CCA-trimming mechanism may also regulate initiation in the mitochondrial compartment.

4 Translational Elongation

Once initiation has been accomplished, the polypeptide needs to be extended by the mitoribosome reading through the entire open reading frame. This phase of translation is more evolutionarily conserved than either initiation or termination. The mature transcript now needs to be translated by the complex molecular machine, the mitoribosome, together with a cluster of elongation factors and charged tRNAs. Details of the mitoribosome and the extensive differences between the machinery and its assembly in different organisms are given in separate chapters of this series and so will not be reiterated here. Suffice to say, that evolution of the RNA content of the mitoribosomes has had to accommodate changes in the mtDNA sequences, ability to import RNA species and also changes to codon usage. The standard codon usage is retained for the most part across mitochondria from a wide spectrum of organisms; however, as in all biological systems, there are exceptions, some of which are given Table 2. The tRNAs used in translation may be entirely mtDNA encoded as in mammals,

Table 2
Alternative termination codons, changes to standard codons

Organism	Standard stop codons			Alternative codon usages							
	UAA	UAG	UGA	AGA	AGG	AAA	CUU	CUC	CUA	CUG	UCA
Vertebrate			Trp	Not assigned	Not assigned						
Invertebrates			Trp	Ser	Ser Absent in *Drosophila*						
Yeast			Trp				Thr	Thr	Thr	Thr	
Starfishessea urchins			Trp	Ser	Ser	Asn					
Mold, protozoan, coelenterate			Trp								
Ascidian			Trp	Gly	Gly						
Flatworms, roundworms	Tyr		Trp	Ser	Ser	Asn					
Chlorophycean		Leu									
Trematode			Trp	Ser	Ser	Asn					
Rhabdopleuridae			Trp	Ser	Lys						
Cephalodiscidae	Tyr		Trp	Ser	Lys						
Green alga *Scenedesmus obliquus*		Leu									STOP

entirely nuclear encoded as in trypanosomes or a mixture of the two as in many plant mitochondria. All need to be correctly conjugated to the cognate amino acid to retain fidelity in translation and to support cellular viability. It is a remarkably accurate process with aminoacylation mischarging at a frequency of only one in 10^4 to 10^5 events [73]. This is done in large part by a battery of specific aminoacyl-tRNA synthetases (aaRSs), all of which are nuclear encoded [74, 75]. One exception to this rule is the absence of a gene encoding a mammalian mitochondrial glutaminyl tRNA synthetase, necessitating the formation of glutaminylated mt-$tRNA^{Gln}$ by an indirect mechanism. Production of mt-Glu-$tRNA^{Gln}$ occurs through misaminoacylation of mt-$tRNA^{Gln}$ by the mt glutamyl tRNA synthetase (mtGluRS), with the correctly aminoacylated mt-Gln-$tRNA^{Gln}$ being reconstituted by a Glu-$tRNA^{Gln}$ amidotransferase (GATCAB) [76]. The genes for aaRSs may encode both the cytosolic and mitochondrial variant allowing for dual localization, or the proteins for the two different compartments may be encoded by different genes. This is in part dictated by where the tRNAs are encoded. In mammals, all the mt-tRNAs are encoded on the mtDNA and differ structurally from their cytosolic counterparts [77]. Thus, each of the mitochondrial aaRSs are encoded by individual genes that differ from those for the aaRs for cytosolic tRNAs, and are specific for their cognate mitochondrial tRNAs, with the exception of two (*GARS* and *KARS*, which encode the glycyl- (GlyRS) and lysyl- (LysRS) tRNA synthetases) where cytosolic or mitochondrial localization of the synthetase is effected by different translation initiation sites in the same transcript; yeast encode 5 dually localized aaRSs; Arabidopsis encodes an almost complete set of dually localized aaRSs [78]; while for trypanosomes where all the tRNAs are nuclear encoded, and therefore essentially identical to the cytosolic pool, most of the aaRSs are encoded by a single gene with only some demonstrating dual localization [79]. Thus, it is curious as to why trypanosomes retain two genes for TrpRS, AspRS, and LysRS, each encoding either a cytosolic or a mitochondrial variant. Studies in *Drosophila* revealed that in addition to the mitochondrial SerRS (DmSerRS2), gene duplication has resulted in a paralog, SLIMP, that is also present in arthropods and echinoderms. Although this protein has retained structural similarity and has lost all tRNA aminoacylation activity, it retains an essential role in protein synthesis [80]. SLIMP interacts with DmSerRS2 to form a heterodimer that is essential for mt-tRNA-aminoacylation [81]. Intriguingly, it also interacts with the substrate-binding domain of the ATP-dependent serine protease, LonP, stimulating TFAM degradation and thereby coordinating mtDNA copy number to mt-protein synthesis [81].

Since these synthetases play a fundamental role in translation, it is intriguing that pathogenic mutations that have been identified in these proteins can cause clinical manifestation in patients with a very defined tissue specific phenotype that closely aligns with the affected individual aaRSs (reviewed in [82]).

In the elongation phase, three elongation factors (mtEFTu, mtEFTs, and mtEFG1) play key roles in mitochondrial translation. Mitochondrial EFTU:GTP is responsible for the transfer of the -aminoacyl-tRNA to the empty A site. It promotes the GTP-dependent binding of aminoacyl-tRNA to the A-site of ribosomes and coordinates specific codon:anticodon pairing between the mRNA and tRNA through a process of kinetic proofreading, a decoding that proceeds at 10^8 M^{-1} s^{-} [83, 84], allowing an elongation speed of 20 amino acid additions/second [85].

Not all mammalian mt-tRNAs fold to produce the canonical cloverleaf structure, but while there are species that lack the tRNA dihydrouridine (D)-arm such as mt-tRNA$^{Ser(AGY),}$ they are still all recognized by a single mtEFTu. For various species of nematode mt-tRNAs, however, there are even more unusual noncanonical forms of mt-tRNA which require more than one mtEFTu for ribosomal tRNA loading and are encoded by two separate genes [86]. This is also the case for *Drosophila* [87], which similarly encode D-armless mitochondrial tRNA species although one isoform, mEFTu1, binds well to both the canonical and D-armless mt-tRNA species. It is unclear whether the second isoform, mtEFTu2, plays any additional role in *Drosophila* mitochondrial translation.

After delivery of the charged mt-tRNA and the hydrolysis of GTP, the newly formed mtEFTu:GDP is released from the mitoribosome and needs the intervention of the guanine exchange factor, mtEFTs, to reassociate with GTP restoring mtEFTu:GTP, ready to deliver the next aminoacylated tRNA. Subsequent peptide bond formation occurs joining the P-site peptidyl-tRNA to the A-site mt-tRNA. The mitoribosome now needs to move along the transcript, a process that requires high fidelity to avoid frame-shifting and mis-reading the message. This process is catalyzed by the third elongation factor, mtEF-G1 [88, 89]. This now binds to the A-site and promotes translocation of the mitoribosome along the mt-mRNA by promoting movement of A-site tRNAs and P-site tRNAs to the P-site and E-site respectively. The tRNA that is now shunted to the E-site leaves the monomer permitting this cycle of tRNA delivery, peptide bond formation and translocation to continue until the polypeptide is completed and a stop codon appears within the A-site (reviewed in [90]). Interestingly, although these factors are required for most mitochondrial translation, budding yeast *Saccharomyces cerevisiae* do not require the guanine exchange factor, mtEFTs [91].

There is a fourth elongation factor, believed to be involved in mitochondrial translation, mtEF4 (also known as GUF1). It is highly conserved in eukaryotic mitochondria, (present in yeast, *C. elegans,* humans) as well as being found in chloroplasts. The bacterial homolog, LepA, promotes backward movement of tRNAs on the ribosome in a process called back-translocation, and mtEF4 becomes especially important under stress conditions when it serves to increase fidelity of OXPHOS polypeptide synthesis [92, 93] (reviewed in [94]), although its exact molecular mechanism has not been elucidated.

5 How Is Mitochondrial Translation Termination Effected?

Based on the standard codon usage, there are three RNA triplets that do not have a tRNA that recognizes them and therefore act as regular stop codons. These are UAA, UAG, and UGA that are recognized instead by *trans*-acting protein elements termed polypeptide release factors (RF). These release factor proteins are part of a large family some of which have specificity to individual stop codons, while others do not recognize the triplets of the mRNA but play other roles in the termination process [95]. Eubacteria have two such elements, RF1 and RF2, that recognize either UAA/UAG or UAA/UGA, respectively. In the eukaryotic cytosol, there is a single factor, eRF1, that recognizes all three triplets. In addition to these codon-specific RFs there is a second class that is codon nonspecific, RF3 in eubacteria helps dissociate RF1/RF2 from the termination complex while eRF3 interacts directly with eRF1 and GTP and on hydrolysis of the latter, mediates the release of the completed polypeptide [96]. In both cases, the posttermination complex is then poised for recycling.

As is ever the case with mitochondria, the situation is not quite so straightforward. Virtually, no mitochondria from any organism use the standard three stop codons. The most common but not completely universal change among organisms is the recoding of UGA to be recognized as a tryptophan codon, with the retention of UAA and UAG as stops by most mitochondrial systems. However, in a limited number of cases, UAA and UAG have also be recoded as Tyr and Leu, respectively [97]. Another common reassignment is that of the arginine codons that are now recognized in different organisms by tRNAs charged with Ser, Lys, Asn, or Gly. In the case of mammalian mitochondria, however, there is no mt-tRNA that will make a codon:anticodon pairing with either of the arginine codons AGA or AGG (Table 2) [97]. Other codon reassignments can occur from one amino acid to another but currently there are limited other changes to a stop codon, these include UUA by *Thraustochytrium aureum* and UCA by the green alga *Scenedesmus obliquus* [97].

The sampling of the A-site by tRNAs to determine if there is a perfect codon:anticodon match is very fast. By comparison, the sampling and recognition by release factor proteins that have evolved to recognize the stop codons is a slower process but more accurate with release factors recognizing sense codons at a frequency of ~1 in 10^5 [98]. The ability of these proteins to bind to the ribosomal A-site is partly based on the structural resemblance between a folded domain within the RF and tRNAs and represents a novel concept of molecular mimicry between nucleic acids and proteins [99]. Once a stop codon arrives at the A-site of a translating ribosome, it is sampled by charged tRNAs as well as by release factors. Only when it is recognized by the appropriate release factor does this binding event cause a structural rearrangement positioning the conserved GGQ motif into the peptidyl transferase center within the large subunit. This induced fit mechanism allows the discrimination of stop codons from sense codons. Matching between the termination codon and the codon recognition sequence in the RF triggers hydrolysis of the ester bond linking the peptide and the terminal tRNA in the P-site, facilitating the release of a completed translation product. The equivalent of an RF3 has not yet been identified in mitochondria, either to assist in the hydrolysis or to facilitate escape of the RF from the complex.

The number of RFs present in mitochondria differs by organism. Yeasts use a single release factor [100, 101], as do trypanosomes [102], while humans have a family of 4 mtRFs based on the conservation of the GGQ motif that confers ribosome dependent peptidyl-tRNA hydrolysis [103]. In addition to this critical domain, release factors have two highly conserved domains that show sequence specificity and are responsible for decoding the A-site RNA triplet. In two of the human mtRFs, ICT1 (renamed mL62) and C12orf65, these decoding domains are absent. Intriguingly, ICT1/mL62 has become a fully integrated structural component of the large subunit of the mitoribosome [103]; however, a subset that remains free may be acting as a rescue factor of stalled translation complexes [104–106]. Due to its lack of codon specificity C12orf65 (renamed mtRF-R) was also initially thought to be involved in rescuing stalled mitoribosomes, or those that had premature drop off after initiation. It has recently been revealed to partake in a new rescue pathway releasing mitoribosomes that experience frequent stalling [107]. To do this, it acts in concert with an RNA-binding protein, MTRES1 (previously C6orf203), to release the incomplete polypeptide still attached to the mt-tRNA from the large mitoribosomal subunit, indicating that this happens after the dissociation of the 39S from the 28S small subunit [108]. Interestingly, *Drosophila* have orthologs of C12orf65 and ICT1 mL62 but only a single copy of a protein with homology to mtRF1/mtRF1a (reviewed in [55]).

A third RF, mtRF1, was originally identified in silico from an alignment of Expressed Sequence Tags (ESTs) and described as a mitochondrial protein due to the similarity with bacteria and yeast peptide chain release factors [109]. Although it had retained the two decoding domains these sequences diverged from the consensus, and this was thought to enable recognition of the AGG and AGA triplets at the end of the open reading frame within human mt-mRNAs *MTND6* and *MTCO1*, respectively, that were initially presumed to be recoded as stop codons [109]. Since then various theories have been put forward to explain the function of mtRF1 including acting as a rescue factor for mitoribosomes that have no RNA in the A-site. This was based on structural models showing that the extra bulk and new interactions would not be compatible with an RNA-occupied A-site [110, 111]. This leaves mtRF1a (mtRF1L), the fourth of the mt-RF family to act as the main release factor [112]. This protein demonstrates recognition of standard UAG and UAA codons. In the case of seven of the UAA termination signals in humans, these are only completed upon oligo/polyadenylation of the processed transcripts [113]. Therefore, translation termination of 11 of the 13 mammalian mitochondrial open reading frames is facilitated by mtRF1a. How does termination occur in the remaining two open reading frames within *MT-ND6* and *MT-CO1*? Originally, it was suggested that two codons AGA and AGG, unassigned for mt-tRNA recognition, somehow mediated translation termination, possibly through A-site recruitment of one of the translation release factor family members [113, 114]. Intriguingly, however, both codons are preceded by a U, such that a −1 ribosomal frameshift would position a standard UAG in the A-site, allowing termination by mtRF1a. Judicious use of a sequence-specific bacterial endoribonuclease supports this mechanism [112], but the putative roles of other release factors in AGA/AGG termination are also still possible. As mentioned earlier, codon recognition by RFs is slower than by charged tRNAs, allowing for the prospect of spontaneous hydrolysis to cleave the ester bond between the nascent peptide and the P-site tRNA. This may also give the opportunity for free ICT1 (mL62) to enter the A-site and due to its lack of sequence specificity, promote peptide release as has been previously suggested [105, 115]. However, the prospect of free ICT1 being able to function *in vivo* in nonspecific translation termination is potentially concerning.

6 How Are Mitoribosomes Recycled?

Upon completion of the encoded polypeptide and its release from the mitoribosome, the remaining elements of the machinery, namely the transcript, deacylated mt-tRNA and the 2 mitoribosomal subunits, need to be disassembled in readiness for the next cycle of translation. This was thought to be effected by a single protein, the mitochondrial ribosome recycling factor (mtRRF), first identified by the Spremulli group in 1998 [109]. It is now clear that other factors are also involved in promoting the separation of the mitoribosomal subunits. Somewhat unusually, mitochondria have evolved two paralogs of the elongation factor EFG. The first has already been mentioned, mtEF-G1, which operates exclusively in the elongation phase, and a second mtEF-G2 that has evolved a divergent function, participating only in mitoribosome recycling [116, 117], and is also present in *Drosophila* [118]. To avoid confusion, this is now termed mtRRF2, and, acting in concert with mtRRF1, they bring about recycling of the mitoribosome at the end of polypeptide synthesis. An alternative route to mitoribosome splitting particularly under stress conditions is effected by GTP-binding protein 6 (GTPBP6), a homolog of the bacterial ribosome recycling factor HflX [119, 120]. Unlike the bacterial counterpart, GTPBP6 has an additional role in biogenesis of the large subunit of the mitoribosome, acting at a late stage of assembly [119]. In addition to the roles described above for mtIF3, it has also been implicated in this final phase of the translation process. It has been suggested that it too can facilitate the dissociation of mitochondrial 55S monosomes [56]. Characterization of the conditional knockout of *Mtif3* in mice confirmed the role of mtIF3 in initiation as discussed above, but there was no supporting evidence that mtIF3 is required for mitoribosome dissociation, as the ratio of mitoribosomal subunits whether free, assembled or translating was unaltered in these knockout mice [62]. What is clear is that following the separation of the mt-SSU from the mt-LSU, mtIF3 combines with the small subunit. This acts to prevent the binding of any mt-tRNA or -mRNA which could cause unscheduled association of the large and small subunits, leaving all the required components ready for a new round of translation [47, 62, 121].

7 Final Comments

Our understanding of mitochondrial translation in a wide variety of species has been revolutionized by the recent advances in cryoelectron microscopy. We now have structures of mitoribosomes as complete monosomes or as partially assembled subcomplexes and with various permutations of bound translation factors and mt-mRNAs/tRNAs. We are at the cusp of another step change in our understanding of mitochondrial gene expression as we can now begin to manipulate the mitochondrial genome using zinc finger/TALEN technologies and mitochondrially targeted base editors. The remaining major challenge in understanding mitochondrial translation is to reconstitute a workable and meaningful *in vitro* translation system. Although the challenges there are vast, surely it cannot be too long before we see real progress in this area as well. Taken together, it is an exciting time to be a mitochondrial biologist.

Acknowledgments

This work was supported by The Wellcome Trust [203105/Z/16/Z] RNL and ZCL.

References

1. Neupert W (1977) Mitochondrial ribosomes. Horiz Biochem Biophys 3:257–296
2. O'Brien TW, Kalf GF (1967) Ribosomes from rat liver mitochondria. I. Isolation procedure and contamination studies. J Biol Chem 242(9):2172–2179
3. O'Brien TW, Kalf GF (1967) Ribosomes from rat liver mitochondira. II. Partial characterization. J Biol Chem 242(9):2180–2185
4. Hamilton MG, O'Brien TW (1974) Ultracentrifugal characterization of the mitochondrial ribosome and subribosomal particles of bovine liver: molecular size and composition. Biochemistry 13(26):5400–5403
5. Gulikova OM, Zaitseva GN, Pakhomova MV (1979) Mitochondrial ribosomes of the phytoflagellata Astasia longa. Biokhimiia 44(11): 2013–2020
6. Lambowitz AM, Chua NH, Luck DJ (1976) Mitochondrial ribosome assembly in Neurospora. Preparation of mitochondrial ribosomal precursor particles, site of synthesis of mitochondrial ribosomal proteins and studies on the poky mutant. J Mol Biol 107(3):223–253
7. Begueret J, Picard-Bennoun M (1979) Ribosomes of the lower eukaryotes. Biochimie 61(7):VII–XVIII
8. Kuntzel H, Noll H (1967) Mitochondrial and cytoplasmic polysomes from Neurospora crassa. Nature 215(5108):1340–1345
9. Dure LS, Epler JL, Barnett WE (1967) Sedimentation properties of mitochondrial and cytoplasmic ribosomal RNA's from Neurospora. Proc Natl Acad Sci U S A 58(5): 1883–1887
10. Borst P, Grivell LA (1971) Mitochondrial ribosomes. FEBS Lett 13(2):73–88
11. Lederman M, Attardi G (1970) In vitro protein synthesis in a mitochondrial fraction from HeLa cells: sensitivity to antibiotic and ethidium bromide. Biochem Biophys Res Commun 40(6):1492–1500
12. Matthews DE, Hessler RA, Denslow ND et al (1982) Protein composition of the bovine mitochondrial ribosome. J Biol Chem 275(15):8788–8794
13. Pinel C, Douce R, Mache R (1986) A study of mitochondrial ribosomes from the higher plant Solanum tuberosum L. Mol Biol Rep 11(2):93–97
14. O'Brien TW, Fiesler SE, Denslow ND et al (1999) Mammalian mitochondrial ribosomal proteins (2). Amino acid sequencing, characterization, and identification of corresponding gene sequences. J Biol Chem 274(51):36043–36051
15. Goldschmidt-Reisin S, Kitakawa M, Herfurth E et al (1998) Mammalian mitochondrial ribosomal proteins. N-terminal amino acid sequencing, characterization, and identification of corresponding gene sequences. J Biol Chem 273(52):34828–34836
16. Koc EC, Burkhart W, Blackburn K et al (2000) A proteomics approach to the identification of mammalian mitochondrial small subunit ribosomal proteins. J Biol Chem 275(42):32585–32591
17. Graack HR, Wittmann-Liebold B (1998) Mitochondrial ribosomal proteins (MRPs) of yeast. Biochem J 329(Pt 3):433–448
18. Koc EC, Burkhart W, Blackburn K et al (2001) Identification of four proteins from the small subunit of the mammalian mitochondrial ribosome using a proteomics approach. Protein Sci 10(3):471–481
19. Jourdain AA, Koppen M, Wydro M et al (2013) GRSF1 regulates RNA processing in mitochondrial RNA granules. Cell Metab 17(3):399–410
20. Antonicka H, Sasarman F, Nishimura T et al (2013) The mitochondrial RNA-binding protein GRSF1 localizes to RNA granules and is required for posttranscriptional mitochondrial gene expression. Cell Metab 17(3): 386–398
21. Ojala D, Montoya J, Attardi G (1981) tRNA punctuation model of RNA processing in human mitochondria. Nature 290:470–474
22. Holzmann J, Frank P, Loffler E et al (2008) RNase P without RNA: identification and functional reconstitution of the human mitochondrial tRNA processing enzyme. Cell 135(3):462–474
23. Brzezniak LK, Bijata M, Szczesny RJ et al (2011) Involvement of human ELAC2 gene

product in 3′ end processing of mitochondrial tRNAs. RNA Biol 8(4):616–626
24. Chang JH, Tong L (2012) Mitochondrial poly(A) polymerase and polyadenylation. Biochim Biophys Acta 1819(9–10):992–997
25. Wilson WC, Hornig-Do HT, Bruni F et al (2014) A human mitochondrial poly (A) polymerase mutation reveals the complexities of post-transcriptional mitochondrial gene expression. Hum Mol Genet 23(23): 6345–6355
26. Temperley RJ, Wydro M, Lightowlers RN et al (2010) Human mitochondrial mRNAs--like members of all families, similar but different. Biochim Biophys Acta 1797(6–7):1081–1085
27. Binder S, Stoll K, Stoll B (2016) Maturation of 5′ ends of plant mitochondrial RNAs. Physiol Plant 157(3):280–288
28. Cahoon AB, Nauss JA, Stanley CD et al (2017) Deep transcriptome sequencing of two green algae, Chara vulgaris and Chlamydomonas reinhardtii, provides no evidence of organellar RNA editing. Genes (Basel) 8(2): 80
29. Gualberto JM, Lamattina L, Bonnard G et al (1989) RNA editing in wheat mitochondria results in the conservation of protein sequences. Nature 341(6243):660–662
30. Rovira AG, Smith AG (2019) PPR proteins – orchestrators of organelle RNA metabolism. Physiol Plant 166(1):451–459
31. Alverson AJ, Wei X, Rice DW et al (2010) Insights into the evolution of mitochondrial genome size from complete sequences of Citrullus lanatus and Cucurbita pepo (Cucurbitaceae). Mol Biol Evol 27(6):1436–1448
32. Groth-Malonek M, Pruchner D, Grewe F et al (2005) Ancestors of trans-splicing mitochondrial introns support serial sister group relationships of hornworts and mosses with vascular plants. Mol Biol Evol 22(1):117–125
33. Knoop V (2004) The mitochondrial DNA of land plants: peculiarities in phylogenetic perspective. Curr Genet 46(3):123–139
34. Salinas-Giege T, Cavaiuolo M, Cognat V et al (2017) Polycytidylation of mitochondrial mRNAs in Chlamydomonas reinhardtii. Nucleic Acids Res 45(22):12963–12973
35. Gallaher SD, Fitz-Gibbon ST, Strenkert D et al (2018) High-throughput sequencing of the chloroplast and mitochondrion of Chlamydomonas reinhardtii to generate improved de novo assemblies, analyze expression patterns and transcript speciation, and evaluate diversity among laboratory strains and wild isolates. Plant J 93(3):545–565
36. Kuhn J, Tengler U, Binder S (2001) Transcript lifetime is balanced between stabilizing stem-loop structures and degradation-promoting polyadenylation in plant mitochondria. Mol Cell Biol 21(3):731–742
37. Hirayama T (2021) PARN-like proteins regulate gene expression in land plant mitochondria by modulating mRNA polyadenylation. Int J Mol Sci 22(19):10776
38. Osinga KA, De Vries E, Van der Horst G et al (1984) Processing of yeast mitochondrial messenger RNAs at a conserved dodecamer sequence. EMBO J 3(4):829–834
39. Butow RA, Zhu H, Perlman P et al (1989) The role of a conserved dodecamer sequence in yeast mitochondrial gene expression. Genome 31(2):757–760
40. Li H, Zassenhaus HP (1999) Purification and characterization of an RNA dodecamer sequence binding protein from mitochondria of Saccharomyces cerevisiae. Biochem Biophys Res Commun 261(3):740–745
41. Aphasizhev R, Aphasizheva I (2011) Mitochondrial RNA processing in trypanosomes. Res Microbiol 162(7):655–663
42. Aphasizhev R, Aphasizheva I (2014) Mitochondrial RNA editing in trypanosomes: small RNAs in control. Biochimie 100:125–131
43. Aphasizheva I, Alfonzo J, Carnes J et al (2020) Lexis and grammar of mitochondrial RNA processing in trypanosomes. Trends Parasitol 36(4):337–355
44. Denslow ND, Michaels GS, Montoya J et al (1989) Mechanism of mRNA binding to bovine mitochondrial ribosomes. J Biol Chem 264(14):8328–8338
45. Jones CN, Wilkinson KA, Hung KT et al (2008) Lack of secondary structure characterizes the 5′ ends of mammalian mitochondrial mRNAs. RNA 14(5):862–871
46. Christian BE, Spremulli LL (2010) Preferential selection of the 5′-terminal start codon on leaderless mRNAs by mammalian mitochondrial ribosomes. J Biol Chem 285(36): 28379–28386
47. Khawaja A, Itoh Y, Remes C et al (2020) Distinct pre-initiation steps in human mitochondrial translation. Nat Commun 11(1): 2932
48. Takemoto C, Spremulli LL, Benkowski LA et al (2009) Unconventional decoding of the AUA codon as methionine by mitochondrial tRNAMet with the anticodon f5CAU as revealed with a mitochondrial in vitro translation system. Nucleic Acids Res 37(5): 1616–1627

49. Hinttala R, Sasarman F, Nishimura T et al (2015) An N-terminal formyl methionine on COX 1 is required for the assembly of cytochrome c oxidase. Hum Mol Genet 24(14): 4103–4113
50. Haack TB, Gorza M, Danhauser K et al (2014) Phenotypic spectrum of eleven patients and five novel MTFMT mutations identified by exome sequencing and candidate gene screening. Mol Genet Metab 111(3): 342–352
51. Sasarman F, Brunel-Guitton C, Antonicka H et al (2010) LRPPRC and SLIRP interact in a ribonucleoprotein complex that regulates posttranscriptional gene expression in mitochondria. Mol Biol Cell 21(8):1315–1323
52. Ruzzenente B, Metodiev MD, Wredenberg A et al (2012) LRPPRC is necessary for polyadenylation and coordination of translation of mitochondrial mRNAs. EMBO J 31(2): 443–456
53. Singh V, Itoh Y, Huynen MA et al (2022) Activation mechanism of mitochondrial translation by LRPPRC-SLIRP. bioRxiv
54. Koc EC, Spremulli LL (2002) Identification of mammalian mitochondrial translational initiation factor 3 and examination of its role in initiation complex formation with natural mRNAs. J Biol Chem 277(38):35541–35549
55. Marygold SJ, Attrill H, Lasko P (2017) The translation factors of Drosophila melanogaster. Fly (Austin) 11(1):65–74
56. Christian BE, Spremulli LL (2009) Evidence for an active role of IF3mt in the initiation of translation in mammalian mitochondria. Biochemistry 48(15):3269–3278
57. Gaur R, Grasso D, Datta PP et al (2008) A single mammalian mitochondrial translation initiation factor functionally replaces two bacterial factors. Mol Cell 29(2):180–190
58. Haque ME, Koc H, Cimen H et al (2011) Contacts between mammalian mitochondrial translational initiation factor 3 and ribosomal proteins in the small subunit. Biochim Biophys Acta 1814(12):1779–1784
59. Haque ME, Grasso D, Spremulli LL (2008) The interaction of mammalian mitochondrial translational initiation factor 3 with ribosomes: evolution of terminal extensions in IF3mt. Nucleic Acids Res 36(2):589–597
60. Bhargava K, Spremulli LL (2005) Role of the N- and C-terminal extensions on the activity of mammalian mitochondrial translational initiation factor 3. Nucleic Acids Res 33(22): 7011–7018
61. Derbikova K, Kuzmenko A, Levitskii S et al (2018) Biological and evolutionary significance of terminal extensions of mitochondrial translation initiation factor 3. Int J Mol Sci 19(12):3861
62. Rudler DL, Hughes LA, Perks KL et al (2019) Fidelity of translation initiation is required for coordinated respiratory complex assembly. Sci Adv 5(12):eaay2118
63. Herbert CJ, Labarre-Mariotte S, Cornu D et al (2021) Translational activators and mitoribosomal isoforms cooperate to mediate mRNA-specific translation in Schizosaccharomyces pombe mitochondria. Nucleic Acids Res 49(19):11145–11166
64. Weraarpachai W, Antonicka H, Sasarman F et al (2009) Mutation in TACO1, encoding a translational activator of COX I, results in cytochrome c oxidase deficiency and late-onset Leigh syndrome. Nat Genet 41(7): 833–837
65. Richman TR, Spahr H, Ermer JA et al (2016) Loss of the RNA-binding protein TACO1 causes late-onset mitochondrial dysfunction in mice. Nat Commun 7:11884
66. Hammani K, Giege P (2014) RNA metabolism in plant mitochondria. Trends Plant Sci 19(6):380–389
67. Gao G, Kapushoc ST, Simpson AM et al (2001) Guide RNAs of the recently isolated LEM125 strain of Leishmania tarentolae: an unexpected complexity. RNA 7(9): 1335–1347
68. Fox TD (ed) (1996) Genetics of mitochondrial translation. In: Hershey JWB, Matthews MB, Sonnenberg N (eds) Translational control. Cold Spring Harbor, New York, pp 733–758
69. Aphasizheva I, Maslov DA, Qian Y et al (2015) Ribosome-associated pentatricopeptide repeat proteins function as translational activators in mitochondria of trypanosomes. Mol Microbiol 99(6):1043–1058
70. Sloan DB, Warren JM, Williams AM et al (2018) Cytonuclear integration and co-evolution. Nat Rev Genet 19(10): 635–648
71. Cristodero M, Brogli R, Joss O et al (2021) tRNA 3′ shortening by LCCR4 as a response to stress in Trypanosoma brucei. Nucleic Acids Res 49(3):1647–1661
72. Clayton C (2019) Regulation of gene expression in trypanosomatids: living with polycistronic transcription. Open Biol 9(6):190072
73. Francklyn CS (2008) DNA polymerases and aminoacyl-tRNA synthetases: shared mechanisms for ensuring the fidelity of gene expression. Biochemistry 47(45):11695–11703

74. Garin S, Levi O, Cohen B et al (2020) Localization and RNA binding of mitochondrial aminoacyl tRNA synthetases. Genes (Basel) 11(10):1185
75. Rubio Gomez MA, Ibba M (2020) Aminoacyl-tRNA synthetases. RNA 26(8): 910–936
76. Nagao A, Suzuki T, Katoh T et al (2009) Biogenesis of glutaminyl-mt tRNAGln in human mitochondria. Proc Natl Acad Sci U S A 106(38):16209–16214
77. Suzuki T, Yashiro Y, Kikuchi I et al (2020) Complete chemical structures of human mitochondrial tRNAs. Nat Commun 11(1):4269
78. Duchene AM, Giritch A, Hoffmann B et al (2005) Dual targeting is the rule for organellar aminoacyl-tRNA synthetases in Arabidopsis thaliana. Proc Natl Acad Sci U S A 102(45):16484–16489
79. Rinehart J, Horn EK, Wei D et al (2004) Non-canonical eukaryotic glutaminyl- and glutamyl-tRNA synthetases form mitochondrial aminoacyl-tRNA in Trypanosoma brucei. J Biol Chem 279(2):1161–1166
80. Guitart T, Leon Bernardo T, Sagales J et al (2010) New aminoacyl-tRNA synthetase-like protein in insecta with an essential mitochondrial function. J Biol Chem 285(49): 38157–38166
81. Picchioni D, Antolin-Fontes A, Camacho N et al (2019) Mitochondrial protein synthesis and mtDNA levels coordinated through an aminoacyl-tRNA synthetase subunit. Cell Rep 27(1):40–47 e45
82. Gonzalez-Serrano LE, Chihade JW, Sissler M (2019) When a common biological role does not imply common disease outcomes: disparate pathology linked to human mitochondrial aminoacyl-tRNA synthetases. J Biol Chem 294(14):5309–5320
83. Blanchard SC, Gonzalez RL, Kim HD et al (2004) tRNA selection and kinetic proofreading in translation. Nat Struct Mol Biol 11(10):1008–1014
84. Rodnina MV, Pape T, Fricke R et al (1996) Initial binding of the elongation factor Tu. GTP.aminoacyl-tRNA complex preceding codon recognition on the ribosome. J Biol Chem 271(2):646–652
85. Hopfield JJ (1974) Kinetic proofreading: a new mechanism for reducing errors in biosynthetic processes requiring high specificity. Proc Natl Acad Sci U S A 71(10):4135–4139
86. Ohtsuki T, Watanabe Y, Takemoto C et al (2001) An "elongated" translation elongation factor Tu for truncated tRNAs in nematode mitochondria. J Biol Chem 276(24):21571–21577
87. Sato A, Suematsu T, Aihara KK et al (2017) Duplication of Drosophila melanogaster mitochondrial EF-Tu: pre-adaptation to T-arm truncation and exclusion of bulky aminoacyl residues. Biochem J 474(6):957–969
88. Kummer E, Ban N (2020) Structural insights into mammalian mitochondrial translation elongation catalyzed by mtEFG1. EMBO J 39(15):e104820
89. Koripella RK, Sharma MR, Bhargava K et al (2020) Structures of the human mitochondrial ribosome bound to EF-G1 reveal distinct features of mitochondrial translation elongation. Nat Commun 11(1):3830
90. Kummer E, Ban N (2021) Mechanisms and regulation of protein synthesis in mitochondria. Nat Rev Mol Cell Biol 22(5):307–325
91. Chiron S, Suleau A, Bonnefoy N (2005) Mitochondrial translation: elongation factor tu is essential in fission yeast and depends on an exchange factor conserved in humans but not in budding yeast. Genetics 169(4): 1891–1901
92. Bauerschmitt H, Funes S, Herrmann JM (2008) The membrane-bound GTPase Gufl promotes mitochondrial protein synthesis under suboptimal conditions. J Biol Chem 283(25):17139–17146
93. Yang F, Gao Y, Li Z et al (2014) Mitochondrial EF4 links respiratory dysfunction and cytoplasmic translation in Caenorhabditis elegans. Biochim Biophys Acta 1837(10): 1674–1683
94. Wang F, Zhang D, Zhang D et al (2021) Mitochondrial protein translation: emerging roles and clinical significance in disease. Front Cell Dev Biol 9:675465
95. Burroughs AM, Aravind L (2019) The origin and evolution of release factors: implications for translation termination, ribosome rescue, and quality control pathways. Int J Mol Sci 20(8):1981
96. Hellen CUT (2018) Translation termination and ribosome recycling in eukaryotes. Cold Spring Harb Perspect Biol 10(10):a032656
97. Elzanowski A, Ostell J. http://www.ncbi.nlm.nih.gov/Taxonomy/taxonomyhome.html/index.cgi?chapter=tgencodes#SG1. NCBI
98. Jorgensen F, Adamski FM, Tate WP et al (1993) Release factor-dependent false stops are infrequent in Escherichia coli. J Mol Biol 230(1):41–50
99. Ito K, Ebihara K, Uno M et al (1996) Conserved motifs in prokaryotic and eukaryotic polypeptide release factors: tRNA-protein

mimicry hypothesis. Proc Natl Acad Sci U S A 93(11):5443–5448
100. Pel HJ, Maat C, Rep M et al (1992) The yeast nuclear gene MRF1 encodes a mitochondrial peptide chain release factor and cures several mitochondrial RNA splicing defects. Nucleic Acids Res 20(23):6339–6346
101. Askarian-Amiri ME, Pel HJ, Guevremont D et al (2000) Functional characterization of yeast mitochondrial release factor 1. J Biol Chem 275(23):17241–17248
102. Cristodero M, Mani J, Oeljeklaus S et al (2013) Mitochondrial translation factors of Trypanosoma brucei: elongation factor-Tu has a unique subdomain that is essential for its function. Mol Microbiol 90(4):744–755
103. Richter R, Rorbach J, Pajak A et al (2010) A functional peptidyl-tRNA hydrolase, ICT1, has been recruited into the human mitochondrial ribosome. EMBO J 29(6):1116–1125
104. Feaga HA, Quickel MD, Hankey-Giblin PA et al (2016) Human cells require non-stop ribosome rescue activity in mitochondria. PLoS Genet 12(3):e1005964
105. Akabane S, Ueda T, Nierhaus KH et al (2014) Ribosome rescue and translation termination at non-standard stop codons by ICT1 in mammalian mitochondria. PLoS Genet 10(9):e1004616
106. Chicherin IV, Dukhalin SV, Khannanov RA et al (2021) Functional diversity of mitochondrial peptidyl-tRNA hydrolase ICT1 in human cells. Front Mol Biosci 8:716885
107. Desai N, Yang H, Chandrasekaran V et al (2020) Elongational stalling activates mitoribosome-associated quality control. Science 370(6520):1105–1110
108. Gopalakrishna S, Pearce SF, Dinan AM et al (2019) C6orf203 is an RNA-binding protein involved in mitochondrial protein synthesis. Nucleic Acids Res 47(17):9386–9399
109. Zhang Y, Spremulli LL (1998) Identification and cloning of human mitochondrial translational release factor 1 and the ribosome recycling factor. Biochim Biophys Acta 1443(1–2):245–250
110. Lind C, Sund J, Aqvist J (2013) Codon-reading specificities of mitochondrial release factors and translation termination at non-standard stop codons. Nat Commun 4:2940
111. Huynen MA, Duarte I, Chrzanowska-Lightowlers ZM et al (2012) Structure based hypothesis of a mitochondrial ribosome rescue mechanism. Biol Direct 7:14
112. Temperley R, Richter R, Dennerlein S et al (2010) Hungry codons promote frameshifting in human mitochondrial ribosomes. Science 327(5963):301
113. Anderson S, Bankier AT, Barrell BG et al (1981) Sequence and organization of the human mitochondrial genome. Nature 290(5806):457–465
114. Lee CC, Timms KM, Trotman CN et al (1987) Isolation of a rat mitochondrial release factor. Accommodation of the changed genetic code for termination. J Biol Chem 262(8):3548–3552
115. Kummer E, Schubert KN, Schoenhut T et al (2021) Structural basis of translation termination, rescue, and recycling in mammalian mitochondria. Mol Cell 81(12):2566–2582.e2566
116. Hammarsund M, Wilson W, Corcoran M et al (2001) Identification and characterization of two novel human mitochondrial elongation factor genes, hEFG2 and hEFG1, phylogenetically conserved through evolution. Hum Genet 109(5):542–550
117. Tsuboi M, Morita H, Nozaki Y et al (2009) EF-G2mt is an exclusive recycling factor in mammalian mitochondrial protein synthesis. Mol Cell 35(4):502–510
118. Trivigno C, Haerry TE (2011) The Drosophila mitochondrial translation elongation factor G1 contains a nuclear localization signal and inhibits growth and DPP signaling. PLoS One 6(2):e16799
119. Lavdovskaia E, Denks K, Nadler F et al (2020) Dual function of GTPBP6 in biogenesis and recycling of human mitochondrial ribosomes. Nucleic Acids Res 48(22):12929–12942
120. Zhang Y, Mandava CS, Cao W et al (2015) HflX is a ribosome-splitting factor rescuing stalled ribosomes under stress conditions. Nat Struct Mol Biol 22(11):906–913
121. Koripella RK, Sharma MR, Risteff P et al (2019) Structural insights into unique features of the human mitochondrial ribosome recycling. Proc Natl Acad Sci U S A 116(17):8283–8288

Part II

Methods to Study Mitoribosome Structure, Function, and Biogenesis

Chapter 5

Sample Preparation of Isolated Mitochondria for Cryoelectron Tomography and In Situ Studies of Translation

Lena Thärichen, Robert Englmeier, and Friedrich Förster

Abstract

Cryoelectron tomography is a method to image biological samples three-dimensionally at molecular resolution. This modality provides insights into intracellular processes in their physiological settings. Obtaining a high-quality sample for cryoelectron tomography on mitochondria, however, can be challenging. In this chapter, we describe the crucial steps from sample preparation to data acquisition enabling studies of mitochondrial translation in situ by cryoelectron tomography. We provide detailed protocols for yeast and human mitochondria preparations yielding a high concentration of intact mitochondrial vesicles on cryo-EM grids. In addition, we describe a workflow for particle identification and spatial mapping in context of the organelle.

Key words Mitochondria, Cryoelectron tomography, Translation, Ribosome

1 Introduction

Cryoelectron tomography (cryo-ET) has extended the scope of structural biology studies from isolated molecules to molecules in context of their physiological organellar environment and even native cellular settings allowing to study their molecular sociology [1]. Cryo-ET provides 3D density volumes of an object of interest, typically referred to as tomograms. They are calculated from a set of 2D projection images of the sequentially rotated object, which is called a tilt series. These projections are acquired using a transmission electron microscope (TEM), which provides parallel projections of the sample along the electron beam to very good approximation [2].

Prior to imaging the sample is vitrified. The transition into a vitreous phase of matter is achieved by rapid cooling via plunge freezing into a suitable cryogen like liquid ethane [3]. Importantly, cryo-ET does not involve dehydration of the sample and does not

Antoni Barrientos and Flavia Fontanesi (eds.), *The Mitoribosome: Methods and Protocols*,
Methods in Molecular Biology, vol. 2661, https://doi.org/10.1007/978-1-0716-3171-3_5,

require any staining or fixation agents, enabling studies under near native aqueous conditions and avoiding preparation artifacts of EM methods relying on resin embedding. Cryo-ET relies on the contrast generated by intrinsic density differences of the sample, which is relatively weak and gives rise to a low signal-to-noise ratio (SNR) of the data [2]. A further limitation of cryo-ET is that sample must not exceed few hundred nanometers in thickness to remain transparent to the electron beam.

The potential of cryo-ET to retain the organelle environment during structural studies makes it attractive for mitochondrial research. Due to the requirement of relatively thin samples, cryo-ET has typically been applied to isolated mitochondria. Early studies revealed the ultrastructure of such isolated mitochondria with relatively modest TEM equipment [4]. High-voltage TEMs (typically operated at 200–300 kV) and advanced direct electron detectors allow insights into the organization of specific macromolecules. Due to the low SNR of tomograms, high-molecular-weight complexes are easier to distinguish in the data due to their larger volume. Thus, cryo-ET has primarily provided information on the intricate spatial organization of large mitochondrial complexes such as the ATP synthetase [5, 6].

Cryo-ET also enables a comprehensive understanding of mitochondrial translation with the large molecular weight mitochondrial ribosome (mitoribosome) as its centerpiece. Two routes are typically pursued to analyze specific particles depicted in tomograms. On the one hand, features of interest such as mitoribosomes can be automatically identified and traced, resulting in a segmentation, as exemplified by Pfeffer et al. and Englmeier et al. for mitochondrial polysomes *in situ* [7, 8]. On the other hand, particle volumes or subtomograms depicting the same type of macromolecule can be aligned in 3D and added up during subtomogram averaging [9]. This step significantly increases the electron density SNR, which manifests in higher-resolution reconstructions of the macromolecule of interest. Thus, subtomogram averaging of yeast and human mitoribosomes *in organello* revealed details of membrane tethering, protein insertion, and translation regulation [7, 8].

Efficient collection of high-quality data relies on a high-quality mitochondria sample. Mitochondria have been isolated from a variety of sources for decades and used in functional studies, examining respiration [10], mitochondrial translation [11], or mitochondrial import [12–14], respectively. However, these protocols include harsh centrifugation steps. For yeast mitochondria preparations, additionally involve extended biochemical treatment for cell wall digestion [15]. Cryo-EM grid preparation from these samples results in a relatively low number of structurally intact mitochondria per grid area, which inevitably leads to extended search time during data collection setup at the cost of expensive and often closely calculated electron microscope time. In this chapter, we describe gentle and fast protocols for mitochondria preparations from human (HEK) and yeast (*S. cerevisiae*) cells for the purpose of cryo-ET. Furthermore, we complement the sample preparation protocols with a data processing workflow for membrane segmentation and particle mapping in tomograms.

2 Materials

2.1 Isolation of Mitochondria from HEK Suspension Cells

This protocol is focused on obtaining a highly enriched, intact mitochondria preparation for the purpose of cryo-ET and will contain other subcellular vesicle contaminations, such as endoplasmic reticulum (ER) vesicles. The underlying procedure is differential centrifugation for separation of different subcellular vesicles after cell lysis (Fig. 1a) [10]. The centrifugation speeds have been optimized for HEK suspension cells (Fig. 1b) and might need to be optimized for other cell lines (*see* **Note 1**). Since Cytochrome C resides in the intermembrane space, its enrichment in the pellet fraction indicates outer membrane intactness (Fig. 1b). For buffer preparations, use sterile filtered (ø 0.2 μm) Hepes and Sucrose stock solutions.

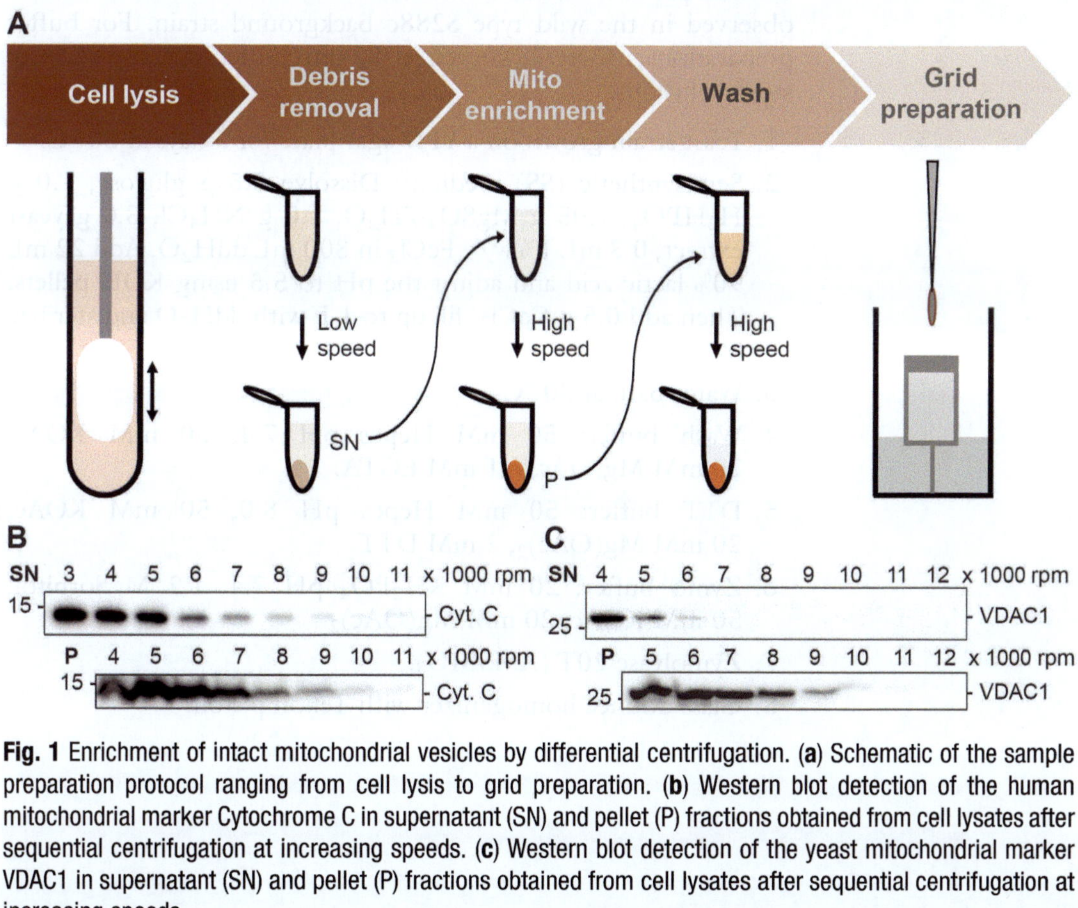

Fig. 1 Enrichment of intact mitochondrial vesicles by differential centrifugation. (**a**) Schematic of the sample preparation protocol ranging from cell lysis to grid preparation. (**b**) Western blot detection of the human mitochondrial marker Cytochrome C in supernatant (SN) and pellet (P) fractions obtained from cell lysates after sequential centrifugation at increasing speeds. (**c**) Western blot detection of the yeast mitochondrial marker VDAC1 in supernatant (SN) and pellet (P) fractions obtained from cell lysates after sequential centrifugation at increasing speeds

1. HEK suspension-adapted cells (e.g., FreeStyle™ 293-F Cells from Thermo Fisher) at 1–1.5 mio cells/mL grown in FreeStyle™ 293 Expression Medium (Thermo Fisher) for >2 days.
2. Phosphate-buffered saline (PBS) w/o magnesium and calcium.
3. Glass dounce homogenizer with Teflon piston.
4. Homogenization buffer HEK: 20 mM Hepes pH 7.4, 250 mM Sucrose, 1 mM EGTA, 1 mM PMSF, 0.2% fatty acid-free BSA in ddH_2O.
5. Mitochondria buffer HEK: 20 mM Hepes pH 7.4, 250 mM Sucrose.
6. UV/Vis spectrophotometer for OD_{280} measurement (Nanodrop™).

2.2 Isolation of Mitochondria from Yeast Cells

Like the protocol for mitochondria preparations from HEK cells (Fig. 1a), this protocol has been optimized for a high abundance of intact mitochondrial vesicles for cryo-ET and will still contain other subcellular vesicles. The protocol was based on Pfeffer et al. [7], and buffers, cell wall digestion, and centrifugation speeds have been optimized for *S. cerevisiae* spheroplasts (BY4742 [16]) (Fig. 1c). This yeast strain carries modifications to maximize mitochondrial biomass production and reduce loss of mitochondrial DNA that is observed in the wild type S288c background strain. For buffer preparations, use sterile filtered (ø 0.2 μm) buffer, salt, and Sorbitol stock solutions.

1. Yeast strain grown on a YPD agar plate for 3 days at 30 °C.
2. Semisynthetic (SS) medium: Dissolve 0.5 g glucose, 1.0 g H_2HPO_4, 1.05 g $MgSO_4{\cdot}7H_2O$, 1.0 g NH_4Cl, 3.0 g yeast extract, 0.3 mL 1% w/v $FeCl_3$ in 800 mL ddH_2O. Add 22 mL 90% lactic acid and adjust the pH to 5.5 using KOH pellets. Then add 0.5 g $CaCl_2$, fill up to 1 L with ddH_2O and sterilize by autoclaving.
3. Water bath at 30 °C.
4. Wash buffer: 50 mM Hepes pH 7.4, 50 mM KOAc, 20 mM $Mg(OAc)_2$, 1 mM EGTA.
5. DTT buffer: 50 mM Hepes pH 8.0, 50 mM KOAc, 20 mM $Mg(OAc)_2$, 2 mM DTT.
6. Zymo buffer: 20 mM KH_2PO_4 pH 7.4, 1.2 M Sorbitol, 50 mM KOAc, 20 mM $Mg(OAc)_2$.
7. Zymolyase 20T (AMSBIO).
8. Glass dounce homogenizer with Teflon piston.

9. Homogenization buffer yeast: 20 mM Hepes pH 7.4, 600 mM Sorbitol, 50 mM KOAc, 20 mM $Mg(OAc)_2$, 1 mM PMSF, 1 mM EGTA.
10. 2 mL Eppendorf tubes.
11. Mitochondria buffer yeast: 20 mM Hepes pH 7.4, 600 mM Sorbitol, 50 mM KOAc, 20 mM $Mg(OAc)_2$.
12. UV/Vis spectrophotometer for OD_{280} measurement (Nanodrop™).

2.3 Grid Preparation

To reduce sample viscosity and beam sensitivity while ensuring membrane intactness, mitochondria samples are diluted with the respective plunging buffer on the mounted grid immediately prior to blotting. Low concentrations of Sucrose or Sorbitol in the plunging buffer will ensure vesicle intactness. Manual one-sided blotting is used to reduce mechanical force on mitochondrial vesicles. BSA-coated gold fiducials are added for tilt series alignment during the tomogram reconstruction process [17].

1. Plunging buffer HEK: 120 mM Sucrose, 20 mM Hepes pH 7.4, 10 nm BSA-coated gold beads (e.g., UMC Utrecht diluted 1:30–1:40).
2. Plunging buffer yeast: 20 mM Hepes pH 7.4, 200 mM Sorbitol, 50 mM KOAc, 20 mM $Mg(OAc)_2$, 10 nm BSA-coated gold beads (e.g., UMC Utrecht diluted 1:30–1:40).
3. Protective goggles and gloves.
4. Manual plunger for single-sided blotting.
5. Ice-free liquid Nitrogen.
6. Pressurized ethane gas.
7. Grid storage boxes for grids.
8. Glow discharger.
9. Quantifoil® Cu 200 Mesh Holey Carbon R2/1 or Cu 200 Mesh Lacey Carbon grids.
10. Blotting filter paper Whatman® 595.

2.4 Image Analysis and Visualization Software

The workflow described here requires already reconstructed tomograms. The software packages required to perform the described segmentation protocol are the following:

1. ChimeraX 1.2.5 [18] for template generation, visualization, and segmentation clean-up.
2. Relion image handler [19] to prepare a template for template matching.

3. PyTOM v0.993 [20] for tomogram deconvolution (`deconv.py`), template matching, and motl file generation (`pytom_xml_to_motl.py`) (https://github.com/FridoF/PyTom).
4. EMAN 2.31 [21] for convolutional neural network (CNN) segmentation [22].
5. TOM toolbox [23].
6. MATLAB vR2018a (The MathWorks Inc.): `plot_particles.m` and `tom_classmask.m` (https://bitbucket.org/FridoF/av3/src/master/utils/).

3 Methods

3.1 Mitochondria Isolation Human

To isolate mitochondria from HEK cells, these need to be grown to logarithmic growth phase (*see* **Notes 2** and **3**). Lysis is performed mechanically while ensuring organelle intactness using a dounce-homogenizer. Gentle differential centrifugation ensures vesicle intactness during the enrichment step (Fig. 1). All buffers should be prepared and cooled prior to the lysis step and the centrifuge and other equipment should be cooled as well to ensure fast execution of the protocol (*see* **Note 4**).

1. Grow suspension-adapted HEK cells in FreeStyle™ medium to a density of approximately 1.5×10^6 cells/mL. 50 mio cells are sufficient for a small-scale purification and preparation of more than 10 grids.
2. Harvest the cells by centrifugation at $400 \times g$ for 5 min.
3. Discard the supernatant, resuspend the pellet in 25 mL ice-cold PBS per 50 mio cells, and repeat **step 2**.
4. Discard the supernatant, resuspend the pellet in 3.5 mL ice-cold homogenization buffer per 50 mio cells, and transfer the suspension to the precooled dounce homogenizer.
5. Lyse the cells by douncing three times with 30 strokes. Shortly cool the lysate on ice in between sets.
6. Transfer the lysate to 2 mL Eppendorf reaction tubes and centrifuge at $800 \times g$ (Heraeus Biofuge 13, 3000 rpm) for 5 min at 4 °C for removal of unbroken cells, nuclei and other cellular debris.
7. Transfer the supernatant into new 2 mL Eppendorf tubes and repeat the centrifugation as in **step 6**.
8. Transfer the supernatant into new 2 mL Eppendorf tubes and centrifuge at $3000 \times g$ (Heraeus Biofuge 13, 6000 rpm) for 5 min at 4 °C to pellet the mitochondria fraction.

9. Remove the supernatant and carefully resuspend each pellet in 1 mL ice-cold mitochondria buffer. Repeat the centrifugation as in **step 8** to wash.
10. Resuspend the mitochondria pellet carefully in ice-cold homogenization buffer to an OD_{280} of 1–1.4 for grid preparation.

3.2 Mitochondria Isolation Yeast

For isolation of yeast mitochondria, the same requirements apply as described under Subheading 3.1. Additionally, DTT buffer and Zymo buffer need to be prewarmed. DTT treatment and zymolyase digestion is kept as short as possible.

1. Streak out the yeast strain onto a YPD agar plate and grow it for 3 days at 30 °C.
2. Use these cells to inoculate 20 mL of SS medium in the morning with an inoculation loop and grow the culture at 30 °C and 160 rpm for 8–10 h.
3. In the evening, inoculate 1 L of SS medium with the preculture, aiming at an OD_{600} of 0.7–1.0 the next day.
4. Harvest the cells at room temperature (RT) and 4000 × *g* for 10 min.
5. Discard the supernatant and resuspend the cell pellet in 50 mL RT wash buffer. Transfer into a preweighed 50 mL Eppendorf tube and centrifuge at RT, 4000 × *g* for 5 min.
6. Discard the supernatant and weigh the cell pellet.
7. Per 1 g of cell pellet, add 6 mL prewarmed DTT buffer (at 30 °C), resuspend well and incubate at 30 °C for 5 min. Then harvest the cells as in **step 5**.
8. Per 1 g of cell pellet add 17 mg Zymolyase 20T dissolved in prewarmed (30 °C) Zymo buffer and resuspend well. Incubate at 30 °C for 40 min and mix the suspension by inverting the tube every 15 min. Then harvest the cells as in **step 5**.
9. Discard the supernatant, resuspend the cell pellet in 25 mL Zymo buffer without Zymolyase 20T and repeat the centrifugation as in **step 5**.
10. Resuspend the pellet in 2.3 mL ice-cold homogenization buffer per 1 g cell mass, transfer to the precooled dounce homogenizer and lyse the cells three times with 20 strokes. Shortly cool the lysate on ice between sets.
11. Transfer the lysate into 2 mL Eppendorf tubes and centrifuge for 5 min at 4 °C and 1200 × *g* (Heraeus Biofuge 13, 4000 rpm). Repeat this step two more times for the respective supernatant sample.

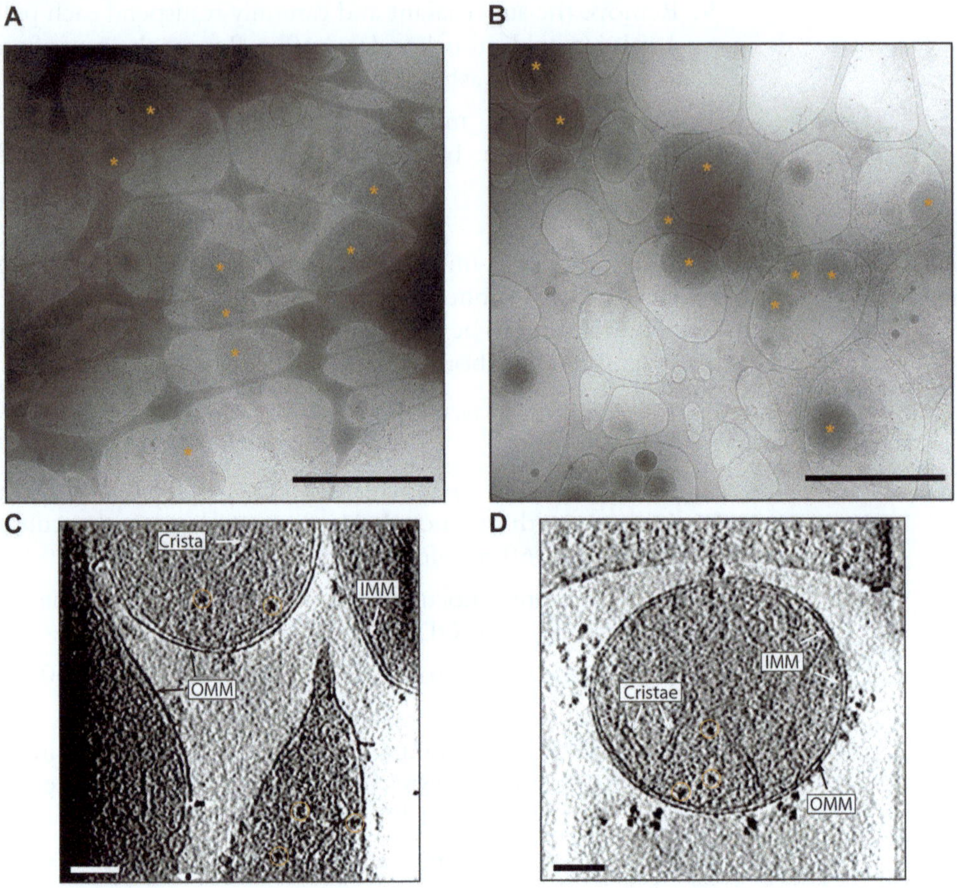

Fig. 2 Cryo-ET sample quality of isolated HEK and yeast cell mitochondria. Electron micrographs of mitochondria samples acquired on a 200 kV Talos Arctica electron microscope from (**a**) HEK cell or (**b**) yeast mitochondria preparations. Intact mitochondrial vesicles are indicated by asterisks and scale bars correspond to 2 μm. (**c**) Example tomogram of HEK cell mitochondria. Intact inner (IMM) and outer (OMM) mitochondrial membranes, Cristae and mitochondrial ribosomes are indicated by arrows or circles, respectively. (**d**) Example tomogram of a yeast mitochondrion with features indicated as in (**c**). Scale bars correspond to 100 nm

12. Pellet the mitochondria fraction for 5 min at 4 °C and 5300 × *g* (Heraeus Biofuge 13, 8000 rpm) and resuspend the pellet in mitochondria buffer to an OD_{280} of 2.5.

3.3 Grid Preparation

The described procedure yields a good distribution (Fig. 2a, b) of largely intact mitochondria (Fig. 2c, d) on grids suitable for efficient cryoelectron tomography data acquisition.

1. Prepare the grid plunger with liquid nitrogen and ethane using goggles and gloves for protection.

2. Glow discharge the grids for 30 s at 20 mA and 0.39 mBar for Holey carbon grids and 25 s at 15 mA and 0.39 mBar for Lacey carbon grids (*see* **Note 5**).
3. Insert a glow discharged grid into the plunging tweezers and add 2 μL mitochondria sample onto the carbon-coated side of the grid. Then insert the tweezer into the plunging device.
4. Add 4 μL gold fiducial buffer (HEK or yeast, respectively) and carefully mix with the sample on the grid (*see* **Note 6**).
5. Immediately blot away excess sample from the back side of the grid using blotting paper and plunge the grid into liquid ethane (*see* **Note 7**).
6. Store grids in grid boxes in liquid nitrogen.

3.4 Tomogram Segmentation

The following protocol describes all necessary steps to segment mitochondrial membranes using the CNN in EMAN2 and to identify molecules of interest by template matching using pyTOM, as well as plotting them back into the tomographic volume using the TOM toolbox and MATLAB. As mentioned in Subheading 2.4, a reconstructed tomogram is required. Furthermore, we recommend working with 6–8 times down-sampled data (pixel size corresponding to approximately 15 Å). An example for a yeast mitochondrion with visualized inner and outer membranes and cytosolic, as well as mitochondrial ribosomes is shown in Fig. 3c.

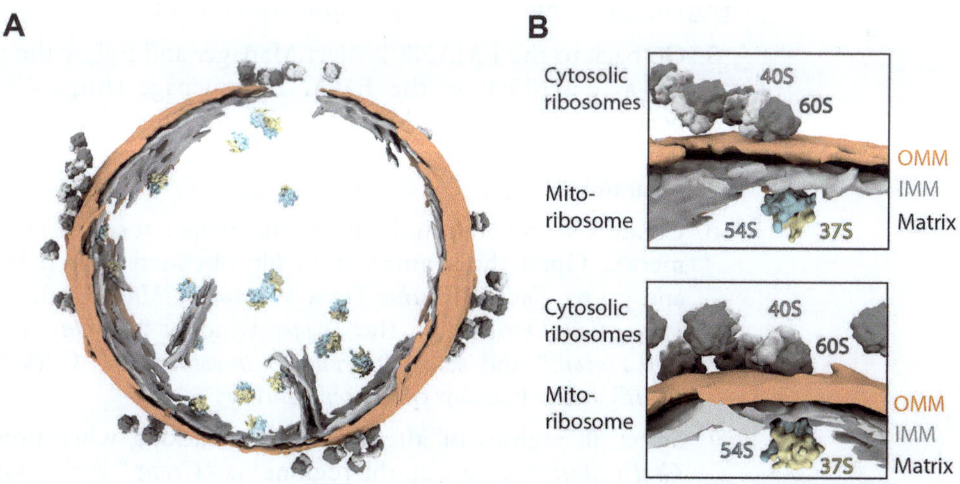

Fig. 3 Visualization of membranes and mitoribosomes in cryoelectron tomograms. (**a**) Tomogram segmentation of a yeast mitochondrion, visualizing cytosolic (40S in light, 60S in dark gray) and mitochondrial (37S in yellow, 54S in blue) ribosomes in the context of the mitochondrial membrane environment (OMM in orange, IMM in gray). (**b**) Cytosolic ribosome clusters coincide with mitochondrial ribosome positions on the IMM

3.4.1 Tomogram Segmentation

The described segmentation workflow is intended to quickly segment apparent features, such as the outer mitochondrial membrane. For features in dense environments or thick parts of a sample, such as cristae in the central mitochondrial regions, additional manual segmentation in, e.g., Amira (Thermo Fisher) is likely necessary.

1. Make sure the tomogram is compatible with EMAN2 (*see* **Note 8**).
2. In a Linux terminal, load pyTOM 0.993 to execute `deconv.py` according to the described usage (`deconv.py --help`). The parameters provided below are an example and should be optimized for each dataset.

```
deconv.py -f <tomogram.mrc> -o <tomogram_deconv.mrc> -s <pixelsize> -z <defocus (positive value in m)> --snrfalloff <1.5> --deconvstrength <0.5> --phaseflipped (in case phase-flipping was used for CTF correction).
```

3. Load EMAN2 and execute `e2projectmanager.py` in a terminal. To prevent automatic pixel addition to the tomographic volume (due to `--clip` option applied during import via EMAN2 Project Manager), create a folder called "tomograms" in the EMAN2 project folder (*see* **Note 9**).
4. Import tomograms into that folder manually as follows:

```
e2proc3d.py <tomogram_deconv.mrc> <tomograms/tomogram_deconv.hdf> --process normalize
```

5. Go back to the EMAN2 Project Manager and follow the tutorial as described on the EMAN2 homepage (https://blake.bcm.edu/emanwiki/EMAN2/Programs/tomoseg, accessed 03/03/2022). Segment inner and outer membranes separately.
6. Create a segmentation file using the Segger tool [24] in ChimeraX: Open the segmentation file obtained from EMAN2 and go to *Tools* > *Volume Data* > *Segment Map*. Under "*Segmentation Options*" in the *Segger* window "*Display at most 300 regions*" and select "*Group by smoothing 6*." Click "*Segment*" at the bottom of *Segger* window.
7. Select all regions of interest by right clicking while pressing *Ctrl* + *Shift*, then group the regions via "*Group*" at the bottom of the *Segger* window.
8. Under Shortcuts Options "*Invert*" the selection, then "*Delete*" all unwanted regions and select the remaining region of interest by *right mouse clicking* + *Ctrl*.

9. Save the segmentation as an mrc file (*File > Save selected regions to .mrc file . . .*).
10. In the command line resample the segmentation onto the original volume dimensions using "vop resample #<ID_segmentation> onGrid #<ID_CNN_hdf_file>" and save the resampled map using "save <filename.mrc> model #<ID_segmentation_resampled.mrc>."
11. Clean up the segmentation in ChimeraX: Hide small volumes using the Hide Dust Tool under *Tools > Volume Data > Hide Dust*. Then open *Tools > Volume Data > Map Eraser*. Remove noise or falsely segmented features from the map and save it as a new file.

3.4.2 Mapping Particles in the Segmentation

1. Prepare templates for template matching using the Relion image handler as follows:

```
relion_image_handler --i <EMDBfile.mrc> --o <EMDB-
file_temp.mrc> --angpix <Input pixel size(é)> --
lowpass <resolution (é)> --rescale_angpix <Desired
pixel size(é)>
```

The pixel size needs to match the tomogram used for segmentation.

2. Open the templates in Chimera and, if necessary, scale the density to match the tomogram density values using "vop scale #<IDEMDBfile_temp.mrc> factor -1."
3. Generate a spherical mask around the template in pyTOM (gen_mask.py) with the same box size as the template. To adjust the mask position with respect to the template, open both files in in Chimera, center them on each other and use the "vop resample" command (*see* Subheading 3.4.1, **step 10**).
4. Perform template matching and extract particles to an .xml file using pyTOM (https://github.com/FridoF/PyTom/wiki).
5. Clean up the particle lists via the pyTOM GUI (pytomGUI.py): Create a new project folder and under "*Enable Stage*" select "*Particle Picking*." In the "*Manual Picking*" tab, open the tomogram and click "*Pick*." Apply a Gaussian filter to the tomogram, select a suitable particle size, and then load the particle list extracted in **step 4**. Use the left and right keyboard arrows to navigate and right click on particles to remove them. Then save the particle list as a new file.
6. Prepare motl files from the pyTOM particle lists as follows with pyTOM:

```
pytom_xml_to_motl.py <input file name>
```

7. In MATLAB add the path for the TOM toolbox (`addpath (genpath(<path>'));`), then open and execute `tom_-classmask.m`. Set paths or add to path if required.
8. Open `plot_particles.m` for plotting molecule references to their positions in a tomogram, enter motl file name, the template file name, the tomogram dimensions and run the script for each molecule type to be displayed.
9. Visualize the segmentation, as well as the plotted particles in ChimeraX.

4 Notes

1. We recommend optimizing the sedimentation settings for differential centrifugation when applying the protocols to other cell lines, yeast strains or other sources of mitochondria.
2. Culture cells for at least 2 days after thawing and do not freeze cells before mitochondria isolation. The freezing process significantly reduces the yield of intact mitochondrial vesicles.
3. Only harvest cells in logarithmic growth phase; otherwise, the yield of mitochondrial vesicles is very low.
4. Working at 4 °C and fast preparation is crucial to yield structurally and functionally intact mitochondrial vesicles.
5. The use of Lacey carbon grids improves the vesicle distribution compared to Holey carbon grids due to a smaller carbon surface. However, holey carbon grids provide more stability during data acquisition.
6. The ratio of mitochondria sample to fiducial buffer can be varied to sample different mitochondria, sucrose/sorbitol or gold fiducial concentrations.
7. Since the sample thickness is one main limiting factor for high-resolution subtomogram averaging, but mitochondrial vesicles are up to 500 nm in diameter, a balance needs to be found between vesicle intactness and sample thickness. Different sample thickness values can be sampled by varying the blotting parameters or by choosing acquisition points depending on ice thickness or vesicle size, respectively.
8. When using tomograms reconstructed with Etomo in IMOD, make sure to untick the "*convert to bytes*" option in both "*Coarse Alignment*" and "*Post-processing*" prior to reconstruction.
9. Coordinate systems can be defined differently in different software packages and need to be considered for combining, e.g., membrane segmentations and back-mapped particles. Especially after a segmentation clean-up in Chimera, the original

tomogram dimensions (box size) are not saved automatically and need to be manually applied prior to saving. Otherwise, Chimera automatically sets the volume dimensions depending on the cleaned segmentation, which might have a smaller box size compared to the initial tomogram.

Acknowledgments

We would like to acknowledge Stuart Howes and the whole EM Square facility at Utrecht University for technical support in electron microscopy as well as Mihajlo Vanevic for computational technical support. We would also like to thank Gijs van der Schot and Marten Chaillet for providing pyTOM scripts.

This work was supported by funding from Nederlandse Organisatie voor Wetenschappelijke Onderzoek (Vici 724.016.001). The electron microscopy within this work is part of the research program National Roadmap for Large-Scale Research Infrastructure (NEMI), project number 184.034.014, which is financed by the Dutch Research Council (NWO).

References

1. Mahamid J, Pfeffer S, Schaffer M et al (2016) Visualizing the molecular sociology at the HeLa cell nuclear periphery. Science (1979) 351:969–972
2. Lučić V, Förster F, Baumeister W (2005) Structural studies by electron tomography: from cells to molecules. Annu Rev Biochem 74:833–865
3. Dubochet J, Adrian M, Chang J-J et al (1988) Cryo-electron microscopy of vitrified specimens. Q Rev Biophys 21:129–228
4. Nicastro D, Frangakis AS, Typke D, Baumeister W (2000) Cryo-electron tomography of Neurospora mitochondria. J Struct Biol 129: 48–56
5. Strauss M, Hofhaus G, Schröder RR, Kühlbrandt W (2008) Dimer ribbons of ATP synthase shape the inner mitochondrial membrane. EMBO J 27:1154–1160
6. Davies KM, Strauss M, Daum B et al (2011) Macromolecular organization of ATP synthase and complex I in whole mitochondria. Proc Natl Acad Sci U S A 108:14121–14126
7. Pfeffer S, Woellhaf MW, Herrmann JM, Förster F (2015) Organization of the mitochondrial translation machinery studied in situ by cryoelectron tomography. Nat Commun 6:1–8
8. Englmeier R, Pfeffer S, Förster F (2017) Structure of the human mitochondrial ribosome studied in situ by cryoelectron tomography. Structure 25:1574–1581
9. Förster F, Hegerl R (2007) Structure determination in situ by averaging of tomograms. Methods Cell Biol 2007:741–767
10. Frezza C, Cipolat S, Scorrano L (2007) Organelle isolation: functional mitochondria from mouse liver, muscle and cultured filroblasts. Nat Protoc 2:287–295
11. Fernández-Silva P, Acín-Pérez R, Fernández-Vizarra E et al (2007) In vivo and in organello analyses of mitochondrial translation. Methods Cell Biol 80:571–588
12. Maccecchini M-L, Rudin Y, Blobelt G, Schatz G (1979) Import of proteins into mitochondria: precursor forms of the extramitochondrially made F1-ATPase subunits in yeast. Proc Natl Acad Sci U S A 76:343–347
13. Lazarou M, Smith SM, Thorburn DR et al (2009) Assembly of nuclear DNA-encoded subunits into mitochondrial complex IV, and their preferential integration into supercomplex forms in patient mitochondria. FEBS J 276:6701–6713
14. Bihlmaier K, Bien M, Herrmann JM (2008) In vitro import of proteins into isolated mitochondria. In: Methods in molecular biology: membrane trafficking. Humana Press, pp 85–94

15. Meisinger C, Pfanner N, Truscott KN (2006) Isolation of yeast mitochondria. Yeast Protoc:33–39
16. Couvillion MT, Soto IC, Shipkovenska G, Churchman LS (2016) Synchronized mitochondrial and cytosolic translation programs. Nature 533:499–503
17. Ress D, Harlow ML, Schwarz M et al (1999) Automatic acquisition of fiducial markers and alignment of images in tilt series for electron tomography. Microscopy 48(3):277–287
18. Pettersen EF, Goddard TD, Huang CC et al (2021) UCSF ChimeraX: structure visualization for researchers, educators, and developers. Protein Sci 30:70–82
19. Scheres SHW (2012) RELION: implementation of a Bayesian approach to cryo-EM structure determination. J Struct Biol 180:519–530
20. Hrabe T, Chen Y, Pfeffer S et al (2012) PyTom: a python-based toolbox for localization of macromolecules in cryo-electron tomograms and subtomogram analysis. J Struct Biol 178:177–188
21. Tang G, Peng L, Baldwin PR et al (2007) EMAN2: an extensible image processing suite for electron microscopy. J Struct Biol 157:38–46
22. Chen M, Dai W, Sun SY et al (2017) Convolutional neural networks for automated annotation of cellular cryo-electron tomograms. Nat Methods 14:983–985
23. Nickell S, Förster F, Linaroudis A et al (2005) TOM software toolbox: acquisition and analysis for electron tomography. J Struct Biol 149: 227–234
24. Pintilie GD, Zhang J, Goddard TD et al (2010) Quantitative analysis of cryo-EM density map segmentation by watershed and scale-space filtering, and fitting of structures by alignment to regions. J Struct Biol 170:427–438

Chapter 6

Cryo-EM for Structure Determination of Mitochondrial Ribosome Samples

Hauke S. Hillen

Abstract

Single-particle cryoelectron microscopy (cryo-EM) allows structure determination of large macromolecular complexes from conformationally and compositionally heterogeneous mixtures of particles. This technique has been used to reveal the architecture of the mitochondrial ribosome and to visualize transient states that occur during the translation cycle or during mitoribosome biogenesis. Here, we outline an exemplary workflow for the analysis of single-particle cryo-EM data of human mitoribosome samples. In addition, we provide an example dataset which can be used for training purposes alongside the protocol.

Key words Mitochondrial ribosome, single-particle cryo-EM, image processing, sample preparation

1 Introduction

The structural analysis of the mitochondrial ribosome (mitoribosome) has led to considerable advances in our understanding of mitochondrial translation [1–6]. These insights have been achieved through the use of single-particle cryoelectron microscopy (cryo-EM), which has undergone a "resolution revolution" driven by technical advances in both microscope hardware and data analysis software [7]. As a result, single-particle cryo-EM now routinely allows structure determination of macromolecular complexes at near-atomic resolution. Within just a few years, single-particle cryo-EM has produced detailed insights into the architecture of the human mitochondrial ribosome [1, 2, 5, 8, 9] and into the mechanisms of its biogenesis [6, 11–14] and mitochondrial translation [3, 4, 15–17].

The structural analysis of mitochondrial ribosomes by single-particle cryo-EM can be achieved using standard state-of-the-art approaches and protocols for single-particle analysis. Several software packages for data acquisition [18] and image processing [19–22] as well as model building [23] and refinement [24] are

Antoni Barrientos and Flavia Fontanesi (eds.), *The Mitoribosome: Methods and Protocols*,
Methods in Molecular Biology, vol. 2661, https://doi.org/10.1007/978-1-0716-3171-3_6,

available, which are becoming increasingly automated. The image processing and 3D reconstruction steps are facilitated by the large size and high nucleic acid content of the mitochondrial ribosome, which usually results in good contrast and easily discernable particles. However, as mitoribosome samples are usually endogenously purified from their native source organisms, they are often heterogeneous both in terms of their composition and the conformational landscape of their constituent particles. For example, they may contain mitoribosome particles in different functional states that occur during the mitochondrial translation cycle or during mitoribosome biogenesis, as well as additionally bound protein factors that mediate these steps. Moreover, the mitochondrial ribosome is a highly dynamic molecular machine, which undergoes structural transitions during its catalytic cycle. As a result, different conformational states of the mitochondrial ribosome are usually present in the sample, which can complicate structural analysis.

Single-particle cryo-EM is uniquely capable of revealing such heterogeneity and produce reconstructions of different states of the mitochondrial ribosome within a single sample. For this, in silico approaches that classify particles based on compositional and conformational differences are used [25]. Despite the increasing automation of single-particle analysis, these approaches often require intervention and decision making by the experimenter to obtain optimal results and identify rare particle subpopulations. In addition, the ideal strategy for image analysis and data processing often depends on the specific properties of the sample. Nevertheless, a basic workflow consisting of a step-wise clean-up followed by heterogeneity analysis of particles is common to all these cryo-EM data processing approaches.

Here, we provide a step-by-step protocol for cryo-EM analysis of a compositionally heterogeneous human mitochondrial ribosome sample. As a general workflow for single-particle cryo-EM of mitochondrial monosomes has been described in a previous edition of this series [26], we here describe the analysis of a dataset of biogenesis intermediates of the large subunit of the human mitochondrial ribosome (mtLSU) which we have previously used to visualize late-stage assembly of the mtLSU [10]. This dataset contains several distinct states of the mtLSU with various protein factors bound, and thus serves as an exemplary case for the analysis of a compositionally heterogeneous mitochondrial ribosome sample. We therefore highlight strategies that can be used to dissect such heterogeneous samples and obtain high-resolution reconstructions of individual particle subpopulations. The outlined experimental and computational steps serve as an example that can be adapted and applied to other types of mitochondrial ribosome samples. Moreover, the workflow presented here can serve as a starting point for the adaption of other strategies and software packages, which may be necessary and more appropriate for other datasets. To facilitate this, we have made the cryo-EM dataset used here as an example available for download from the public EMPIAR database, which can be used for training purposes and to recapitulate the described steps.

2 Materials

2.1 For Sample Preparation and Cryo-EM Imaging

1. Purified mitochondrial ribosome sample (*see* **Note 1**).
2. TEM grids, for example, Quantifoil R3.5/1 Holey Carbon Films on Cu 200 mesh.
3. Mica sheets.
4. High vacuum sputter coater to deposit a thin carbon film onto Mica sheets. We use a Leica ACE600.
5. Glow discharger to clean TEM grids and make them hydrophilic. We use a Pelco EasiGlow.

2.2 For Image Processing and 3D Reconstruction

2.2.1 Computer Hardware Requirements

1. A computer with a modern CPU with at least 8 cores. We typically use a computer with a 24-core AMD EPYC ROME 7402P CPU at 2.8 GHz.
2. An NVIDIA graphics processing unit (GPU) with at least 6 GB of memory. We used a computer with 4 NVIDIA RTX3090 GPUs, each with 24 GB of memory.
3. Sufficient storage space for the cryo-EM dataset. We recommend a local solid-state scratch disk to which the dataset can be copied for fast processing. We use a 4 TB SSD.

2.2.2 Computer Software Requirements

1. A Linux-based operating system.
2. RELION [19, 25, 27, 28]. https://www3.mrc-lmb.cam.ac.uk/relion. Compilation and installation of RELION require several additional software packages. For more information on this, we refer to the RELION documentation at https://relion.readthedocs.io/en/latest/Installation.html.
3. A molecular graphics viewer. We use UCSF Chimera [29] and ChimeraX [30]. Download from https://www.cgl.ucsf.edu/chimera.
4. COOT for molecular modeling [23].
5. PHENIX for structure refinement [24, 31]. The PHENIX suite includes an implementation of MolProbity [32] used for assessment of model quality.

2.3 Example Data

An example dataset can be downloaded from EMPIAR (EMPIAR ID: EMPIAR-11317). This comprises cryo-EM data of a purified assembly intermediate of the large subunit (mtLSU) of the human mitochondrial ribosome (dataset 2 in [10]).

3 Methods

The first and last steps (Subheadings 3.1, 3.2, and 3.4) are common practices in single-particle cryo-EM and are not specific to the analysis of mitochondrial ribosomes. As an in-depth description of these methods would be beyond the scope of this work, we refer the reader to published protocols and tutorials for these steps (e.g., [33]), and focus here on experimental details crucial to the sample preparation and cryo-EM data acquisition of mitochondrial ribosome samples.

3.1 Cryo-EM Sample Preparation and Data Collection

1. Prepare cryo-EM grids with a thin amorphous carbon film. A detailed protocol for this can be found elsewhere [34]. We typically aim for a target thickness of 2.5 nm (*see* **Note 2**).
2. Glow discharge carbon-coated grids. We typically use 15 mA for 60 s at 0.4 mbar in a Pelco EasiGlow (*see* **Note 3**).
3. Plunge-freeze samples. We use a Vitrobot (Thermo Fisher) and incubate the blotting papers for at least 1 h in the humidifier chamber at 4 °C and 100% humidity. We use mitoribosome samples with an absorbance of ~3 units at 260 nm (10 mm path length) and incubate 3.5 μL of sample on the grid for 30 s before blotting and plunge-freezing in liquid ethane (*see* **Notes 4** and **5**).
4. Collect a cryo-EM dataset. Detailed protocols for single-particle cryo-EM data acquisition have been described elsewhere (e.g., in [33]), and mitoribosome-specific considerations have been outlined in a previous edition of this series [26]. We usually collect data using a Titan Krios 300 kV transmission electron microscope equipped with an energy filter and a direct electron detection camera, such as the K3 (Gatan) or the Falcon 4 (Thermo Fisher). We typically use a magnification that corresponds to a pixel size of 1.05 Å/pix, set the energy filter to a slit width of 20 eV and illuminate with a total dose of around 40 e/A^2 fractionated into approx. 40 movie frames (*see* **Note 6**).

3.2 Image Preprocessing and Particle Extraction

1. Preprocess raw movie stacks. This step comprises gain correction (if raw images were not stored as gain-corrected movies), CTF and defocus estimation, and motion correction. We typically use Warp [22] for preprocessing. The workflow for on-the-fly processing of single-particle cryo-EM data with Warp has been described previously [22], and we refer to this work for details (*see* **Note 7**).
2. Pick and extract particles. We typically use the particle picker implemented in Warp, which employs a machine-learning-based approach. For mammalian mitochondrial ribosomes (55S as well as small and large subunits), the generic BoxNet of Warp performs well (*see* **Notes 8**, **9**, and **10**).

3.3 Image Processing, Classification, and Reconstruction

We typically perform all of the steps described here in RELION, and if recommended settings are provided, they refer to RELION. However, these steps can also be performed with other software packages (e.g., cryoSPARC), which should result in similar outcomes.

1. Perform 2D classification. Select classes that represent different views of mitoribosome particles (Fig. 1) (*see* **Note 11**).
2. Create a reference 3D volume for 3D classification and refinement. This can be done either using the "3D initial model" job in RELION, or by rescaling an appropriate map from the EMDB to the pixel size and box size used (*see* **Notes 12** and **13**).

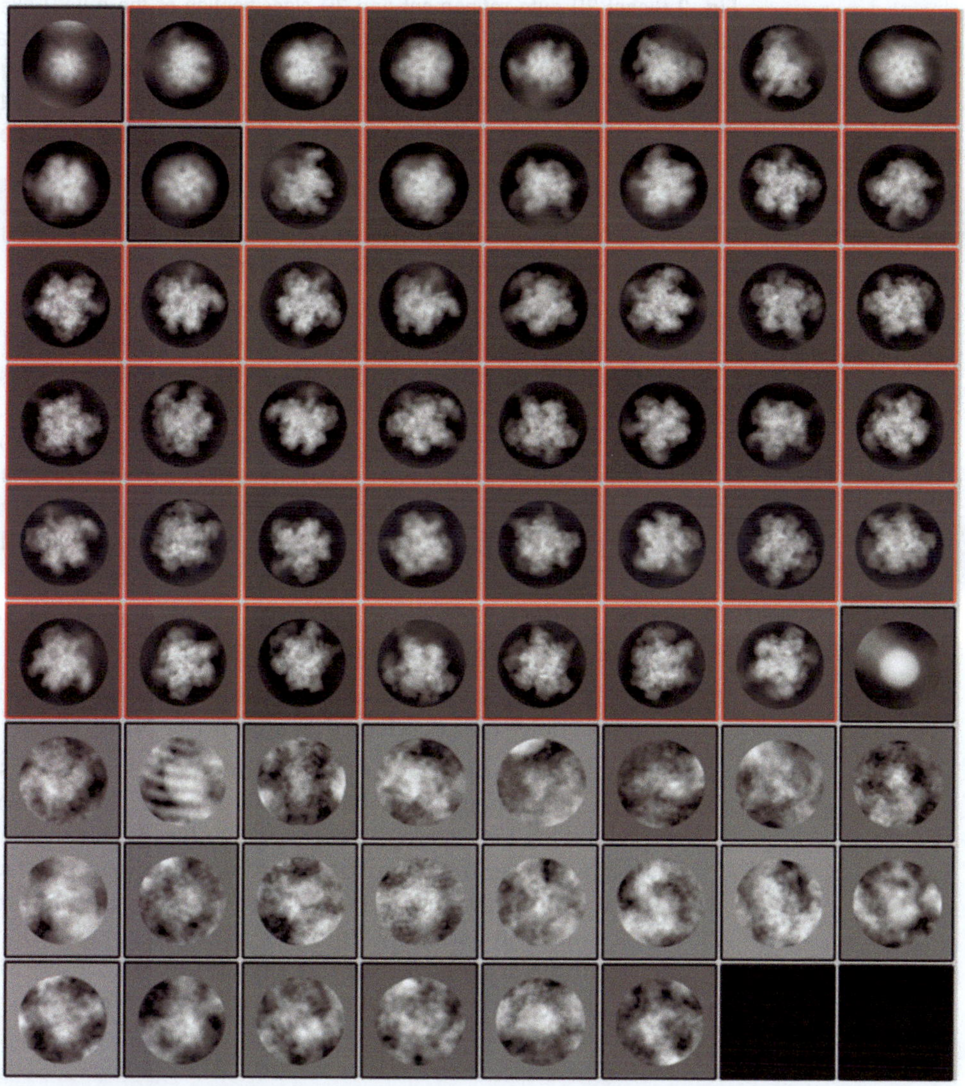

Fig. 1 Example results of 2D classification of the mtLSU example dataset. 2D class averages from RELION are shown. Selected particles are marked by a red box

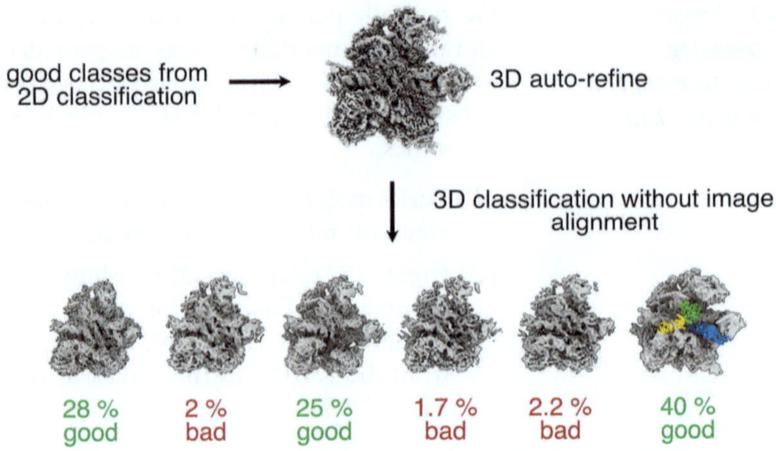

Fig. 2 Global 3D auto-refinement and 3D classification of the mtLSU example dataset. After global 3D auto-refinement, a 2D classification with 6 classes ($K = 6$) without image alignment reveals three good classes (marked in green) and three bad classes (marked in red), which show less high-resolution features. The sixth class contains additional ribosome-binding factors (indicated by coloring)

3. Perform 3D auto-refinement of selected good particles from **step 1** with the reference created in **step 2**.
4. Perform a 3D classification using the output particles from **step 3** to sort out low-quality particles as well as compositional and conformational heterogeneity (Fig. 2). As the particles are already aligned after refinement in **step 3**, this can be done without further image alignment to speed up the calculation. If the focus of research lies on a specific part of the mitoribosome (e.g., the small- or large subunit), a soft solvent mask around the respective region can be used (*see* **Notes 14**, **15**, and **16** as well as [26]).
5. Select particles corresponding to classes that show high-resolution features for 3D refinement.
6. Perform 3D auto-refinement and postprocessing of the individual particle subsets selected in **step 5** (*see* **Note 17**).
7. Frequently, one or more of the resulting reconstructions show residual heterogeneity within specific regions that cannot be resolved with the strategy described in **step 4**, which is often caused by transient binding of mitoribosome-associated factors or local conformational flexibility. In this case, perform further 3D classification with a solvent mask around the region of these regions interest (Fig. 3). Repeat this step until compositional heterogeneity is resolved as far as possible (*see* **Note 18**).

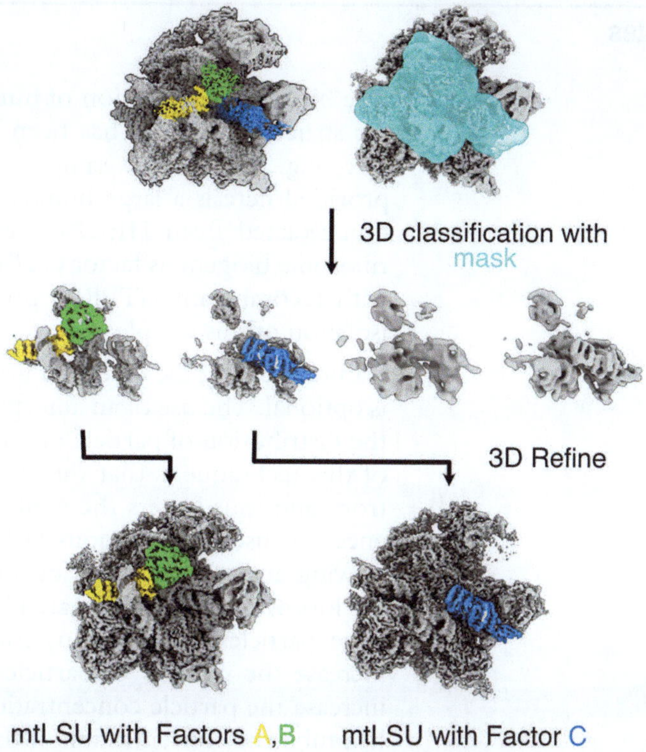

Fig. 3 Local 3D classification of the mtLSU example dataset. Focused 3D classification of class 6 from the initial global 3D classification (Fig. 2) with a soft mask around the additional ribosome-binding factors separates particles with different factors bound (Factor A + B or Factor C). The resulting particle subpopulations can be used for global and focused 3D auto-refinement to yield high-resolution maps

8. Perform 3D auto-refinement and postprocessing of compositionally homogenous particle populations. Perform focused 3D refinement to improve the local resolution of conformationally heterogeneous regions of the map.

3.4 Model Building and Refinement

1. Fit an appropriate starting model from the PDB into your map of interest from 3D auto-refine and postprocessing. This can be done in UCSF Chimera (*see* **Note 19**).
2. Refine and rebuild the structure manually in real-space in COOT. Regions not present in the starting model can either be built de novo, or by fitting and rebuilding additional starting models (*see* **Notes 20** and **21**).
3. Refine the structural model against the experimental cryo-EM map(s) to obtain a stereochemically plausible model. We typically use phenix.real_space_refine for this (*see* **Note 22**).

4 Notes

1. The biochemical isolation of human mitochondrial ribosomes for structural analysis has been described in detail elsewhere (see, e.g., [26]). The sample used for the example dataset provided here is a large human mitochondrial ribosome subunit isolated from HEK293 cells lacking the mitochondrial ribosome biogenesis factor GTPBP6, which was supplemented with recombinant GTPBP6 prior to structural analysis. The isolation of this sample is described in detail in [10].
2. Carbon-coating the cryo-EM grids prior to sample application is optional. The use of an amorphous carbon layer can improve the distribution of particle orientation [35]. The disadvantage of this technique is that the amorphous carbon absorbs electrons and thus lowers the contrast of particles. In our experience, the use of continuous carbon-coated grids improves the viewing angle distribution of mitochondrial ribosomes, while the loss of contrast is mitigated by the strong signal from these large particles. In addition, continuous carbon support can increase the number of particles visible per micrograph (i.e., increase the particle concentration on the grid), which can be helpful if the concentration of the sample is low.
3. When using grids without continuous carbon support film, a longer glow discharging protocol can be used (e.g., 100 s).
4. The absorbance can be estimated using a microliter-scale spectrophotometer. We typically use a NanoDrop One UV-Vis Spectrophotometer.
5. The waiting time may be increased (for example to 60, 120, or 240 s) for samples with low concentration.
6. The magnification should be chosen according to the resolution aimed for. Higher magnifications come at the cost of fewer particles per field of view and thus requires longer data acquisition times and more data storage space to obtain an identical number of particles.
7. A detailed documentation for Warp can be found at http://www.warpem.com. Preprocessing can also be carried out using other software packages, for example, RELION [19] or cryoSPARC [20].
8. Particle picking and/or extraction can also be carried out using other software packages, for example RELION [19, 28], cryoSPARC [20], or others. In our experience, most particle picking approaches perform well on mitoribosome cryo-EM data due to the large size and high contrast of particles.

9. Make sure to use an appropriate box size. A common recommendation is to use a box that is 1.5–2.0× the longest axis of the particle of interest.
10. Particles can be binned during extraction to speed up downstream computing steps. We recommend a binning factor of 3.
11. The number of classes, circular mask diameter, and number of iterations may need to be optimized to obtain best results. For the 2D classification shown in Fig. 1, we used 200 classes, $T = 2$, 25 iterations, 280 Å mask diameter, angular sampling 6, offset range 5, offset **step 1**.
12. The "3D initial model" job in RELION may produce an initial model with wrong handedness. It is advisable to check the handedness of the map by attempting to fit a mitoribosome coordinate file from the PDB into the map. The handedness of the map can be flipped using the command "vop zflip" in Chimera. Use a map with correct handedness as reference for the following steps.
13. The command-line program "relion_image_handler" can be used for rescaling.
14. We recommend testing 3D classification with and without image alignment at this step to compare which strategy leads to better classification results. For the 3D classification shown in Fig. 2, we used 6 classes ($K = 6$), $T = 4$, 25 iterations, 280 Å mask diameter and no image alignment. This 3D classification step may be repeated iteratively until no further heterogeneity can be classified.
15. 3D classification may also be performed with a soft solvent mask encompassing the particle, as described in [26].
16. Solvent masks should not include high-resolution features and should have a soft edge of 6–10 pixels [26].
17. If particles were binned during extraction, the particle subsets selected in **step 5** should be re-extracted without binning prior to 3D auto-refinement.
18. In the 3D classification shown in Fig. 3, we used 4 classes ($K = 4$), $T = 4$, 25 iterations, 280 Å mask diameter, and no image alignment. If heterogeneity within the masked region cannot be resolved using this approach, it can sometimes help to increase the regularization parameter T in RELION. In our experience, values up to $T = 100$ can yield useful results [10].
19. PDB files can be roughly fit to the map through manual rotation and translation, followed by accurate fitting using the "fit in map" tool.

20. Additional starting models can be obtained either from previous high-resolution structures in the PDB or from AlphaFold predictions [36].
21. For regions with low resolution or poor density, other real space modeling and refinement tools such as ISOLDE may be useful [37].
22. For regions with low resolution or poor density, it may be necessary to restrain refinement to preserve plausible stereochemistry. This can be done with secondary structure restraints as well as with reference model restraints if high-resolution reference models are available.

Acknowledgments

This work was supported by the Deutsche Forschungsgemeinschaft (FOR2848, SFB1190, EXC 2067/1-390729940). The work was funded by the Deutsche Forschungsgemeinschaft (DFG, German Research Foundation) under Germany's Excellence Strategy—EXC 2067/1-390729940.

References

1. Amunts A, Brown A, Toots J et al (2015) Ribosome. The structure of the human mitochondrial ribosome. Science 348:95–98. https://doi.org/10.1126/science.aaa1193
2. Greber BJ, Bieri P, Leibundgut M et al (2015) Ribosome. The complete structure of the 55S mammalian mitochondrial ribosome. Science 348:303–308. https://doi.org/10.1126/science.aaa3872
3. Aibara S, Singh V, Modelska A, Amunts A (2020) Structural basis of mitochondrial translation. elife 9:e58362. https://doi.org/10.7554/elife.58362
4. Kummer E, Leibundgut M, Rackham O et al (2018) Unique features of mammalian mitochondrial translation initiation revealed by cryo-EM. Nature 560:263–267. https://doi.org/10.1038/s41586-018-0373-y
5. Itoh Y, Andréll J, Choi A et al (2021) Mechanism of membrane-tethered mitochondrial protein synthesis. Science 371:846–849. https://doi.org/10.1126/science.abe0763
6. Itoh Y, Khawaja A, Laptev I et al (2022) Mechanism of mitoribosomal small subunit biogenesis and preinitiation. Nature 606:603–608. https://doi.org/10.1038/s41586-022-04795-x
7. Kühlbrandt W (2014) Biochemistry. The resolution revolution. Science 343:1443–1444. https://doi.org/10.1126/science.1251652
8. Brown A, Amunts A, Bai X et al (2014) Structure of the large ribosomal subunit from human mitochondria. Science 346:718–722. https://doi.org/10.1126/science.1258026
9. Greber BJ, Boehringer D, Leibundgut M et al (2014) The complete structure of the large subunit of the mammalian mitochondrial ribosome. Nature 515:283. https://doi.org/10.1038/nature13895
10. Hillen HS, Lavdovskaia E, Nadler F et al (2021) Structural basis of GTPase-mediated mitochondrial ribosome biogenesis and recycling. Nat Commun 12:3672. https://doi.org/10.1038/s41467-021-23702-y
11. Brown A, Rathore S, Kimanius D et al (2017) Structures of the human mitochondrial ribosome in native states of assembly. Nat Struct Mol Biol 24:866–869. https://doi.org/10.1038/nsmb.3464
12. Harper NJ, Burnside C, Klinge S (2023) Principles of mitoribosomal small subunit assembly in eukaryotes. Abstract Nature 614(7946): 175–181. https://doi.org/10.1038/s41586-022-05621-0
13. Chandrasekaran V, Desai N, Burton NO et al (2021) Visualizing formation of the active site

in the mitochondrial ribosome. elife 10: e68806. https://doi.org/10.7554/elife.68806

14. Cheng J, Berninghausen O, Beckmann R (2021) A distinct assembly pathway of the human 39S late pre-mitoribosome. Nat Commun 12:4544. https://doi.org/10.1038/s41467-021-24818-x
15. Kummer E, Ban N (2021) Mechanisms and regulation of protein synthesis in mitochondria. Nat Rev Mol Cell Biol 22:307–325. https://doi.org/10.1038/s41580-021-00332-2
16. Singh V, Itoh Y, Huynen MA, Amunts A (2022) Activation mechanism of mitochondrial translation by LRPPRC-SLIRP. Biorxiv 2022.06.20.496763. https://doi.org/10.1101/2022.06.20.496763
17. Desai N, Yang H, Chandrasekaran V et al (2020) Elongational stalling activates mitoribosome-associated quality control. Science 370:1105–1110. https://doi.org/10.1126/science.abc7782
18. Mastronarde DN (2018) Advanced data acquisition from electron microscopes with SerialEM. Microsc Microanal 24:864–865. https://doi.org/10.1017/s1431927618004816
19. Zivanov J, Nakane T, Forsberg BO et al (2018) New tools for automated high-resolution cryo-EM structure determination in RELION-3. elife 7:163. https://doi.org/10.7554/elife.42166
20. Punjani A, Rubinstein JL, Fleet DJ, Brubaker MA (2017) cryoSPARC: algorithms for rapid unsupervised cryo-EM structure determination. Nat Methods 14:290–296. https://doi.org/10.1038/nmeth.4169
21. Wagner T, Merino F, Stabrin M et al (2019) SPHIRE-crYOLO is a fast and accurate fully automated particle picker for cryo-EM. Commun Biol 2:218. https://doi.org/10.1038/s42003-019-0437-z
22. Tegunov D, Cramer P (2019) Real-time cryo-electron microscopy data preprocessing with Warp. Nat Methods 16:1146–1152. https://doi.org/10.1038/s41592-019-0580-y
23. Emsley P, Lohkamp B, Scott WG, Cowtan K (2010) Features and development of Coot. Acta Crystallogr D Biol Crystallogr 66:486–501. https://doi.org/10.1107/s0907444910007493
24. Afonine PV, Poon BK, Read RJ et al (2018) Real-space refinement in PHENIX for cryo-EM and crystallography. Acta Crystallogr Sect D Struct Biol 74:531–544. https://doi.org/10.1107/s2059798318006551
25. Scheres SHW (2012) RELION: implementation of a Bayesian approach to cryo-EM structure determination. J Struct Biol 180:519–530. https://doi.org/10.1016/j.jsb.2012.09.006
26. Singh V, Amunts A (2020) Application of Cryo-EM for visualization of mitoribosomes. Methods Mol Biol Clifton NJ 2192:197–210. https://doi.org/10.1007/978-1-0716-0834-0_15
27. Scheres SHW (2016) Processing of structurally heterogeneous Cryo-EM data in RELION. Methods Enzymol 579:125–157. https://doi.org/10.1016/bs.mie.2016.04.012
28. Kimanius D, Dong L, Sharov G, Nakane T, Scheres SHW (2021) New tools for automated cryo-EM single-particle analysis in RELION-4.0. Biochem J 478(24):4169–4185. https://doi.org/10.1042/BCJ20210708
29. Pettersen EF, Goddard TD, Huang CC et al (2004) UCSF Chimera–a visualization system for exploratory research and analysis. J Comput Chem 25:1605–1612. https://doi.org/10.1002/jcc.20084
30. Pettersen EF, Goddard TD, Huang CC et al (2021) UCSF ChimeraX: structure visualization for researchers, educators, and developers. Protein Sci 30:70–82. https://doi.org/10.1002/pro.3943
31. Adams PD, Afonine PV, Bunkóczi G et al (2010) PHENIX: a comprehensive Python-based system for macromolecular structure solution. Acta Crystallogr D Biol Crystallogr 66:213–221. https://doi.org/10.1107/s0907444909052925
32. Williams CJ, Headd JJ, Moriarty NW et al (2018) MolProbity: more and better reference data for improved all-atom structure validation. Protein Sci 27:293–315. https://doi.org/10.1002/pro.3330
33. Thompson RF, Iadanza MG, Hesketh EL et al (2019) Collection, pre-processing and on-the-fly analysis of data for high-resolution, single-particle cryo-electron microscopy. Nat Protoc 14:100–118. https://doi.org/10.1038/s41596-018-0084-8
34. Chang L, Barford D (2018) Methods for preparing Cryo-EM grids of large macromolecular complexes. Methods Mol Biol Clifton NJ 1844:209–215. https://doi.org/10.1007/978-1-4939-8706-1_14
35. Drulyte I, Johnson RM, Hesketh EL et al (2018) Approaches to altering particle distributions in cryo-electron microscopy sample preparation. Acta Crystallogr Sect D 74:560–571. https://doi.org/10.1107/s2059798318006496

36. Jumper J, Evans R, Pritzel A et al (2021) Highly accurate protein structure prediction with AlphaFold. Nature 596:583–589. https://doi.org/10.1038/s41586-021-03819-2

37. Croll TI (2018) ISOLDE: a physically realistic environment for model building into low-resolution electron-density maps. Acta Crystallogr Sect D Struct Biol 74:519–530. https://doi.org/10.1107/s2059798318002425

Chapter 7

Sucrose Gradient Analysis of Human Mitochondrial Ribosomes and RNA

Kah Ying Ng and Brendan J. Battersby

Abstract

Faithful expression of the mitochondrial genome is required for the synthesis of the oxidative phosphorylation complexes and cell fitness. In humans, mitochondrial DNA (mtDNA) encodes 13 essential subunits of four oxidative phosphorylation complexes along with tRNAs and rRNAs needed for the translation of these proteins. Protein synthesis occurs on unique ribosomes within the organelle. Over the last decade, the revolution in genetic diagnostics has identified disruptions to the faithful synthesis of these 13 mitochondrial proteins as the largest group of inherited human mitochondrial pathologies. All of the molecular steps required for mitochondrial protein synthesis can be affected, from the genome to protein, including cotranslational quality control. Here, we describe methodologies for the biochemical separation of mitochondrial ribosomes from cultured human cells for RNA and protein analysis. Our method has been optimized to facilitate analysis for low-level sample material and thus does not require prior organelle enrichment.

Key words Mitochondria, Ribosomes, RNA, Sucrose gradient, Human disease, Mitochondrial disease

1 Introduction

Mitochondria are fundamental for cellular metabolism and functionally require coordinated gene expression (transcription and translation) from two genomes. Although 99% of the human mitochondrial proteome is encoded in the nucleus, the faithful expression of the mitochondrial genome is essential for oxidative phosphorylation. Mitochondrial DNA (mtDNA) is a multicopy maternally inherited intracellular genome that encodes 13 essential proteins of four oxidative phosphorylation complexes along with tRNAs and rRNAs needed for the translation of these proteins [1]. Regulation of mtDNA expression is coordinated by dedicated factors encoded in the nuclear genome, which are synthesized in the cytosol and then imported across the two mitochondrial lipid bilayers [2].

Antoni Barrientos and Flavia Fontanesi (eds.), *The Mitoribosome: Methods and Protocols*,
Methods in Molecular Biology, vol. 2661, https://doi.org/10.1007/978-1-0716-3171-3_7,

Protein synthesis within mitochondria occurs on unique ribosomes [3, 4] by mechanisms more similar to those in bacteria than to cytoplasmic ribosomes [2]. Mitochondrial ribosomes require rRNA and tRNA components transcribed from mtDNA to assemble with ribosomal proteins that are encoded in the nucleus, synthesized in the cytosol and then imported across the two mitochondrial lipid bilayers to assemble the large and small ribosomal subunits. Recent progress has been made to elucidate the mechanisms for the step-wise assembly of human mitochondrial ribosomes [5]. Over the last decade, the revolution in genetic diagnostics has identified disruptions to the faithful synthesis of these 13 mitochondrial proteins as the largest group of inherited human mitochondrial pathologies [6]. It appears that all of the molecular steps required for mitochondrial protein synthesis can be affected, from the genome to protein, including cotranslational quality control. Despite the importance of oxidative phosphorylation to cell fitness, this group of diseases manifest with an exceptional diversity in clinical presentations, which demonstrates that dysregulation of mitochondrial protein synthesis can have cell and tissue-specific consequences. The biochemical defect in ATP synthesis alone does not account for the clinical presentations or disease severity, suggesting the importance of other factors critical to the molecular pathogenesis. Thus, investigating the mechanisms by which a primary molecular defect disrupts the synthesis of mitochondrial proteins is a critical step to elucidating the molecular pathogenesis of these disorders.

An important tool in the study of human mitochondrial protein synthesis requires the biochemical separation of mitochondrial ribosomes and analysis of the associated RNA. One of the most common sources of patient-derived sample material is cultured primary dermal fibroblast. In some cases, myoblasts can be also derived from skeletal muscle biopsies. Here, we provide a robust methodological paradigm to facilitate the analysis of mitochondrial ribosomes and RNA from cultured fibroblasts that can be readily applied to any cultured cell type and or small tissue biopsy material. Importantly, our method does not require any prior enrichment of mitochondria and thus is well suited when the input material can be limiting. This pipeline can then be used for traditional northern blotting analysis or for next-generation deep sequencing analysis. Our preferred approaches will be described in detail herein. We also provide a complementary approach for the analysis of the mitochondrial ribosomal proteins to identify potential interacting factors.

2 Materials

All solutions are prepared in ultrapure water (Merck Millipore Milli-Q) unless indicated otherwise.

2.1 Cell Culture

1. Culture medium for human dermal fibroblasts from healthy controls or patient-derived: High glucose Dulbecco's Modified Eagle's Medium (DMEM), supplemented with 10% fetal bovine serum, 1× glutamax (Gibco, 35050-038), and 50 μg/mL uridine. Store at 4 °C (*see* **Note 1**).
2. Culture medium for human myoblast cultures: skeletal muscle cell growth medium (Sigma, 151-500) supplemented with 50 μg/mL uridine. Store at 4 °C (*see* **Note 1**).
3. 1× phosphate-buffered saline (PBS): 137 mM NaCl, 2.7 mM KCl, 10 mM Na_2HPO_4, 1.8 mM KH_2PO_4.

2.2 Cell Lysis Buffer for Mitochondrial Ribosome Isolation

1. Mitochondrial ribosome buffer (MRB): 50 mM Tris–HCl pH 7.2, 10 mM magnesium acetate, 40 mM ammonium chloride (NH_4Cl), 100 mM potassium chloride (KCl) in 1 L of ultrapure water. Filter the solution through 0.45 μM by vacuum filtration. Store at 4 °C.
2. 10 mM Adenosine 5′ triphosphate (ATP) stock solution. Dissolve 55.1 mg of ATP in 10 mL of MRB buffer. Filter through a 0.45 μM syringe filter. Prepare fresh and store at 4 °C.
3. Dodecyl-maltoside (DDM).
4. Phenylmethylsulfonyl fluoride solution (PMSF).
5. 67 mg/mL chloramphenicol (Sigma, C3175-100MG). Stock solution prepared in 1× PBS and stored at −20 °C.
6. Prepare lysis buffer fresh by mixing 0.02 g DDM (1% final), 20 μL PMSF (1 mM final), 200 μL ATP stock solution (1 mM final), and 12 μL of chloramphenicol solution (400 μg/mL final) in 2 mL of MRB.

2.3 Preparation of 10–30% Linear Sucrose Gradient

1. 10% sucrose solution: Dissolve 2 g sucrose in 13 mL MRB. Add 200 μL PMSF (1 mM final), 2 mL of 10 mM ATP stock solution and fill up to 20 mL with MRB. Filter through a 0.45 μM syringe filter. This volume is sufficient for two sample gradient preparations. Prepare fresh.
2. 30% sucrose solution: Dissolve 6 g sucrose in 13 mL MRB. Add 200 μL PMSF, 2 mL of 10 mM ATP stock solution and fill up to 20 mL. Filter through a 0.45 μM syringe filter. This volume is sufficient for two sample gradient preparations. Prepare fresh.

3. Ultraclear centrifuge tube (16 × 102 mm, 17 mL) (Beckman Coulter, 344061).
4. Beckman Coulter SW 32.1 Ti rotor (swinging bucket).
5. Beckman Coulter ultracentrifuge Optima L series.

2.4 RNA Isolation from Sucrose Gradients

1. TRIzol LS reagent (Thermo Fisher, 10296028). Store at 4 °C.
2. Chloroform.
3. GlycoBlue™ Coprecipitant (Thermo Fisher, AM9516).
4. Analytical grade isopropanol.
5. Pure ethanol.
6. 3 M Sodium acetate pH 4.8.
7. Microsep™ centrifugal filter (MWCO 3K) (VWR, 516-0370).

2.5 Bioanalyzer Analysis of Purified RNA to Assess Quality

1. Agilent 2100 Bioanalyzer system (Agilent Technologies).
2. Agilent RNA 600 Nano kit (Agilent Technologies, 5067-1511).
3. Syringe kit (supplied with Agilent RNA 600 Nano kit).
4. RNA ladder (supplied with Agilent RNA 600 Nano kit).
5. Chip priming station (supplied with Agilent 2100 Bioanalyzer).
6. IKA vortex mixer (supplied with Agilent 2100 Bioanalyzer).

2.6 Northern Blotting

2.6.1 Preparation of RNA Sample

1. 10× MOPS (3-(N-morpholino)propanesulfonic acid): Dissolve 83.7 g MOPS, 13.6 g sodium acetate trihydrate, 3.7 g EDTA disodium dihydrate in 800 mL of ultrapure water. Adjust to pH 7.0 with sodium hydroxide (NaOH). Fill up to 1 L with ultrapure water. Autoclave to sterilize and store at 4 °C. Protect the solution from light.
2. RNA sample buffer stock solution (1.325 mL): 150 μL 10× MOPS, 300 μL 36.5–38% formaldehyde, 850 μL formamide, and 25 μL EtBr. The stock solution is sufficient for 50 samples. Store at −20 °C.
3. RNA loading buffer: 50 μL glycerol, 2 μL 500 mM EDTA pH 8, 4 mg bromophenol blue, add ultrapure water to 1 mL. Store at −20 °C.
4. RNase AWAY® decontaminating solution (Molecular BioProducts™, 6227799).

2.6.2 Preparation of Denaturing Gel

1. Agarose, molecular grade.
2. 10× MOPS.
3. 36.5–38% formaldehyde.

2.6.3 Membrane Transfer

1. Saline sodium citrate (SSC) solution (10×, pH 7.0) (300 mM sodium citrate, 1 M NaCl).
2. Hybond-N+ membrane (Cytiva, RPN203S).
3. UV crosslinker.

2.6.4 5′ End Labeling of Oligonucleotide with ^{32}P

1. T4 Polynucleotide kinase (PNK) (NEB, M0201L) and 10× T4 PNK reaction buffer.
2. 500 mM EDTA, pH 8.0.
3. Antisense DNA oligonucleotide probe between 24 and 26 nucleotides.
4. ATP, [γ-^{32}P]- 3000 Ci/mmol 10 mCi/mL (Perkin Elmer, BLU002A250UC).
5. ProbeQuant™ G-50 micro columns (Cytiva, 28-9034-08).

2.6.5 Oligonucleotide Hybridization

1. 1 M sodium phosphate buffer, pH 7.2: 138 g $NaH_2PO_4{\cdot}H_2O$ in 1 L ultrapure water, 142 g Na_2HPO_4 in 1 L ultrapure water. Mix 280 mL $NaH_2PO_4{\cdot}H_2O$ with 720 mL Na_2HPO_4.
2. 10% bovine serum albumin (BSA) in ultrapure water.
3. Hybridization buffer (20 mL): 5 mL phosphate buffer pH 7.2, 960 μL 5 M NaCl, 40 μL 500 mM EDTA pH 8.0, 7 mL 20% SDS, and 5 mL formamide. Store at −20 °C.
4. Wash solution 1: 2× SSC, 0.2% SDS, 1 mM EDTA pH 8.0.
5. Wash solution 2: 0.5× SSC, 0.2% SDS, 1 mM EDTA pH 8.0.
6. Wash solution 3: 0.2× SSC, 0.1% SDS, 1 mM EDTA pH 8.0.

2.7 Next-Generation RNA Sequencing

2.7.1 RNA Library Preparation with InGex TGIRT®-III Reverse Transcriptase

1. Monarch total RNA Miniprep kit (NEB, T2010).
2. NEBNext® magnesium RNA fragmentation module (NEB, E6150S).
3. T4 PNK (NEB, M0201L) and 10× T4 PNK reaction buffer.
4. Zymo RNA clean & concentrator-5 (Zymo Research, R1013).
5. Agilent 2100 Bioanalyzer system (Agilent Technologies).
6. Agilent RNA 600 Nano kit (Agilent Technologies, 5067-1511).
7. TGIRT™ improved Modular Template-switching RNA-seq kit (InGex): include TGIRT®-III reverse transcriptase, 10× primer mix (1 μM R2 RNA and 1 μM R2 DNA in 10 mM Tris–HCl, pH 7.5, 1 mM EDTA), 10× DTT (5 mM final) and 5× reaction buffer (2.25 M NaCl, 25 mM $MgCl_2$, 100 mM Tris–HCl, pH 7.5).
8. R1R DNA adaptor: 5′/5Phos/GAT CGT CGG ACT GTA GAA CTC TGA ACG TGT AG/3SpC3/.
9. Illumina multiplex PCR primer: 5′ AAT GAT ACG GCG ACC ACC GAG ATC TAC ACG TTC AGA GTT CTA CAG TCC GAC GAT C 3′.

10. Illumina barcode PCR primer: 5′ CAA GCA GAA GAC GGC ATA CGA GAT BARCODE GTG ACT GGA GTT CAG ACG TGT GCT CTT CCG ATC T 3′.
11. MinElute Reaction Cleanup kit (QIAGEN, 28206).
12. 5′ DNA Adenylation kit (NEB, E2610S).
13. 5′ App DNA/RNA Ligase (NEB, M0319S).
14. Phusion™ High-Fidelity DNA Polymerase (2 U/μL) (Thermo Fisher, F530S).
15. MagSi-NGS^{PREP} Plus (Magtivio BV, MDKT00010005).
16. 10 mM dNTPs.
17. 5 M NaOH.
18. 5 M HCl.

2.8 Sodium Dodecyl Sulfate-Polyacrylamide Gel Electrophoresis (SDS-PAGE)

1. 30% Acrylamide/Bis solution, 29:1 (Bio-rad, 1610156).
2. 10% ammonium persulfate (APS) solution in ultrapure water. Store at 4 °C.
3. N,N,N^1,N^1 -Tetramethylethane-1,2-diamine (TEMED).
4. 12% separating gel, 10 mL for one Criterion™ empty cassette: 3.3 mL ultrapure water, 4 mL 30% Acrylamide/Bis solution, 2.5 mL 1.5 M Tris–HCl, pH 8, 100 μL 10% SDS, 100 μL 10% APS, 4 μL TEMED.
5. 5% stacking gel, 3 mL for one Criterion™ empty cassette: 1.05 mL ultrapure water, 250 μL 30% Acrylamide/Bis solution, 190 μL 1 M Tris–HCl, pH 6.8, 15 μL 10% SDS, 15 μL 10% APS, 1.5 μL TEMED.
6. Running buffer (10×): 60.6 g Tris–HCl, 288.4 g glycine and 20 g SDS in 1 L of ultrapure water, adjust pH to 8.3 and fill up to 2 L. Prepare 1× running buffer with ultrapure water. Store at room temperature (RT).
7. Semidry transfer buffer: 2.9 g glycine, 5.8 g Tris base and 0.75 g SDS in 800 mL of ultrapure water and add 200 mL of methanol. Store at RT.
8. Tris-buffered saline (TBS) (10×): 24.2 g Tris base and 80 g sodium chloride (NaCl) in 1 L of ultrapure water. Store at RT.
9. TBST wash buffer: 1× TBS supplemented with 0.1% Tween-20. Store at RT.
10. 5% BSA (IgG-free, protease-free) (Jackson ImmunoResearch, 001-000-162) in TBST.
11. Blocking solution: 1% nonfat dry milk in TBST wash buffer. Prepare fresh.

12. 2× Laemmli sample buffer: 120 mM Tris–HCl (pH 6.8), 20% glycerol, 4% SDS, 0.02% (w/v) bromophenol blue, 5% β-mercaptoethanol (β-ME). Store at RT. Add β-ME fresh just prior to use.
13. Primary antibodies against mitochondrial ribosomal proteins and RNA-binding proteins: From Proteintech Group: uL11m (15543-1-AP, 1:20,000); LRPPRC (21175-1-AP, 1:8000); SLIRP (26006-1-AP, 1:1000); and mS35 (16457-1-AP, 1: 5000). From Sigma-Aldrich: MTRES1 (HPA049535, 1: 1000). All antibodies are diluted in 5% BSA (IgG-free, protease-free, Jackson ImmunoResearch, 001-000-162) and can be reused multiple times following storage at −20 °C.
14. Horseradish peroxidase (HRP)-conjugated secondary antibodies: rabbit HRP (1: 20,000 in TBST) (Jackson Immunoresearch, 111-035-144) and mouse HRP (1: 10,000 in TBST) (Jackson ImmunoResearch, 115-035-146).
15. SuperSignalTM West Femto Maximum Sensitivity Substrate (Thermo Fisher, 34095).
16. Trichloroacetic acid solution (TCA), 6.1 N.
17. Acetone. Store at −20 °C.
18. Protein molecular weight ladder.
19. Criterion™ Cell (Bio-rad, 1656001).
20. Criterion™ empty cassette with 26-well comb (Bio-rad, 3459903).
21. Nitrocellulose hybridization transfer membrane.
22. Whatman gel blotting paper.
23. Owl™ HEP series semidry electro blotting systems. Alternative semidry or wet transfer systems can be used.

3 Methods

3.1 Cell Culture

1. Culture human fibroblasts cells in 145 mm plate.
2. Culture until cell density reaches 80–90% confluency. The number of plates per sample preparation will be determined by the experimental analysis (*see* **Note 2**).
3. Precool benchtop centrifuge to 4 °C.
4. Pour off the media quickly, rinse 1× with ice cold PBS and pour off then place culture plate on dry ice to rapidly freeze the cells. Transfer to ice, add cold 1× PBS and scrape the cell suspension and transfer to a 15 mL conical tube. Centrifuge at 7500 × *g* for 5 min, at 4 °C. Aspirate the PBS and store the cell pellet at − 80 °C.

3.2 Preparation of 10–30% Linear Sucrose Gradient

1. Gradients are made fresh on the day of centrifugation according to the protocol described [7]. Pipette 8 mL of 30% sucrose solution into ultraclear centrifuge tube (16 × 102 mm). Carefully layer 8 mL of 10% sucrose solution on top of the 30% sucrose solution. A visible interface should form between the layers (Fig. 1) (*see* **Note 3**).
2. Seal the tube with parafilm and *slowly* invert horizontally.
3. Incubate for 1 h at RT.
4. Slowly invert the tube back to a vertical position and place in tube holder. Store at 4 °C until ready for loading of clarified lysate.

3.3 Cell Lysis

1. Thaw frozen cell pellet on ice.
2. Resuspend cell pellet in freshly prepared lysis buffer and incubate on ice for 30 min (*see* **Note 4**).
3. Clarify cell lysate with centrifugation at 18,000 × *g* for 20 min at 4 °C.
4. Transfer clarified cell lysate to a new microcentrifuge tube on ice (*see* **Note 5**).
5. Determine protein concentration.

3.4 Sucrose Gradient Sedimentation

1. Precool ultracentrifuge (Beckman Coulter Optima L series), SW 32.1 Ti rotor and tube holder at 4 °C at least 1 h before run.
2. Add clarified lysate on top of the prepared 10–30% sucrose solution (Fig. 1).
3. Gently place tube into the precooled tube holder and cap tightly, and then place into precooled rotor.

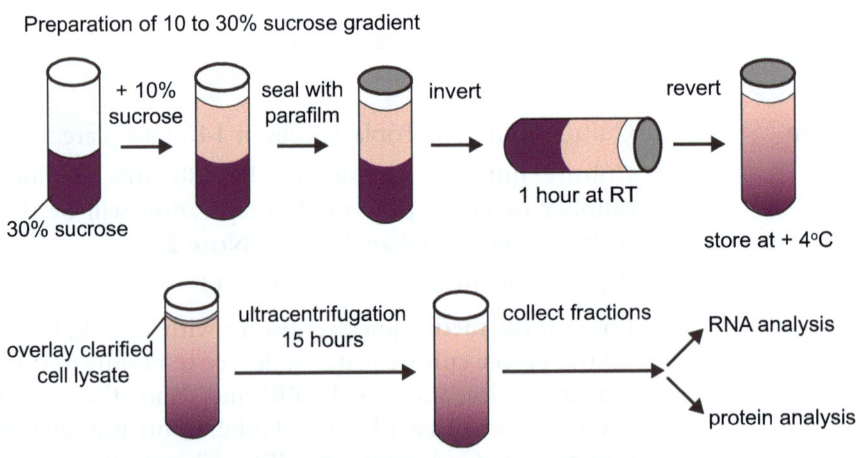

Fig. 1 Schematic illustrating the workflow to prepare a 10–30% sucrose gradient for biochemical isolation of human mitochondrial ribosomes for analysis

4. Ultracentrifuge settings: 74,400 × *g*, 15 h run time, running temperature at 4 °C, maximum acceleration and "no brake" for deceleration. Start run.
5. Following the run, collect 24 equal volume fractions into 1.5 mL tubes on ice for RNA isolation and analysis (Subheadings 3.5, 3.6, 3.7, and 3.8) or for protein analysis (Subheading 3.9). Samples can be stored at −80 °C or proceed directly to Subheading 3.5 or 3.9 (*see* **Note 6**).

3.5 RNA Isolation

1. Clean surface area and equipment with RNase AWAY® decontaminating solution.
2. From the 24 collected fractions, combine fractions 1–4, 5–7, 9–11, 13–15, 16–18, and 19–24 (Fig. 2a). Transfer into Microsep™ centrifugal filter (MWCO 3K).
3. Centrifuge tubes of combined fractions at 7500 × *g* for 90 min at 4 °C. This step concentrates the sample and reduces the volume for RNA isolation.
4. Transfer concentrated fraction from the centrifugal filter into a microcentrifuge tube, with a maximum volume of 250 μL in each tube.
5. For every 250 μL sample, add 750 μL TRIzol LS reagent and mix by inverting the tube. Adjust volume of TRIzol LS reagent according to sample volume. Incubate at RT for 5 min.
6. Add 200 μL of chloroform and shake vigorously for 15 s. Incubate at RT for 3 min.
7. Centrifuge at 12,000 × *g* for 15 min at 4 °C.
8. Carefully transfer the top aqueous phase containing the RNA into a new microcentrifuge tube containing 2 μL of GlycoBlue™ coprecipitant (*see* **Note 7**).
9. Add an equal volume of isopropanol (~550 μL) and store overnight at −20 °C.
10. Centrifuge for 30 min at 18,000 × *g* at 4 °C.
11. Wash RNA pellet with 75% ethanol for 15 min at 18,000 × *g* at 4 °C.
12. Remove ethanol and air-dry pellet for 5 min at RT.
13. Resuspend pellet in 20 μL nuclease-free water.
14. To remove any potential residual phenol-chloroform, add 1/10 volume of 3 M sodium acetate (pH 4.8) and 2.5 volume of ethanol to the resuspended RNA sample. Incubate at −80 ° C for 2 h or overnight.
15. Repeat **steps 10–13**.
16. Measure RNA concentration and integrity using Bioanalyzer.

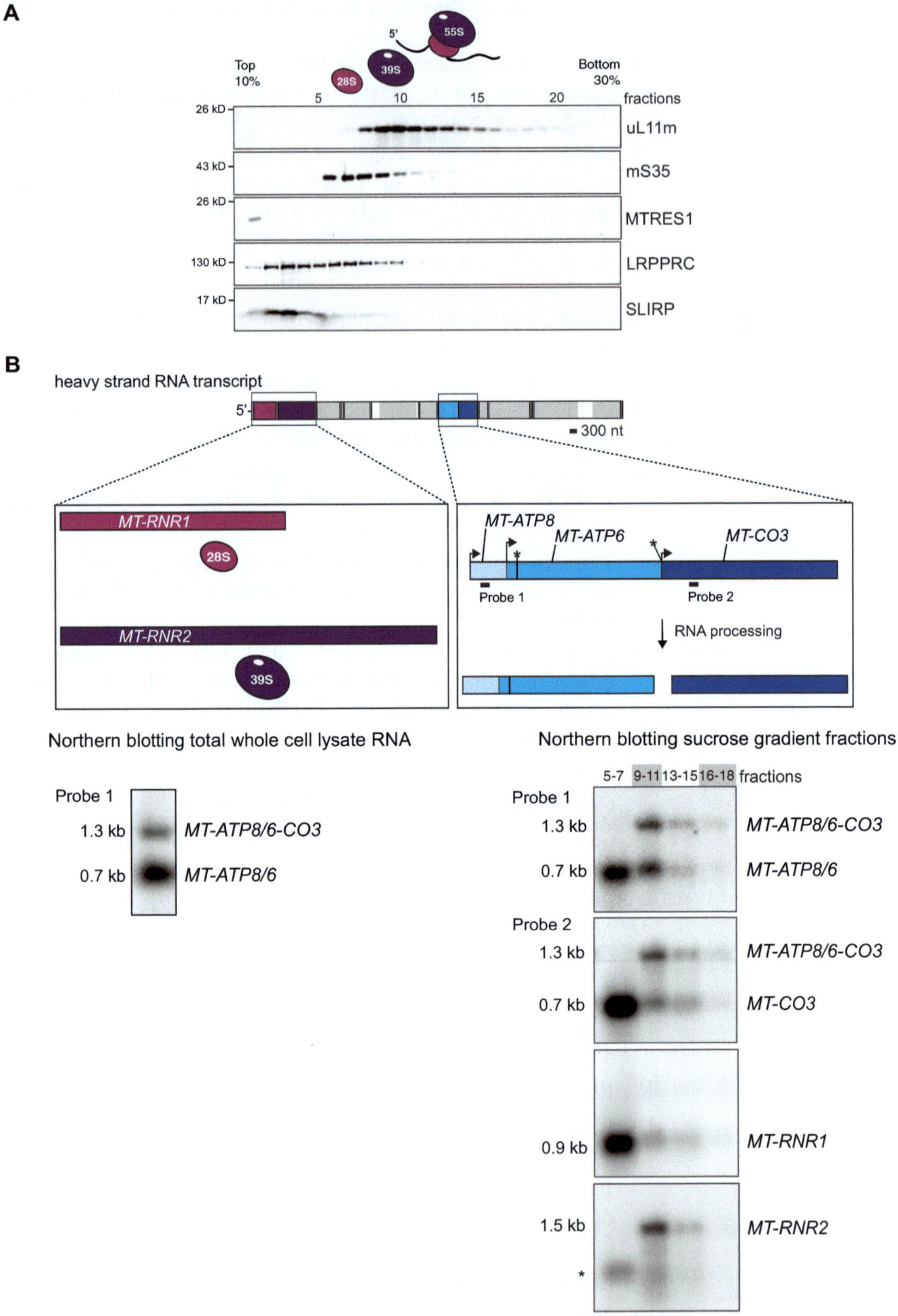

Fig. 2 (**a**) Sucrose gradient sedimentation profile of mitochondrial ribosomal proteins from whole cell lysates of cultured human myoblasts. Immunoblotting of mitochondrial ribosomal proteins (uL11m and mS35) and RNA-binding proteins (MTRES1, LRPPRC, and SLIRP) analyzed on 12% SDS-PAGE. (**b**) Northern blot analysis of

3.6 Bioanalyzer Analysis of Purified RNA

1. Thaw RNA ladder on ice.
2. Equilibrate all reagents to RT for 30 min before use.
3. Prepare gel-dye by mixing 1 μL of RNA 6000 Nano dye concentrate to 65 μL of filtered RNA 6000 Nano gel matrix.
4. Vortex and centrifuge for 10 min at 13,000 × *g* at RT. Use the gel-dye mix within 1 day.
5. Place RNA 6000 Nano chip on priming station.
6. Pipette 9 μL gel-dye mix into the well marker "G."
7. Press the plunger of the priming station and wait for 30 s.
8. Release the plunger and pipette 9 μL gel-dye mix into remaining two marked wells.
9. Pipette 5 μL RNA 6000 Nano marker into the ladder and sample wells. Pipette 1 μL of thawed ladder into the ladder well.
10. Pipette 1 μL of RNA sample into the sample wells.
11. Vortex the chip on IKA vortex mixer for 1 min at 2400 rpm.
12. Insert the chip in Agilent 2100 Bioanalyzer and start the run within 5 min.
13. For detailed protocols, please refer to https://www.agilent.com/cs/library/usermanuals/Public/G2938-90034_RNA6000Nano_KG.pdf.

3.7 Northern Blotting

3.7.1 Preparation of Denaturing Gel

1. Prepare a 1.2% agarose gel, adding 1.2 g of agarose to 70 mL ultrapure water and 10 mL 10× MOPS.
2. Microwave the solution until the agarose dissolves. Cool down to about 50 °C and add 20 mL formaldehyde in the fume hood.
3. Mix by swirling and pour the gel solution into a gel casting system. Allow gel to polymerize for at least 2 h.
4. Prerun agarose gel in 1× MOPS for 10 min at 20 V.
5. Wash wells with 1× MOPS before loading samples.

Fig. 2 (continued) RNA isolated from pooled sucrose gradient fractions. Top, schematic illustrating the mRNAs and rRNAs analyzed by northern blotting. Bottom left, northern blotting of total RNA taken from a whole cell human fibroblast lysate. Bottom right, northern blotting of RNA collected from pooled sucrose gradient fractions as shown in (**a**). Note the sucrose sedimentation profile of the *MT-ATP8/6-CO3* vs. *MT-ATP8/6* and *MT-CO3*, comparing to the analysis of whole cell lysates. The membranes were probed with labeled antisense oligonucleotides in the following sequence: Probe 1; Probe 2; *MT-RNR1* and *MT-RNR2*. * indicates a residual signal from the previous *MT-RNR1* probe following stripping of the membrane

3.7.2 Preparation and Separation of RNA

1. Preheat heating block to 75 °C.
2. Pipette 5 μg of purified RNA into a new microcentrifuge tube.
3. Add 10 μL of RNA sample buffer.
4. Incubate at 75 °C for 15 min.
5. Place tube on ice and add 3 μL of RNA loading buffer.
6. Load all RNA sample into the agarose gel and run overnight (18 h) at 10 V in a fume hood.

3.7.3 RNA Transfer onto a Hybridization Membrane

1. Evaluate the separation and integrity of the RNA using nucleic acid imaging system. Two distinct bands corresponding to the 28S and 18S rRNA should be visible (*see* **Note 8**).
2. Place the gel in ultrapure water for 15 min with gentle agitation at RT.
3. Replace with 10× SSC. Repeat once.
4. Transfer the RNA from agarose gel onto the Hybond-N+ nylon membrane overnight for 18 h with 10× SSC using capillary transfer approach [8].
5. Wash the membrane in 10× SSC for 5 min.
6. UV crosslink at 120 mJoule for 1 min. Store membrane at 4 °C until hybridization.

3.7.4 5′ End Labeling of Oligonucleotide Probes

1. Preheat heating block to 37 °C.
2. Mix 5 μL 10 μM primer, 2 μL T4 PNK buffer, 2 μL T4 PNK in a microcentrifuge tube and place on ice.
3. Working in a radioisotope laboratory and following local safety precautions, add 5 μL [γ-^{32}P]-ATP (3000 Ci/mmol) to the mixture. Caution: wear double gloves and work behind a plexiglass shield when handling radioisotopes.
4. Incubate at 37 °C for at least 20 min and up to 1 h.
5. Add 2 μL 500 mM EDTA to terminate the reaction.
6. Add 30 μL of ultrapure water.
7. Vortex ProbeQuant™ G-50 microcolumn for 15 s.
8. Centrifuge for 1 min at 735 × *g* at RT to remove storage buffer.
9. Remove unincorporated [γ-^{32}P]-ATP by adding the reaction mixture into ProbeQuant™ G-50 micro column.
10. Centrifuge for 2 min at 735 × *g* at RT. Store labeled oligonucleotide at −20 °C or use immediately.

3.7.5 Hybridization

1. Preheat hybridization oven to 65 °C.
2. Warm hybridization buffer to 50 °C.
3. Place the membrane in a hybridization tube and add 4 mL of hybridization buffer. Place the tube in the hybridization oven for at least 1 h at 65 °C with rotation.

4. Add the labeled oligonucleotide to 4 mL of hybridization buffer to make the probe solution.
5. Pour away hybridization buffer from the hybridization tube without removing the membrane.
6. Slowly add the hybridization buffer with the labeled oligonucleotide probe into the hybridization tube. Close the tube tightly.
7. Place the tube back into the hybridization oven and incubate overnight for 18 h at 37 °C with rotation.
8. Transfer the oligonucleotide probe solution into a 15 mL conical tube and store at −20 °C. Probes can reused multiple times for up to 2 months from receipt of the fresh radioisotope.
9. Wash the membrane with wash buffer 1 at 37 °C for 1 h with rotation inside the hybridization oven. Alternatively, washing can be done in a plastic container with a shaking water bath.
10. Replace with wash buffer 2 and continue washing at 37 °C for 1 h with agitation.
11. Replace with wash buffer 3 and continue washing at 37 °C for 30 min with agitation.
12. Air-dry the membrane for 1 h and expose on storage phosphor screen for 1–3 days then scan with a phosphor imaging station. *See* Fig. 2b for representative data.

3.8 Library Preparation for RNA Sequencing of the Entire Mitochondrial Transcriptome on the Illumina Platform

1. The method is based upon the protocol described by InGex for TGIRT [9].
2. Preheat thermal cycler to 94 °C.
3. In a sterile PCR tube, add 2 μL of 10× fragmentation buffer (NEBNext RNA fragmentation module), 2 μg of total RNA and fill up volume to 20 μL with nuclease-free water. The desired fragment size is between 200 and 300 nucleotides.
4. Incubate reaction at 94 °C for 4 min.
5. Purify fragmented RNA using Zymo RNA Clean & Concentrator-5 kit, elute in 15 μL of nuclease-free water.
6. Add 2 μL of T4 PNK and 2 μL of T4 PNK buffer (10×) to the tube containing the purified fragmented RNA. Incubate the sample at 37 °C for 30 min. This step removes the 3′ phosphate that is known to impede TGIRT template-switching cDNA synthesis.
7. Purify the reaction mixture using Zymo RNA Clean & Concentrator-5 kit, elute in 15 μL of nuclease-free water.
8. Use the Agilent Bioanalyzer (*see* Subheading 3.6) to determine the size range of RNA fragments and concentration.
9. Preanneal R2 RNA/R2R DNA primer mix (10×) by incubating at 85 °C in a thermal cycler for 2 min.

10. Cool down sample to 25 °C with 3% ramp in the thermal cycler.
11. In a sterile PCR tube, add 2 μL of preannealed R2 RNA/R2R DNA heteroduplex, 4 μL of 5× reaction buffer, 2 μL of 10× DTT, 50 ng of RNA and fill up volume to 18 μL with nuclease-free water.
12. Preincubate the reaction mixture at RT for 30 min.
13. Add 2 μL of 10 mM dNTPs and incubate at 60 °C for 1 h in a thermal cycler.
14. Add 1 μL of 5 M NaOH and incubate at 95 °C for 3 min in a thermal cycler.
15. Cool down to RT and neutralize reaction with 1 μL of 5 M HCl. Leave the reaction mixture at RT.
16. Clean up cDNAs with MinElute Reaction Cleanup kit (QIAGEN). Repeat once.
17. Preadenylate the R1R DNA adaptor with the 5′ DNA adenylation kit (NEB).
18. The preadenylated R1R DNA adaptor is ligated to the 3′ end of cDNA using 5′ App DNA/RNA Ligase (NEB) at 65 °C for 2 h.
19. Purify ligated products using MinElute Reaction Cleanup kit (QIAGEN). Repeat once.
20. Set up the PCR reaction mixture by adding 1 μL of 10 μM Illumina multiplex primer, 1 μL of 10 μM Illumina barcode primer, 10 μL of purified ligated products, 5 μL of 5× Phusion High-fidelity buffer, 0.3 μL of 2 U/μL Phusion High-fidelity DNA polymerase and nuclease-free water to a total volume of 25 μL.
21. PCR conditions: 1 cycle at 98 °C for 5 s, 21 cycles at 98 °C 5 s, 60 °C 10 s, 72 °C 15–30 s/kb, hold at 4 °C.
22. Clean up PCR products using magnetic beads (Magsi NGS^{PREP} Plus).
23. Deliver the library to a genomics facility for deep sequencing on an Illumina NextSeq 500 or other platform.

3.9 Protein Analysis of Sucrose Gradient Fractions

3.9.1 TCA Precipitation of Proteins from Collected Fractions

1. Precool the microcentrifuge to 4 °C and preheat the heating block to 95 °C.
2. Add TCA solution to each collected fraction for a final concentration of 13%. Invert tube a few times to mix. Proceed with TCA precipitation (**step 3**) or store at −80 °C.
3. Incubate samples on ice for at least 30 min.
4. Centrifuge samples for 30 min at 18,000 × *g* at 4 °C.
5. Carefully remove the supernatant from the tube by pipetting and discarding into TCA waste container. A white precipitate should be visible.

6. Add 1 mL of cold acetone and centrifuge for 5 min at 18,000 × *g* at 4 °C.
7. Carefully remove all of the acetone from the tube by pipetting. Repeat **steps 6** and 7 once.
8. Air dry the pellet at RT for 5 min.
9. Prepare each fraction for immunoblotting by adding 40 μL of 1× Laemmli. If the sample is yellow, gently add 5 M NaOH 1 μL at a time to restore the pH so that the solution turns blue.
10. Heat the sample at 95 °C for 5 min. Cool down to RT. Proceed to SDS-PAGE (Subheading 3.9.2) for immunoblotting or store samples at −80 °C.

3.9.2 SDS-PAGE and Immunoblotting

1. Prepare a 12% SDS-PAGE and 5% stacking in an empty Criterion™ empty cassette.
2. Remove the comb and wash wells with 1× running buffer.
3. Load 5 μL of prestained protein ladder, followed by 15 μL of protein sample into the gel.
4. Start electrophoresis at 90 V for 30 min so that the protein sample migrates into the separating gel.
5. Increase the voltage to 120 V and continue until the 10 kDa molecular weight marker is near the bottom edge of the gel.
6. Transfer proteins from the gel to a nitrocellulose membrane using a semidry electroblotting system, for example, the Owl™ HEP series. Fix the gel and prewet the nitrocellulose membrane in semidry transfer buffer. Wet a stack of three pieces of Whatman gel blotting paper (7 × 14 cm) and place onto the electroblotting system, carefully place the gel on top, followed by the membrane. Lastly, wet another stack of three pieces of Whatman paper and place on top of the membrane. Carefully roll out bubbles between the sandwich, e.g., with a serological pipette. When using Owl™ system, proteins are transferred to a 6.5 × 13.5 cm nitrocellulose membrane at 75 mA for 1 h and 25 min (*see* **Note 9**).
7. Following the protein transfer, incubate the membrane in a plastic container with blocking solution for 1 h at RT on a platform rocker.
8. Discard the blocking solution and rinse the membrane in TBST wash buffer three times.
9. Incubate the membrane with a primary antibody overnight at 4 °C on a platform rocker.
10. The next day, remove the antibody and return it to −20 °C. Wash the membrane three times in TBST wash buffer, 20 mi each.
11. Incubate the membrane with HRP-conjugated secondary antibodies in TBST at RT for 1 h on a platform rocker.

12. Wash the membrane three times in TBST wash buffer, 20 min each.
13. Use enhanced chemiluminescence (ECL) reagents to detect the proteins using an iBright Imaging system (Thermo Fisher) or with X-ray films and a film developer. *See* Fig. 2a for a representative sedimentation profile for mitochondrial ribosomal proteins and RNA-binding proteins from cultured human myoblasts.

4 Notes

1. The cell culture medium is supplemented with 50 μg/mL uridine to bypass any defects in the respiratory chain that would affect the function of dihydroorotate dehydrogenase, which is required for de novo pyrimidine biosynthesis [10].
2. For cultured fibroblasts and myoblasts, the number of cultured plates of cells required per sample analysis varies depending upon the experimental method pursued. For protein immunoblotting and RNA analysis by NGS RNA-seq, one plate (145 mm) per sample is sufficient for sucrose gradient centrifugation. For northern blotting, however, cells from four culture plates (145 mm) are pooled together for cell lysis and separation by sucrose gradient.
3. Adjust pipette controller to minimum power and avoid bubbles when layering the sucrose solution. This method to form a 10–30% sucrose gradient is highly reproducible [11–15] and advantageous in that a gradient maker is not required.
4. The desired protein concentration of the cell lysate should be 2 mg/mL.
5. Excess lysate can be stored at −80 °C for future use.
6. Optional for protein analysis of sucrose gradient fractions: 1–2 mg of BSA can be added to the empty microcentrifuge tubes before collection of fractions for TCA precipitation to help in the precipitation of proteins from the sucrose fractions.
7. Optional: Addition of GlycoBlue™ coprecipitant during RNA isolation increases the visibility of RNA pellets after centrifugation.
8. It is not necessary to add EtBr to the denaturing gel as it is already added to the RNA sample buffer.
9. Following the protein transfer to the nitrocellulose membrane, the membrane can be cut horizontally into strips using the molecular weight marker as a guide. This step facilitates analysis of individual proteins of differing molecular weights all at once and speeds up the experimentation.

Acknowledgments

We thank Olesia Ignatenko for discussion. This research was supported by funding to BJB from the Academy of Finland (307431 and 314706) and the Sigrid Juselius Foundation Senior Investigator Award. KYN was supported by the Orion Research Foundation and the Finnish Cultural Foundation.

References

1. Anderson S, Bankier AT, Barrell BG et al (1981) Sequence and organization of the human mitochondrial genome. Nature 290: 457–465
2. Kummer E, Ban N (2021) Mechanisms and regulation of protein synthesis in mitochondria. Nat Rev Mol Cell Biol 22:307–325
3. Amunts A, Brown A, Toots J et al (2015) Ribosome. The structure of the human mitochondrial ribosome. Science 348:95–98
4. Greber BJ, Bieri P, Leibundgut M et al (2015) Ribosome. The complete structure of the 55S mammalian mitochondrial ribosome. Science 348:303–308
5. Ferrari A, Del'Olio S, Barrientos A (2021) The diseased mitoribosome. FEBS Lett 595:1025–1061
6. Suomalainen A, Battersby BJ (2018) Mitochondrial diseases: the contribution of organelle stress responses to pathology. Nat Rev Mol Cell Biol 19:77–92
7. Graham J (2001) Biological centrifugation (the basics), 1st edn. BIOS Scientific Publishers Ltd
8. Sambrook J (2001) Molecular cloning: a laboratory manual (3 volume set), 3rd edn. Cold Spring Harbor Laboratory Press
9. Nottingham RM, Wu DC, Qin Y et al (2016) RNA-seq of human reference RNA samples using a thermostable group II intron reverse transcriptase. RNA 22:597–613
10. Rawls J, Knecht W, Diekert K et al (2000) Requirements for the mitochondrial import and localization of dihydroorotate dehydrogenase. Eur J Biochem 267:2079–2087
11. Richter U, Lahtinen T, Marttinen P et al (2013) A mitochondrial ribosomal and RNA decay pathway blocks cell proliferation. Curr Biol 23:535–541
12. Carroll CJ, Isohanni P, Pöyhönen R et al (2013) Whole-exome sequencing identifies a mutation in the mitochondrial ribosome protein MRPL44 to underlie mitochondrial infantile cardiomyopathy. J Med Genet 50:151–159
13. Richter U, Lahtinen T, Marttinen P et al (2015) Quality control of mitochondrial protein synthesis is required for membrane integrity and cell fitness. J Cell Biol 211:373–389
14. Jackson CB, Huemer M, Bolognini R et al (2019) A variant in MRPS14 (uS14m) causes perinatal hypertrophic cardiomyopathy with neonatal lactic acidosis, growth retardation, dysmorphic features and neurological involvement. Hum Mol Genet 28:639–649
15. Ng KY, Richter U, Jackson CB et al (2022) Translation of MT-ATP6 pathogenic variants reveals distinct regulatory consequences from the co-translational quality control of mitochondrial protein synthesis. Hum Mol Genet 31:1230–1241

Chapter 8

Yeast Mitoribosome Purification and Analyses by Sucrose Density Centrifugation and Immunoprecipitation

Andreas Aufschnaiter, Andreas Carlström, and Martin Ott

Abstract

Mitochondrial protein biosynthesis is maintained by an interplay between the mitochondrial ribosome (mitoribosome) and a large set of protein interaction partners. This interactome regulates a diverse set of functions, including mitochondrial gene expression, translation, protein quality control, and respiratory chain assembly. Hence, robust methods to biochemically and structurally analyze this molecular machinery are required to understand the sophisticated regulation of mitochondrial protein biosynthesis. In this chapter, we present detailed protocols for immunoprecipitation, sucrose cushions, and linear sucrose gradients to purify and analyze mitoribosomes and their interaction partners.

Key words Mitoribosome, Sucrose gradient, Sucrose cushion, Immunoprecipitation, Yeast, Mitochondria

1 Introduction

Yeast mitochondrial ribosomes (mitoribosomes) interact with a large set of proteins, which mediate functions related to gene expression, protein biosynthesis, protein quality control, as well as early steps in respiratory chain assembly [1, 2]. The biochemical and structural analysis of these interactions is crucial to understand the molecular principles governing mitochondrial translation. Hence, a reliable and reproducible toolset of methods to determine these interactions is required for a comprehension of mitochondrial function. Immunoprecipitation of mitoribosomes, sucrose cushions, and linear gradients, as well as a combination of these techniques were used in the past as highly reproducible methods to determine the structure and mechanistic principles of mitoribosomes [1–7].

Andreas Aufschnaiter and Andreas Carlström contributed equally with all other contributors.

Antoni Barrientos and Flavia Fontanesi (eds.), *The Mitoribosome: Methods and Protocols*,
Methods in Molecular Biology, vol. 2661, https://doi.org/10.1007/978-1-0716-3171-3_8,

The possibility to combine density gradient centrifugation with various different upstream sample preparations and an equally high range of downstream processes makes them versatile techniques for structural and biochemical applications. Density gradient centrifugation is a quick and reliable method to separate subcellular particles with equipment accessible in most biochemical laboratories. Thereby, solutions with different densities are layered (step gradient) or admixed (linear gradient) in centrifugation tubes to form a gradient with the highest density on the bottom and the lowest density on top. The sample in form of a tissue lysates or cellular/organellar lysates is carefully layered on top of the gradient, and organelles, subcellular particles, or macromolecular structures are separated by centrifugation. It is important to note that the separation of particles depends on both particle density and size. In the experimental set-ups described here, particles separate mainly by their size (with larger particles sedimenting farther), if separation according to density is desirable, other conditions will apply [8].

In this chapter, we provide detailed protocols for the preparation of mitoribosomes via immunoprecipitation, sucrose cushions, and/or linear sucrose gradients for structural and biochemical analysis.

2 Materials

2.1 FLAG Affinity Purification of Mitoribosomes

The following solutions should be prepared by mixing stock solutions. Please note that these recipes represent examples, and the exact composition might need to be adapted according to experimental requirements (*see* **Note 1**). All solutions used for cell lysis and subsequent immunoprecipitation have to be prepared with RNase-free ultrapure water in clean and RNase-free vessels, in order to prevent degradation of the yeast mitoribosome, which is particularly sensitive.

1. We recommend using strains lacking the main unspecific mitochondrial nuclease Nuc1 in order to minimize rRNA degradation and a strain background with high respiratory activity, such as W303 [1].
2. Yeast growth medium: 1% yeast extract, 2% peptone medium supplemented with 2% glycerol or other carbon source. Mix 10 g of yeast extract and 20 g of bacto peptone in 1 L of water. Adjust pH to 5.5 and autoclave.
3. CryoMill (Retsch MM400).
4. Liquid nitrogen (N_2).
5. Lysis buffer: 10 mM Tris–HCl, pH 7.4, 250 mM KCl, 50 mM $MgCl_2$, 1% n-dodecyl-beta-maltoside (DDM; or an alternative detergent (*see* **Note 1**)), 1× cOmplete protease inhibitor (Roche Applied Science).

6. Wash buffer: 10 mM Tris–HCl, pH 7.4, 250 mM KCl, 50 mM $MgCl_2$, 0.05% DDM (or the alternative detergent used in the lysis buffer).
7. 200 mM phenylmethylsulfonyl fluoride (PMSF) stock solution.
8. ANTI-FLAG M2 Affinity Gel (Sigma Aldrich).
9. 3× FLAG Peptide stock (Sigma Aldrich).
10. SDS Sample Buffer: 63 mM Tris–HCl, 2% SDS, 10% glycerol, 0.1% β-mercaptoethanol, and 0.0005% bromophenol blue, pH 6.8.

2.2 Sucrose Cushions for Purification and/or Analysis of Mitoribosomes

All solutions need to be prepared with RNase-free ultrapure water in clean and RNAse-free vessels. The following solutions should be prepared in advance by mixing respective stock solutions (*see* **Note 1**).

1. Lysis buffer: 1% Triton X-100. 50 mM KCl, 0.5 mM $MgCl_2$. 20 mM HEPES/KOH, pH 7.4, 1 mM PMSF, 1× cOmplete protease inhibitor mix (Roche Applied Science).
2. Sucrose cushion solution: 1.2 M sucrose, 50 mM KCl, 0.5 mM $MgCl_2$, 20 mM HEPES/KOH, pH 7.4.

2.3 Linear Sucrose Gradients for the Biochemical Analysis of Mitoribosomes

All solutions need to be prepared with RNase-free ultrapure water and RNase-free tubes. The following solutions should be prepared in advance by mixing respective stock solutions (*see* **Note 1**). Be aware that the depicted recipes only present examples and might need to be adapted according to research questions and experimental requirements. For example, the addition of ethylenediaminetetraacetic acid (EDTA) results in separation of large and small mitoribosomal subunits, allowing to analyze their interactome individually. Please carefully observe **Note 1** for respective ingredient descriptions, alternatives and suggestions for adjustments for certain experimental approaches, and recommended stock solutions.

1. Lysis buffer: 10 mM Tris–HCl, pH 7.4, 10 mM KOAc, 0.5 mM $Mg(OAc)_2$, 10 mM EDTA, 5 mM 2-mercaptoethanol, 1 mM PMSF, 1% DDM or other desired detergent (*see* **Note 1**), 5% glycerol, 1× cOmplete protease inhibitor mix, 0.1 mM spermidine. Adjust volume with RNase-free ultrapure water and note that 285 μL of the lysis buffer are required per sample/gradient.
2. Dilution buffer: 10 mM Tris–HCl (pH 7.4), 10 mM KOAc, 0.5 mM $Mg(OAc)_2$, 10 mM EDTA, 5 mM 2-mercaptoethanol, 1 mM PMSF, 5% glycerol, 0.1 mM spermidine. Adjust volume with RNase-free ultrapure water

and note that 285 μL of the dilution buffer are required per sample/gradient.

3. 1 M sucrose solution: 10 mM Tris–HCl (pH 7.4), 10 mM KOAc, 0.5 mM $Mg(OAc)_2$, 10 mM EDTA, 5 mM 2-mercaptoethanol, 1 mM PMSF, 0.1% DDM or other desired detergent, 0.1 mM spermidine, 1 M sucrose. Adjust volume with RNase-free ultrapure water and note that 2 mL of the 1 M sucrose solution are required per sample/gradient.
4. 0.3 M sucrose solution: 10 mM Tris–HCl (pH 7.4), 10 mM KOAc, 0.5 mM $Mg(OAc)_2$, 10 mM EDTA, 5 mM 2-mercaptoethanol, 1 mM PMSF, 0.1% DDM or other desired detergent, 0.1 mM spermidine, 0.3 M sucrose. Adjust volume with RNase-free ultrapure water and note that 2 mL of the 1 M sucrose solution are required per sample/gradient.
5. 72% trichloroacetic acid (TCA) solution.
6. SDS Sample Buffer: 63 mM Tris–HCl, 2% SDS, 10% glycerol, 0.1% β-mercaptoethanol, and 0.0005% bromophenol blue, pH 6.8.
7. Coomassie G-250 (e.g., NativePAGE™ 5% G-250 Sample Additive, Thermo Fisher Scientific).
8. Blue Native PAGE (e.g., NativePAGE™ 3–12%, Bis-Tris, 1.0 mm, Mini Protein Gels; Mini Gel Tank and Blot Module and respective power supplies).
9. Tris-buffered saline (TBS, 50 mM Tris–HCL, 0.15 M NaCl, pH 7.4).

3 Methods

3.1 FLAG Affinity Purification of Mitoribosomes

Affinity purification of mitochondrial ribosomes can be performed in a variety of ways. Different types of affinity tags can be fused either directly to mitoribosomal proteins or to interaction partners. Furthermore, a purification can either start from a whole-cell lysate or lysates from previously isolated mitochondria. In this section, we are presenting a protocol for performing FLAG affinity purification of mitoribosomes from a whole-cell lysate after flash-freezing cells in liquid nitrogen and subsequent mechanical disruption (Fig. 1). Rapid freezing of live yeast cells has the advantage that interactions with factors and the functional states of mitoribosomes are as close as possible to a physiological state [7]. This protocol can easily be scaled up or down depending on desired application and analyses. As an additional purification step, this protocol can be combined with subsequent sucrose cushions (see protocol below). However, both methods can be also used individually for the analysis of

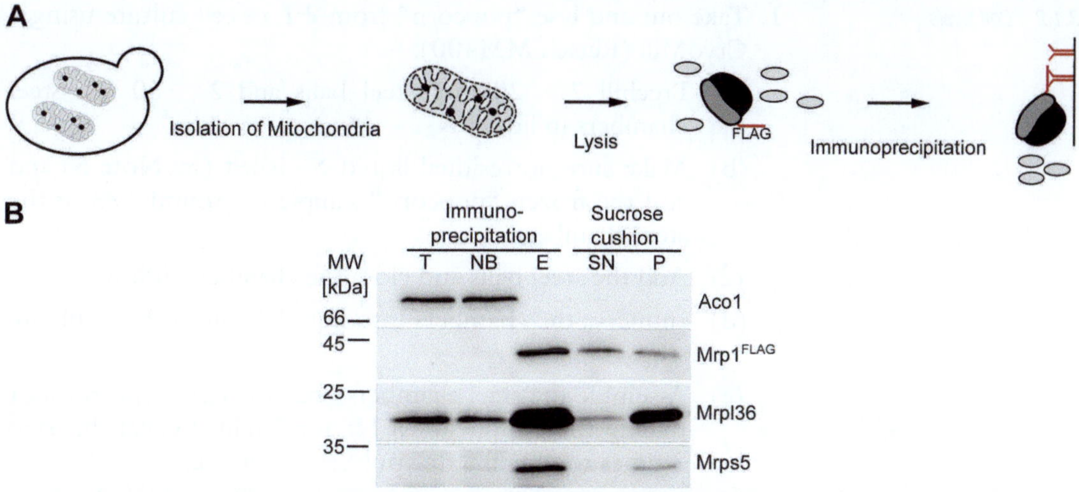

Fig. 1 Immunoprecipitation of mitochondrial ribosomes. (**a**) Scheme of the workflow for purification of mitoribosomes. Short protein tags are fused to mitoribosomal proteins or respective interactors of the mitoribosome (e.g., FLAG epitope). After cell disruption, mitochondria are extracted, lysed and immunoprecipitation is performed. Downstream analyses may for example include immunoblotting to identify interactors, or sucrose cushions (see later sections of this chapter) to exchange buffer or concentrate mitoribosome for structural analysis. Black = large mitochondrial subunit, dark gray = small mitoribosomal subunit, red = FLAG epitope and anti-FLAG antibodies. (**b**) Sample of an immunoblot performed after immunoprecipitation of mitoribosomes. The FLAG epitope was chromosomally fused to the C-terminus of the small mitoribosomal subunit protein Mrp1 ($Mrp1^{FLAG}$) and purification using Anti-FLAG beads was conducted. A total sample (T), the nonbound fraction (NB) and the eluate (E) were TCA precipitated and analyzed via immunoblotting using anti Aco1 (a soluble protein of the mitochondrial matrix), anti Mrpl36 (protein of the large mitoribosomal subunit), Mrps5 (protein of the small mitoribosomal subunit) and anti-FLAG antibodies. The eluate was further applied for sucrose cushions to exchange the buffer and control samples of the supernatant (SN) and the pellet (P) of the sucrose cushion were also analyzed via immunoblotting

mitoribosomes. Hence, we will describe these methods individually in detail below.

3.1.1 Yeast Cell Culturing and Freezing of "Popcorn"

1. Prepare an overnight culture of respective yeast strain with a FLAG-tagged protein in approx. 5 mL of growth medium using a nonfermentative carbon source (e.g., glycerol; *see* **Note 2**).
2. After 16–20 h growth, dilute the culture to an optical density (OD_{600}) of 0.5 in 2–4 L in the same media (*see* **Note 3**).
3. Grow the culture to mid-log phase with an OD_{600} of 0.8–1.5.
4. Harvest the cells by centrifugation at 5000 g, 10 min at 4 °C.
5. Carefully decant the supernatant and resuspend the cells in lysis buffer (use 8 mL lysis buffer per 0.5 L cells).
6. Freeze cells into "popcorn" by drop-wise pipetting the cell suspension into liquid nitrogen (N_2) (*see* **Note 4**).
7. Store cell "popcorn" at −80 °C until further use.

3.1.2 Cell Lysis

1. Take out and lyse "popcorn" from 2 L of cell culture using a CryoMill (Retsch MM400).
   (a) Prechill 2 × 20 mm steel balls and 2 × 50 mL steel chambers in liquid N_2.
   (b) Make sure no residual liquid N_2 is left (*see* **Note 5**) and add the frozen "popcorn" sample to around 1/3 of the steel chambers.
   (c) Add the steel balls and close the chambers tightly.
   (d) Immerse the chambers into liquid N_2 until the bubbling stops.
   (e) Assemble the steel chambers onto the CryoMill and start milling the sample at 15 Hz for 3 min. Repeat this step 3 times to obtain a fine pulverized sample.
   (f) Carefully transfer the pulverized sample with lysed cells to a 50 mL tube.
2. Thaw the lysed cells on ice.
3. Measure and adjust OD_{600} to 25 per mL by adding lysis buffer.
4. Add PMSF to a final concentration of 1 mM (5 μL of 200 mM stock solution per mL lysate).
5. Let the cell membranes solubilize further in lysis buffer for 15 min, tumbling at 4 °C.
6. Perform a clarifying spin at 20,000 g for 10 min at 4 °C to remove cell debris and keep the supernatant.
7. Measure Abs260 (nucleic acid absorbance, i.e., rRNA) and collect a volume corresponding to 1 Abs260 unit as a total sample (T).

3.1.3 Immuno-precipitation of FLAG-Tagged Mitoribosomes

1. Use 10 μL ANTI-FLAG M2 Affinity gel beads (20 μL slurry) per 25 OD_{600} units of desired lysate (*see* **Note 6**).
2. Equilibrate the beads by adding 20 column volumes (cv) of lysis buffer.
3. Pellet the beads by centrifugation at 2000 g for 3 min at 4 °C.
4. Add lysate to the beads and incubate for 2 h, tumbling at 4 °C.
5. Centrifuge the beads and carefully remove the supernatant.
6. Measure Abs260 of the supernatant and take out a volume corresponding to 1 Abs260 unit as a nonbound sample (NB).
7. Wash the beads with 15 cv lysis buffer and carefully remove the supernatant.
8. Wash the beads with 15 cv wash buffer, again carefully removing the supernatant.
9. Dissolve 30 μL 3× FLAG peptide stock (*see* **Note 7**) in 1 mL wash buffer to prepare a FLAG peptide elution buffer.

10. Add 0.5 mL FLAG elution buffer to the beads and incubate for 20 min at 4 °C. Mix every 5 min by gently tapping or shaking the tube.
11. Centrifuge the beads, carefully collect the supernatant and save it as the elution sample (E).
12. Repeat **steps 10** and **11** for a second elution and pool the two eluates.
13. Proceed with downstream applications. These could include SDS-PAGE and immunoblotting to validate the purification (as described in Subheading 3.1.4), for which a small part of the eluate can be applied (Fig. 1). The rest of the eluate can either be directly used for biochemical or structural analyses, or used as starting material for sucrose cushions to remove/exchange the buffer, remove FLAG peptide and concentrate the sample (*see* Subheading 3.2).

3.1.4 TCA Precipitation, SDS-PAGE, and Immunoblot Analysis

1. Add TCA to the total, nonbound and elution samples to a final concentration of 12%.
2. Vortex samples and incubate on ice for 20 min.
3. Centrifuge samples at 25,000 g for 30 min at 4 °C.
4. Remove supernatant, rinse the pellet in 1 mL of precooled 100% acetone.
5. Centrifuge at 25,000 g for 30 min at 4 °C.
6. Remove supernatant completely.
7. Resuspend pellet in 70 μL of SDS sample buffer.
8. Apply 30 μL of each sample for SDS PAGE, followed by immunoblotting according to standard protocols.
9. Immunoblots can be probed with antibodies against the mitochondrial ribosome (we recommend using antibodies against both large and small mitoribosomal subunits, Fig. 1).

3.2 Sucrose Cushions for Purification and/or Analysis of Mitoribosomes

Sucrose cushions are a versatile technique for both, purification and analysis of mitoribosomes [3, 4]. In general, isolated mitochondria are lysed, separated from insoluble components via high-speed centrifugation, optionally pretreated (e.g., by chemical crosslinking [4]) and subsequently loaded on a high-density sucrose cushion in order to sediment mitoribosomes (Fig. 2). Further, sucrose cushions can be applied after immunoprecipitation of mitochondrial ribosomes to exchange buffers and concentrate the elution fraction. After sucrose cushions, the sample can be used for a diverse set of biochemical and/or structural analyses. For the use of mitochondria as starting material, both crude mitochondrial extracts and sucrose step-gradient-purified mitochondria according to Meisinger et al. [9] can be applied. We recommend isolating

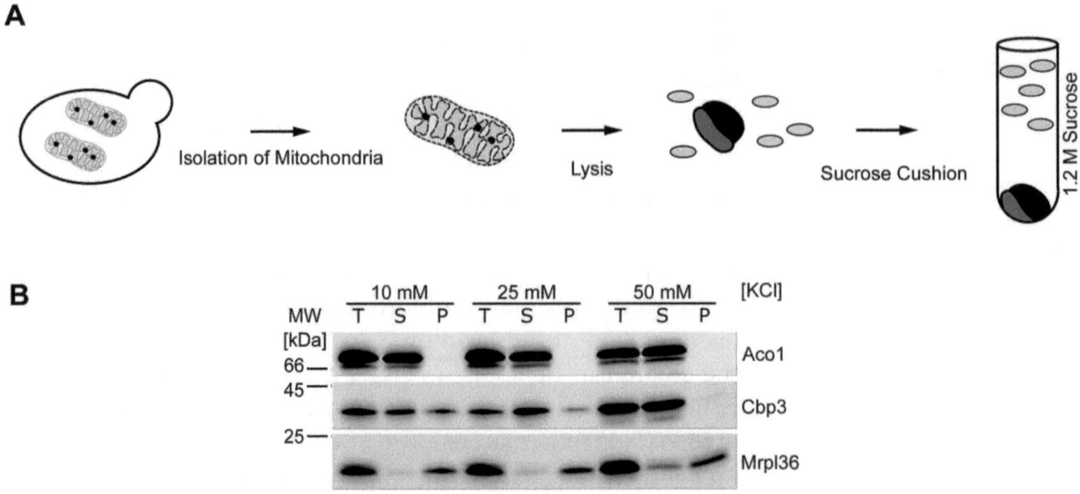

Fig. 2 Sucrose cushion for purification and analysis of mitoribosomes. (**a**) Scheme of the workflow for sucrose cushions of the mitoribosome. Cell disruption and isolation of mitochondria is followed by detergent-based lysis of mitochondria. The mitochondrial lysate is layered on a 1.2 M sucrose solution and the mitoribosome is pelleted by high-speed centrifugation. Black = large mitoribosomal subunit, dark gray = small mitoribosomal subunit. (**b**) Immunoblot to study the interaction of the protein Cbp3 with the mitoribosome. Cbp3 is an assembly factor of cytochrome *b* that is part of a regulatory feedback loop that adjusts synthesis of cytochrome *b* [10, 11]. A total sample (T) after lysis of mitochondria, as well as the supernatant (S) and the pellet (P) of the sucrose cushion were TCA precipitated and applied for immunoblotting. Indicated KCl concentrations were used in the experiment to determine the salt stability of the Cbp3-mitoribosomal interaction. Blots were probed against the soluble matrix protein Aco1, the mitoribosomal protein Mrpl36 (large subunit) and Cbp3. A partial association of Cbp3 with the mitoribosome can be visualized (as seen by Cbp3 signal in both, the pellet and soluble fraction). The association with the mitoribosome is lost upon higher salt concentrations

mitochondria from strains lacking the main unspecific mitochondrial nuclease Nuc1 in order to minimize rRNA degradation [1].

To avoid potential contamination with RNases and subsequent degradation of the mitoribosome, clean the workspace and all instruments in advance with 0.1% SDS solution (or alternative cleaning solutions).

1. Collect mitochondria by centrifugation at 10,000 g, 10 min at 4 °C

   Optional: Perform pretreatments (e.g., chemical crosslinking [4]).
2. Resuspend mitochondria in lysis buffer (1 mL lysis buffer per 10 mg mitochondria) and incubate on ice for 10 min.
3. Clarify the lysate by centrifugation at 20,000 g for 10 min at 4 °C.
4. Harvest the supernatant and repeat centrifugation at 20,000 g for 10 min at 4 °C.

5. Add sucrose cushion solution into centrifugation tubes (suitable for a fixed-angle rotor and centrifugation at 290,000 g).
6. Overlay sucrose cushion solution with mitochondrial lysate (1 mL sucrose cushion solution per 1 mL mitochondrial lysate).
7. Centrifuge samples in a fixed-angle rotor at 245,000 rcf for 5 h at 4 °C.
8. Collect the supernatant and resuspend the pellet (containing mitoribosomes) in 100–200 μL of desired buffer for downstream applications.
9. For direct analysis via immunoblots, resuspend the pellet in 70 μL SDS sample buffer and TCA precipitate the supernatant as described above (Fig. 2).

3.3 Linear Sucrose Gradients for the Biochemical Analysis of Mitoribosomes

In this section, we describe how mitochondrial gradients are used to analyze interactions of proteins with yeast mitoribosomes. Using isolated mitochondria as starting material, we describe how to lyse samples and perform sucrose gradients, and further provide general protocols for analysis via immunoblotting and/or blue native page to analyze the interactome of yeast mitoribosomes, as we have performed previously [1] (Fig. 3).

3.3.1 Lysis of Mitochondria and Sucrose Gradient

To avoid potential contamination with RNases and subsequent degradation of the mitoribosome, clean the workspace and all instruments in advance with 0.1% SDS solution.

1. Use 1 mg of mitochondria per sample and thaw on ice.
2. Spin down the sample at 10,000 g, 4 °C, 10 min.
3. Carefully resuspend the sample in 285 μL of lysis buffer and incubate for 10 min on ice.
4. Add 285 μL of dilution buffer, carefully mix by pipetting.
5. Perform a clarifying spin at 16,000 g, 10 min, 4 °C.
6. Collect 50 μL of the supernatant as a total sample and temporarily store on ice.
7. Prepare sucrose gradients and carefully layer the rest of the supernatant on top of the sucrose gradient.
   (a) Prepare the sucrose gradient by adding 2 mL of the lower concentrated sucrose solution to the right column of the gradient mixer and start the stirrer.
   (b) Use a clamp to prevent the solution from leaking and open the valve so the sucrose solution is distributed within all connections of the gradient mixer.
   (c) Close the valve and transfer the solution back to the right column.

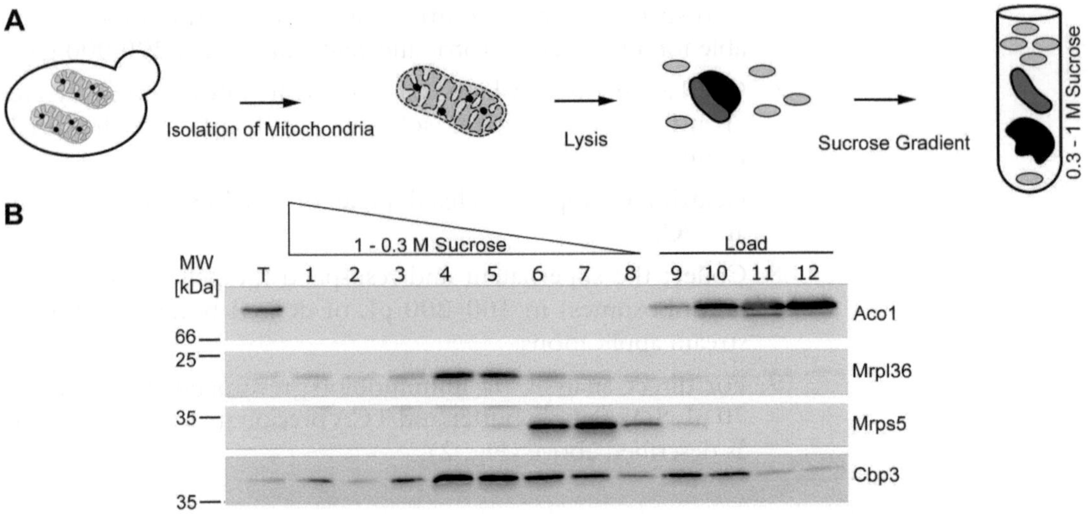

Fig. 3 Linear sucrose gradients for the analysis of mitoribosomes and the investigation of their interactome. (**a**) Scheme of linear sucrose gradients. Cell disruption is followed by mitochondrial extraction and lysis of mitochondria. The mitochondrial lysate is carefully layered on top of a 0.3 M–1 M linear sucrose gradient, which is subsequently applied for high-speed centrifugation. Depending on the buffer composition, mitoribosomal subunits can be dissociated or maintained. (**b**) Example immunoblot to study the interaction of Cbp3 with the mitoribosome. A total sample (T) of the mitochondrial lysate and the 12 harvested fractions after sucrose gradient centrifugation (1–8 corresponding to the sucrose gradient and 9–12 to the volume of the mitochondrial lysate = load) are TCA precipitated and used for immunoblotting. EDTA was added to the lysis buffer to split the large and small mitoribosomal subunits and blots were probed with antibodies against the soluble protein Aco1, large mitoribosomal subunit protein Mrpl36, small mitoribosomal subunit protein Mrps5 and Cbp3. In line with sucrose cushions in Fig. 2, a partial comigration of Cbp3 with the mitoribosome can be confirmed. The majority of Cbp3 comigrates with the large mitoribosomal subunit (fractions 4 + 5)

(d) Add 2 mL of the higher concentrated sucrose solution to the left column.

(e) Keep stirring for approx. 30 s.

(f) Open the clamp and let the mixed solution slowly fill a centrifugation tube.

(g) Carefully layer the supernatant of the sample on top of the sucrose gradient.

8. Place the gradient into a precooled SW 60Ti swinging bucket rotor.
9. Perform ultracentrifugation at 60,000 rpm (average RCF = 370,000 g) for 1 h at 4 °C.
10. Collect 375 μL fractions (12 in total) of the gradient after centrifugation, either using a dedicated fraction collector or a peristaltic pump into fresh sample tubes (*see* **Note 8**).

3.3.2 Downstream Analysis: Preparation for Immunoblotting

1. Bring the volume of each fraction to 1 mL by adding 625 μL ultrapure H_2O, including the previously harvested total sample.
2. Add 200 μL of 72% trichloroacetic acid (TCA) (final concentration of 12%).
3. Vortex samples and incubate on ice for 20 min.
4. Centrifuge samples at 25,000 g for 30 min at 4 °C.
5. Remove supernatant, rinse the pellet in 1 mL of precooled 100% acetone.
6. Centrifuge at 25,000 g for 30 min at 4 °C.
7. Remove supernatant completely.
8. Resuspend pellet in 70 μL of SDS sample buffer.
9. Apply 30 μL of each sample for SDS PAGE, followed by immunoblotting according to standard protocols.
10. Immunoblots can be probed with antibodies against the mitochondrial ribosome (we recommend using antibodies against both, large and small mitoribosomal subunits), against a soluble protein (e.g., Aco1) and proteins of interest (Fig. 3).

3.3.3 Downstream Analysis: Preparation for Blue Native PAGE

1. Make sure to keep gradients and collected fractions on ice after centrifugation.
2. Take a sample of 15 μL from each collected fraction, as well as from the harvested total sample prior to centrifugation.
3. Add Coomassie G-250 to your sample prior to loading Blue Native PAGE gels and ensure that G-250 concentration is one-fourth the detergent concentration.
4. Load 15 μL of each sample on Blue Native PAGE gels and perform electrophoresis.
5. Transfer separated proteins on polyvinylidene difluoride (PVDF) membranes.
6. Destain membranes by washing 5 min in 100% methanol followed by washing the membranes 5× for 5 min in TBS.

4 Notes

1. Here, we list recommended stock solutions for the preparation of working solutions for both sucrose cushions and linear sucrose gradients, as well as immunoprecipitations. As mentioned above, the composition of the buffers used strongly depends on the experimental design. Hence, we also provide a brief information on the role of the used chemicals in order to help finding the appropriate buffer recipe.
   (a) Detergents: Nonionic detergents are required to maintain protein-protein interactions when purifying and analyzing

mitoribosomes. Examples described in above protocols include Triton X-100, DDM, and Digitonin (for all mentioned detergents, a 20% stock solution is recommended). Although mitoribosomes are stable with these detergents, keep in mind that macromolecular structures interacting with mitoribosomes might be destabilized by those detergents. Hence, especially when analyzing mitoribosome interactions, the selection of the detergent is a crucial factor and ideally, experiments should be conducted with several detergents. We recommend heating the detergent solutions during stock preparation to facilitate solubilization.

(b) EDTA: To inhibit RNases and hence maintain mitoribosomal integrity, EDTA can be added as a potent inhibitor. However, as magnesium is important for the stability of ribosomes, chelating magnesium by EDTA will result in the dissociation of the large and small mitoribosomal subunits. Thus, using a knockout of the main unspecific nuclease Nuc1 can be used as an alternative [1]. For EDTA, we recommend a 500 mM stock solution.

(c) $MgCl_2/Mg(OAc)_2$: The concentration of magnesium is important for the stability of ribosomes, as mentioned above. 50 mM stock solutions are convenient for preparation of the buffers used in above protocols.

(d) KCl/KOAc: Mitoribosomes have a large interactome, which can be dissolved in a salt-dependent manner. Titration of salt concentration is important to maintain desired interactions and can also be used to determine the salt stability of interactions. 1 M stock solutions are usually convenient to use a wide range of salt concentrations.

(e) HEPES/KOH pH 7.5: Usually, a 1 M stock solution is convenient for buffer preparation.

(f) PMSF: 200 mM stock solutions can be prepared by dissolving PMSF in ethanol.

(g) Complete protease inhibitor mix: Prepare 50× stock my dissolving one tablet in 1 mL H_2O.

(h) Sucrose: 2 M stock solution can be prepared by dissolving in sucrose in H_2O and heating to facilitate solubilization.

(i) Tris–HCl pH 7.5: Usually, a 1 M stock solution is convenient for buffer preparation.

(j) Glycerol: 50% stock solutions can be prepared in H_2O.

(k) Spermidine: The addition of spermidine stabilizes ribosomal interactions. 100 mM stock solutions can be prepared in H_2O. Avoid freeze/thaw cycles by preparing small aliquots.

2. Growth on a nonfermentative carbon source relieves glucose repression and significantly increases the mitochondrial mass and number of mitoribosomes in the cells. If desired strains are respiratory deficient, a fermentative carbon source such as galactose can be used instead.
3. The culture volume can be scaled up or down to desired volume depending on planned downstream application (e.g., small volumes for immunoprecipitations followed by analytical immunoblotting vs. large volumes for isolation of mitoribosomes for cryo-EM studies).
4. The freezing of cells in liquid N_2 can be performed in several ways. In order for the frozen cells to fit well in the CryoMill, we prefer to make "popcorn," i.e., small frozen balls of cells, by slowly pipetting the cell suspension into a tube immersed in in liquid N_2. Alternative ways of cryofreezing the cells can be just as effective as long as it is compatible with the CryoMill.
5. Liquid N_2 is highly volatile and any residuals left in the steel chambers before cryomilling creates a danger for explosions.
6. The FLAG affinity beads can with preference be collected and equilibrated in a microcentrifuge tube before transfer to a tube or column of desired volume.
7. Prepare the 3× FLAG peptide stock solution according to instructions from the manufacturer (Sigma Aldrich).
8. In case protein complexes should be kept intact (e.g., for subsequent Blue Native Page analysis), keep gradients and fractions cooled after centrifugation (ideally by harvesting fractions in a cold room) and proceed immediately with sample preparation without freezing collected fractions.

Acknowledgment

This work was supported by the Swedish Research Council and the Knut and Alice Wallenberg foundation (to Martin Ott) and the Austrian Science Fund FWF (J4398-B to Andreas Aufschnaiter).

References

1. Kehrein K, Schilling R, Möller-Hergt BV et al (2015) Organization of mitochondrial gene expression in two distinct ribosome-containing assemblies. Cell Rep 10:843–853. https://doi.org/10.1016/J.CELREP.2015.01.012
2. Singh AP, Salvatori R, Aftab W et al (2020) Molecular connectivity of mitochondrial gene expression and OXPHOS biogenesis. Mol Cell 79:1051–1065.e10. https://doi.org/10.1016/j.molcel.2020.07.024
3. Desai N, Brown A, Amunts A, Ramakrishnan V (2017) The structure of the yeast mitochondrial ribosome. Science (80-) 355:528–531. https://doi.org/10.1126/science.aal2415
4. Gruschke S, Gröne K, Heublein M et al (2010) Proteins at the polypeptide tunnel exit of the yeast mitochondrial ribosome. J Biol Chem

285:19022–19028. https://doi.org/10.1074/jbc.M110.113837

5. Salvatori R, Kehrein K, Singh AP et al (2020) Molecular wiring of a mitochondrial translational feedback loop. Mol Cell 77:887–900.e5. https://doi.org/10.1016/J.MOLCEL.2019.11.019
6. Zeng R, Smith E, Barrientos A (2018) Yeast mitoribosome large subunit assembly proceeds by hierarchical incorporation of protein clusters and modules on the inner membrane. Cell Metab 27:645–656.e7. https://doi.org/10.1016/j.cmet.2018.01.012
7. Couvillion MT, Soto IC, Shipkovenska G, Churchman LS (2016) Synchronized mitochondrial and cytosolic translation programs. Nature 533:499–503. https://doi.org/10.1038/nature18015
8. Price CA (1982) Particle abstraction in biology—an introduction. In: Centrifugation in density gradients. Academic Press, pp 1–11
9. Meisinger C, Pfanner N, Truscott KN (2006) Isolation of yeast mitochondria. Methods Mol Biol 313:33–39. https://doi.org/10.1385/1-59259-958-3:033
10. Gruschke S, Römpler K, Hildenbeutel M et al (2012) The Cbp3–Cbp6 complex coordinates cytochrome b synthesis with bc1 complex assembly in yeast mitochondria. J Cell Biol 199:137. https://doi.org/10.1083/JCB.201206040
11. Gruschke S, Kehrein K, Römpler K et al (2011) Cbp3-Cbp6 interacts with the yeast mitochondrial ribosomal tunnel exit and promotes cytochrome b synthesis and assembly. J Cell Biol 193:1101–1114. https://doi.org/10.1083/jcb.201103132

Chapter 9

Rapid Cryopurification of the Yeast Mitochondrial Ribosome

Hong Weng Pang and Antoni Barrientos

Abstract

Cryogenic milling, or cryomilling, involves the use of liquid nitrogen to lower the temperature of the biological material and/or the milling process. When applied to the study of subcellular or suborganellar structures and processes, it allows for their rapid extraction from whole cells frozen in the physiological state of choice. This approach has proven to be useful for the study of yeast mitochondrial ribosomes. Following cryomilling of 100 mL of yeast culture, conveniently tagged mitochondrial ribosomes can be immunoprecipitated and purified in native conditions. These ribosomes are suitable for the application of downstream approaches. These include mitoribosome profiling to analyze the mitochondrial translatome or mass spectrometry analyses to assess the mitoribosome proteome in normal growth conditions or under stress, as described in this method.

Key words Yeast mitoribosome, Cryomilling, Immunoprecipitation, Mass spectrometry, Immunoblotting

1 Introduction

Traditional method of mitochondrial ribosome (mitoribosome) purification requires extensive physical processing of biological samples. This protocol involves a multiday process to obtain a 1–2 L yeast culture that is used for mitochondria preparation and purification. Mitochondrial extracts are then used for linear sucrose gradient sedimentation of the desired mitoribosome fractions containing the small subunit, large subunit, and monosome. Sucrose gradient sedimentation has allowed researchers to separate and study protein complexes such as cytosolic ribosomes, mitoribosome, or ribosome-interacting proteins [1–3]. Alternatively, purification of crude mitoribosome particles from mitochondrial extracts using sucrose cushions has provided samples useful for cryoelectron microscopy studies [4, 5]. However, these methods are not ideal to study, in intact cells, temporal and/or dose-dependent changes of the mitoribosome and its interactome with various treatment conditions (i.e., oxidative stress, heat stress, or antibiotics treatment).

Antoni Barrientos and Flavia Fontanesi (eds.), *The Mitoribosome: Methods and Protocols*,
Methods in Molecular Biology, vol. 2661, https://doi.org/10.1007/978-1-0716-3171-3_9,

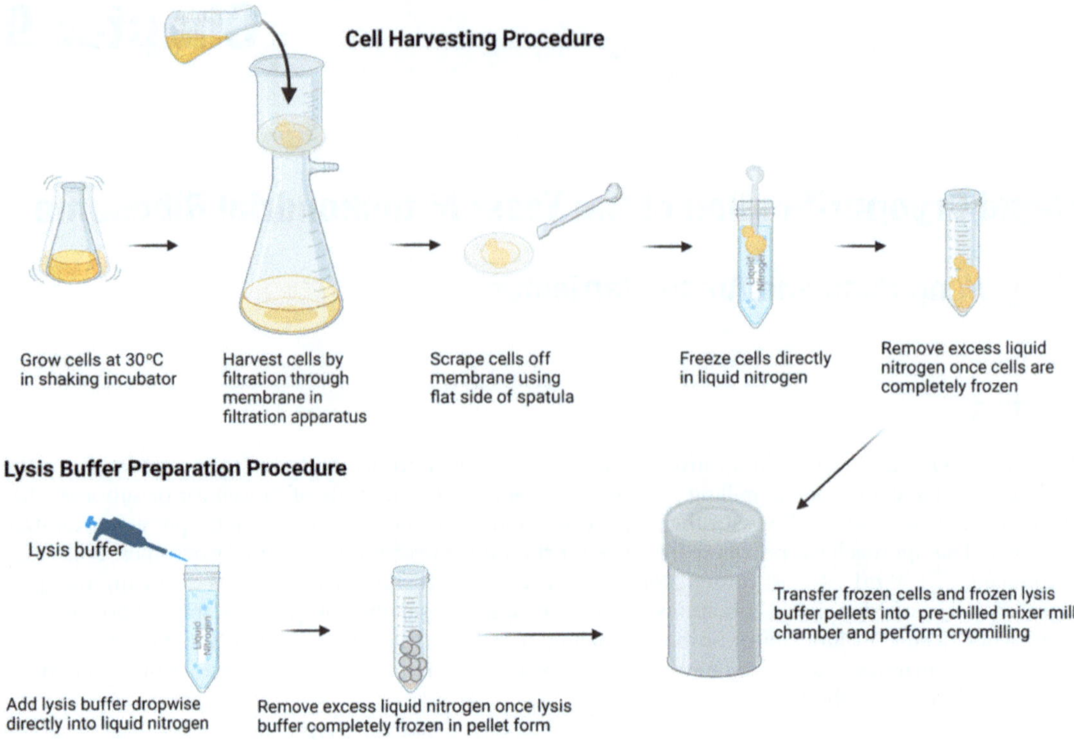

Fig. 1 Workflow diagram. Graphical depiction of how to properly collect and freeze cells and freezing of lysis buffer pellets

For this purpose, we optimized a procedure initially developed for mitoribosome profiling and translation footprinting studies [6], that does not require the isolation of the organelle and allows for a rapid cryopurification of the yeast mitoribosome for proteomic analyses using mass spectrometry approaches. The approach requires the generation of yeast strains expressing a functional epitope-tagged mitoribosome subunit suitable for full mitoribosome immunoprecipitation (i.e., uL29m-FLAG [7]). The basic workflow of the approach (Fig. 1) involves culturing the cells to optimal confluency in control and experimental conditions, rapidly harvesting using the filtration apparatus (Fig. 2a) and flash-freezing them into liquid nitrogen. The frozen cell and lysis buffer pellets are placed in steel chamber containing steel ball (Fig. 2b), which maintains the frozen state of the biological material and mixes and disrupts crude cell wall and membrane structures. The actual process of milling is performed by the cryomiller (Fig. 2c), to obtain a pulverized sample in which all cellular components are extracted. The extract is then suitable for mitoribosome purification by immunoprecipitation. This procedure opens a window into the cells for the analysis of mitoribosomes (Fig. 3) in their native environment.

Many yeast mitoribosome interactors such as assembly factors or translational activators have been identified and characterized by

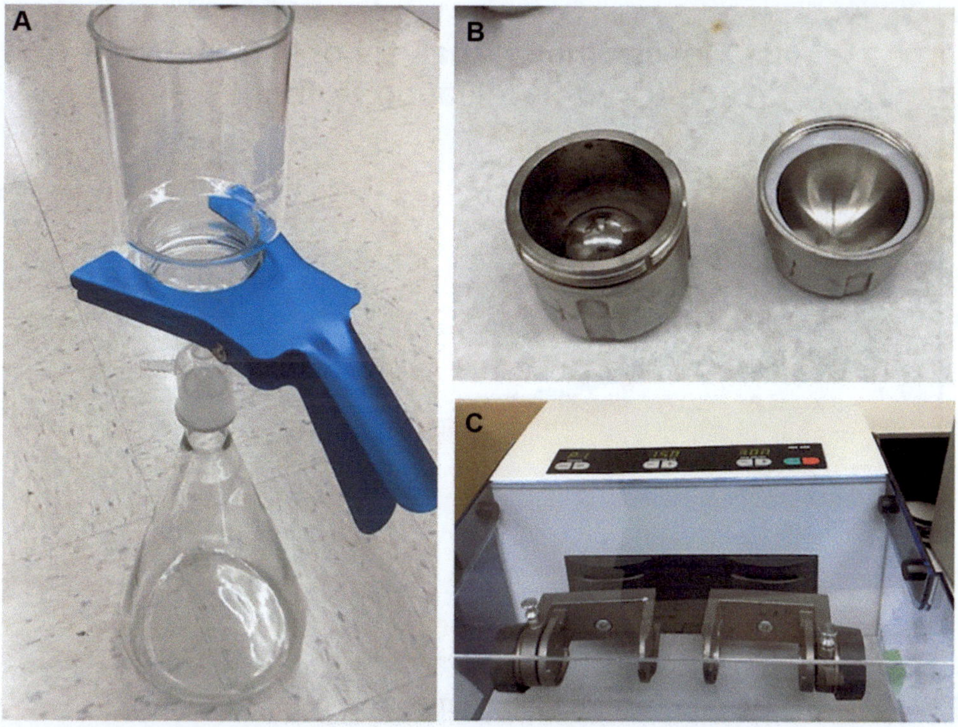

Fig. 2 (**a**) Fully assembled filtration apparatus for cell harvesting; (**b**) Stainless steel chamber and ball; (**c**) Mixer mill

genetic and biochemical approaches using mitochondria isolated from yeast cells deleted for each of these factors [1, 2, 7]. However, it is also important to describe stress- and nutrient-dependent changes in stoichiometric ratios of the various mitoribosome-associated factors in wild-type cells to uncover underlying adaptations that may occur. Examples of such an adaptation have been described in bacteria, where, i.e., the ribosome-associated factor RelA plays a crucial role in signaling stringent response [8]. Given the prokaryotic origin of mitochondria [9], it is indicated to hypothesize that mitochondrial translation and ribosome-associated factors could play a signaling role in eukaryotic stress response. The methodology we are describing to cryopurify mitoribosomes allows for in-depth interrogation of the mitoribosome-associated proteome and its potential role in interorganellar signaling to maintain cellular homeostasis during changing environmental conditions.

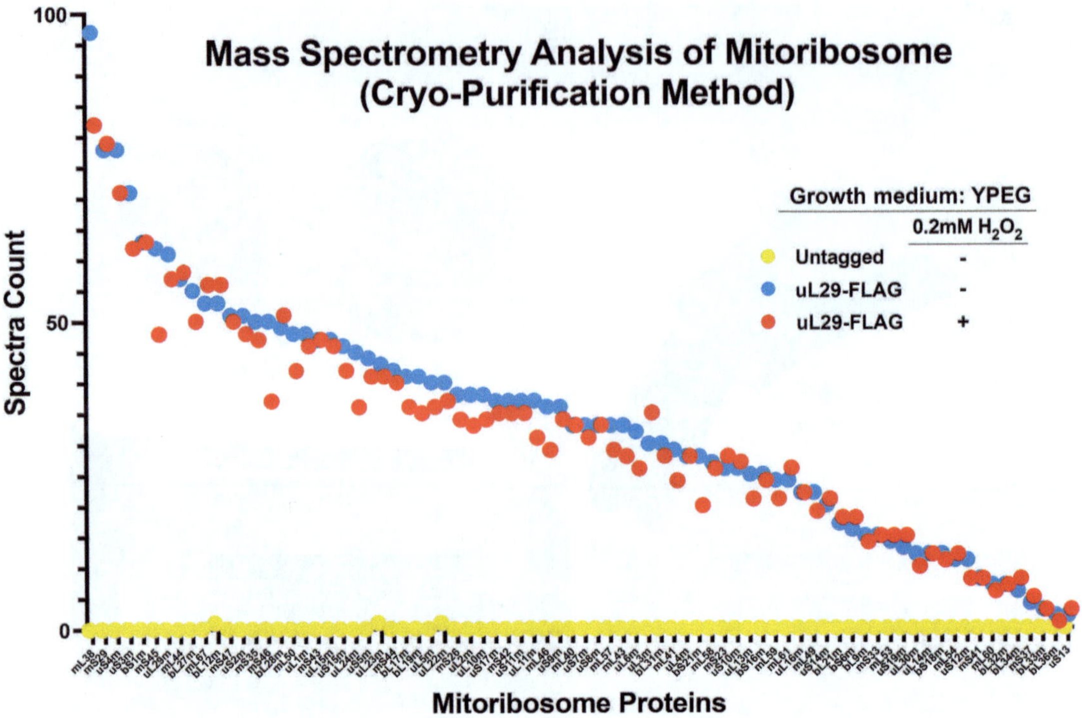

Fig. 3 Mass spectrometry analysis of immunoprecipitated mitoribosome protein subunits. All cells were grown in respiratory complete media. Yellow dots represent WT untagged Negative control. Blue dots represent uL29m-FLAG-tagged strain. Red dots represent uL29m-FLAG-tagged strain treated with 0.2 mM H_2O_2. The X-axis represents all mitoribosome proteins subunits identified in the most recent yeast mitoribosome cryo-EM structure [5]. The Y-axis represents total spectral count for each mitoribosome subunit. WT untagged control is necessary to distinguish between specifically bound proteins from nonspecific binding

2 Materials

All solutions need to be prepared using ultrapure water (*see* **Note 1**) and analytical grade reagents. Prepare and store all reagents at room temperature unless indicated otherwise. Diligently follow all waste disposal regulations when disposing waste materials.

2.1 Preparing Frozen Lysis Buffer

1. Lysis Buffer: 10 mM Tris, pH 8.0, 50 mM NH_4Cl, 20 mM $MgCl_2$, 0.6% Digitonin (EMD Millipore cat# 300410), 0.25 mM DTT, 1× Yeast Protease Inhibitor Cocktail (Sigma P8715), RNase-free water, RNAse inhibitor (Invitrogen AM2694, optional *see* **Note 2**).
2. 50 mL conical tube.
3. Liquid nitrogen (*see* **Note 3**).
4. Styrofoam container.

2.2 Yeast Cell Culture and Collection

1. Yeast strain expressing a C-terminal epitope tagged (FLAG) uL29m grown in medium of choice (*see* **Note 4**).
2. Spectrophotometer.
3. Vacuum source.
4. Liquid nitrogen.
5. Microfiltration assembly (90 mm, ULTRA-WARE) (Fig. 2a):
   (a) Four-liter side-arm flask with a fritted glass support base.
   (b) Glass funnel.
   (c) Anodized aluminum clamp.
   (d) No. 8 silicone stopper.
5. Nitrocellulose (0.45 μm, 90 mm diameter membranes; Whatman).
6. Metal spatulas with one flat end and one curved end.
7. Styrofoam container.

2.3 Cryogenic Cell Lysis Using Mixer Mill (Cryomilling)

1. Mixer mill, 50 mL chambers, and 25 mm stainless steel ball (*see* **Note 5**) (Fig. 2b, c).
2. Metal spatulas with one flat end and one curved end.
3. Styrofoam container.
4. 50 mL conical tube.
5. Tongs.
6. Cryogloves.

2.4 Immuno-precipitate Mitoribosome

1. Floor centrifuge with rotor able to reach 20,000 × *g* (e.g., Sorvall with SS-34 rotor and Oak Ridge tubes).
2. Wash buffer: 10 mM Tris–HCl, pH 8.0, 50 mM NH_4Cl, 20 mM $MgCl_2$, 1× Yeast Protease Inhibitor Cocktail, 0.1% Triton X-100.
3. 5 mg/mL 3X-FLAG peptide in TBS (aliquot and store at −80 °C, avoid freeze-thaw; Sigma F4799).
4. End over end rotator.
5. Refrigerated centrifuge.

3 Methods

3.1 Preparing Frozen Lysis Buffer

1. Prepare 2.6 mL frozen lysis buffer as described in the materials section.
2. Prepare 50 mL conical tubes prechilled and filled with liquid nitrogen.

3. Dropwise, add lysis buffer directly into conical tubes filled with liquid nitrogen (*see* **Note 6**).
4. Allow a few seconds for each lysis buffer drop to completely freeze before adding more. Lysis buffer drops will visibly turn opaque white when completely frozen.
5. Once the appropriate volume of lysis buffer is added into each tube and completely frozen into pellets, cap the tubes and invert to remove any excess liquid nitrogen.
6. Frozen lysis buffer can be stored at −80 °C.

3.2 Yeast Cell Culture and Collection

1. Inoculate yeast cells in desired media (100 mL).
2. Allow cells to grow until mid-logarithm phase for cells harvesting and freezing (*see* **Note 7**).
3. Prior to cell harvest, set up filtration apparatus and prepare 50 mL conical tubes prechilled and filled with liquid nitrogen (*see* **Note 8**).
4. Harvest cells by filtering through the filtration apparatus and cells will collect on the top of 0.45 μm membrane.
5. Once all the media filtered through the membrane, remove the membrane, and use the flat side of the spatula to scrape cells then dip into liquid nitrogen in conical tubes to rapidly freeze these cells (*see* **Note 9**).
6. When all the cells are removed from the membrane and frozen in liquid nitrogen, cap conical tubes and invert to remove excess liquid nitrogen (*see* **Note 10**).
7. Frozen cells can be stored at −80 °C.

3.3 Cryogenic Cell Lysis Using Mixer Mill (Cryomilling)

1. Combine frozen cells and lysis buffer into one tube and tap several times on the benchtop to release any frozen material stuck on the walls of the tube.
2. Prechill mixer mill chamber and steel ball in liquid nitrogen (*see* **Notes 8** and **11**).
3. Pour all the frozen cell and lysis buffer content into mixer mill chamber.
4. Place the chamber into mixer mill and secure tightly. Mixer mill will be run for 3 min cycles at 15 Hz for a total of 6 cycles. Allow the chamber to fully chill in liquid nitrogen between each cycle (*see* **Note 8**).
5. After the mixer mill cycles are completed, open the chambers and scrape the grindate into 50 mL conical tubes prechilled in liquid nitrogen (*see* **Note 12**).
6. The grindate can be stored at −80 °C.

3.4 Immunoprecipitate Mitoribosome

1. Place the grindate in a water bath at room temperature, swirling every 5 min until fully thawed.
2. Clear lysate by centrifugation at 20,000 × *g* at 4 °C for 15 min.
3. Anti-Flag agarose slurry should be washed 3 times with 40 volumes of wash buffer. To wash agarose slurry, add appropriate volume of wash buffer, centrifuge at 1000 × *g* for 1 min, remove supernatant, and repeat 3 times. 12 μL of agarose beads should be used per mL of grindate volume.
4. After agarose slurry is washed, resuspend beads in 4.5 volume of wash buffer.
5. To cleared lysate, add 60 μL of washed and diluted agarose slurry per mL of lysate.
6. Incubate for 3 h at 4 °C rotating end over end.
7. Centrifuge 2 min at 1000 × *g* to pellet beads.
8. Discard supernatant.
9. Wash beads with 50 volume of wash buffer and allow 10 min rotating end over end.
10. Repeat wash **steps** 7–9 for 3 washes.
11. After wash, elute flag-tagged protein by incubating with 3X-FLAG peptide, resuspending in 6 bead slurry volume wash buffer containing 0.2 mg/mL of 3X-FLAG peptide.
12. Allow 40 min of elution rotating end over end at room temperature.
13. Centrifuge 2 min at 1000 × *g* to pellet beads.
14. Collect the eluant without disturbing the beads.
15. Samples can be stored at −20 °C for less than a week or −80 °C for longer durations.

4 Notes

1. Ultrapure water can be prepared by purifying deionized water, to attain a sensitivity of 18 MΩ-cm at 25 °C using a water GenPurePro purifier (Thermo).
2. Many intermolecular interactions with the ribosome are RNA-dependent. Therefore, to maintain important native interactions, it may require the usage of RNAse inhibitors to prevent RNA degradation.
3. ALWAYS use cryogloves when handling items prechilled in liquid nitrogen. Lab coat and protective goggles highly recommended.

4. uL29m (Mrpl4) is located near the polypeptide tunnel exit of the mitoribosome large subunit. Due to its location in the periphery of the mitoribosome, it has been demonstrated to be functioning, with exposed epitope, for immunoprecipitation [7].
5. The Mixer mill from Retsch (model# MM400) is a laboratory ball mill specifically designed for cryogenic grinding. Cooling of grinding jar is maintained by submersion in liquid nitrogen directly and removing for 3 min for each milling intervals. Thus, the sample is embrittled, and volatile components are preserved.
6. Each lysis buffer drop will take between 2 and 3 s to completely freeze. Wait for the lysis buffer to turn opaque before adding more.
7. Harvest the cells when the culture is in the exponential growth phase, at an OD^{600} between 1 and 1.5. Do not have cells growing beyond $OD^{600} = 1.5$ unless interested in studying changes in ribosomes at various growth phases, such as the stationary phase, or under nutrient starvation.
8. Wait for intense bubbling of the liquid nitrogen to stop, which indicates that the tubes or mixer mills have equilibrated to the temperature of liquid nitrogen.
9. Collect cells completely onto the spatula from the membrane before dipping into liquid nitrogen. To have the spatula at room temperature is best for scooping up most cell amount during the collection step.
10. Be careful while pouring out excess liquid nitrogen from conical tubes, rapid expansion of liquid nitrogen may cause gaseous spray. Poking small holes on the top of caps may be advised to allow outflow of gas.
11. Make sure there is no moisture when closing the mixer mill chamber. Any water between the two parts of the mixer mill chamber will cause it to freeze shut and will not loosen until thawed.
12. Be careful not to spill the grindate over the edge of the mill chamber while removing as much grindate as possible by scraping it off the walls with the spatula. Work fast at this step to prevent the samples to thaw.

Acknowledgment

This research was supported by the National Institutes of Health grant R35-GM118141 (AB).

References

1. De Silva D, Poliquin S, Zeng R et al (2017) The DEAD-box helicase Mss116 plays distinct roles in mitochondrial ribogenesis and mRNA-specific translation. Nucleic Acids Res 45(11): 6628–6643
2. De Silva D, Fontanesi F, Barrientos A (2013) The DEAD-box protein Mrh4 functions in the assembly of the mitochondrial large ribosomal subunit. Cell Metab 18:712–725
3. Beckham C, Hilliker A, Cziko AM et al (2008) The DEAD-box RNA helicase Ded1p affects and accumulates in *Saccharomyces cerevisiae* P-bodies. Mol Biol Cell 19(3):984–993. https://doi.org/10.1091/mbc.e07-09-0954
4. Amunts A, Brown A, Bai X et al (2014) Structure of the yeast mitochondrial large ribosomal subunit. Science 343:1485–1489
5. Desai N, Brown A, Amunts A et al (2017) The structure of the yeast mitochondrial ribosome. Science 355(6324):528–531
6. Couvillion MT, Churchman LS (2017) Mitochondrial ribosome (mitoribosome) profiling for monitoring mitochondrial translation in vivo. Curr Protoc Mol Biol 119(4.28):1–4
7. Kehrein K, Schilling R, Moller-Hergt BV et al (2015) Organization of mitochondrial gene expression in two distinct ribosome-containing assemblies. Cell Rep 12(15):S2211–S1247
8. Brown A, Fernández IS, Gordiyenko Y et al (2016) Ribosome-dependent activation of stringent control. Nature 534(7606):277–280. https://doi.org/10.1038/nature17675
9. Gabaldón T (2021) Origin and early evolution of the eukaryotic cell. Annu Rev Microbiol 75: 631–647. https://doi.org/10.1146/annurev-micro-090817-062213

Chapter 10

Methods to Study the Biogenesis of Mitoribosomal Proteins in Yeast

Lea Bertgen, Tamara Flohr, and Johannes M. Herrmann

Abstract

The biogenesis of mitoribosomes is an intricate process that relies on the coordinated synthesis of nuclear-encoded mitoribosomal proteins (MRPs) in the cytosol, their translocation across mitochondrial membranes, the transcription of rRNA molecules in the matrix as well as the assembly of the roughly 80 different constituents of the mitoribosome. Numerous chaperones, translocases, processing peptidases, and assembly factors of the cytosol and in mitochondria support this complex reaction. The budding yeast *Saccharomyces cerevisiae* served as a powerful model organism to unravel the different steps by which MRPs are imported into mitochondria, fold into their native structures, and assemble into functional ribosomes.

In this chapter, we provide established protocols to study these different processes experimentally. In particular, we describe methods to purify mitochondria from yeast cells, to import radiolabeled MRPs into isolated mitochondria, and to elucidate the assembly reaction of MRPs by immunoprecipitation. These protocols and the list of dos and don'ts will enable beginners and experienced scientists to study the import and assembly of MRPs.

Key words Immunoprecipitation, Isolation of mitochondria, Mitochondrial protein import, Mitoribosomal protein (MRP), Sample preparation for mass spectrometry

1 Introduction

The model organism *Saccharomyces cerevisiae* also known as budding yeast provides an easy and controllable system to study the biology of mitoribosomal proteins (MRPs). Cells can be grown under fermentative conditions allowing the cultivation and characterization of respiration-deficient mutants. Yeast mitochondria can easily be purified by differential centrifugation and can be stored frozen until needed [1]. The protein import machinery of yeast mitochondria is well characterized and in vitro, proteins can be easily imported and sorted into the four different mitochondrial

Lea Bertgen and Tamara Flohr contributed equally with all other contributors.

Antoni Barrientos and Flavia Fontanesi (eds.), *The Mitoribosome: Methods and Protocols*,
Methods in Molecular Biology, vol. 2661, https://doi.org/10.1007/978-1-0716-3171-3_10,

subcompartments: the outer membrane, the intermembrane space (IMS), the inner membrane, and the matrix. The mitochondrial import machineries of eukaryotes, in particularly of fungi and animals, are highly conserved, so that yeast serves as excellent model system for the situation in, for example, human mitochondria [2–6]. In yeast, genetic manipulations, such as expression of proteins equipped with affinity purification tags, are straightforward and simple. Using these tags, proteins can be isolated from whole-cell extracts or from isolated mitochondria to identify interaction partners during their import, folding and assembly using mass spectrometry. This chapter provides detailed protocols for the analysis of the import and biogenesis of MRPs in yeast.

Mitochondria are descendants from α-proteobacteria that entered an archaeal host [7]. While most genes were lost due to functional redundancy or transferred to the nucleus, mitochondria of most eukaryotes still contain small residual genomes revealing their prokaryotic ancestry [8]. These largest mitochondrial genomes are found in a class of protists called Jacobids in which, up to 70 proteins can be mitochondrially encoded [9, 10]. However, most mitochondrial genomes are much smaller just coding for a few hydrophobic core subunits of the enzymes of the mitochondrial respiratory chain and the ATPase, as well as a minimalistic set of RNAs, including two or three rRNAs.

Mitochondrial ribosomes are derived from those of the bacterial ancestors, and their sensitivity to antibiotics still reveals this origin. However, during evolution, mitoribosomes were considerably remodeled, lost many RNA segments and protein subunits, and gained others [11]. Thus, present-day mitoribosomes are considerably different from those of bacteria, and mitoribosomes of different groups of eukaryotes often are highly diverse in structure and composition [12–16]. The biology of mitoribosomes is still poorly understood. The impressive success of cryoelectron microscopy revealed exciting details of the structures of mitoribosomes of many organisms [15, 17–24]. However, despite some exciting recent studies [25–27], the assembly of mitoribosomes from proteins and rRNAs is largely elusive [27, 28]. Even less is known about the mechanisms by which mitoribosomes are degraded or by which their activity can be adapted to cellular needs. Recent studies suggest that eukaryotes tightly regulate the amount and activity of their mitochondrial translation machinery in dependence of the import of mitochondrial precursor proteins or in response to stress conditions [29–33]. However, the underlying mechanisms still await to be explored.

In animals and humans, the genes for all MRPs were transferred to the nucleus. In contrast, many fungi including yeast still carry one MRP-coding gene on the mitochondrial genome. This protein, Var1 or uS3m, is a necessary part of the small subunit. Why this protein is mitochondrially encoded is not entirely clear as mutants which express Var1 as fusions with a mitochondrial targeting sequence in the cytosol fully complement cells that lack the *VAR1* gene in mitochondria [34]. However, the retention of *VAR1* in the mtDNA allows for a translational regulatory mechanism by which the *VAR1* mRNA translational activator Sov1 fine-tunes Var1 synthesis with its assembly into the mitoribosome [35]. In addition to Var1, yeast mitoribosomes contain 72 nuclear-encoded proteins, all of which are imported from the cytosol [19].

1.1 Protein Import Routes into Mitochondria

The majority of mitochondrial proteins are synthesized in the cytosol and have to be imported into mitochondria, using a number of distinct import pathways. The import and sorting of these proteins is mediated by protein complexes of the outer and the inner mitochondrial membrane. The process is well studied, and details are described in many review articles [36–40].

The TOM (translocase of the outer membrane) complex of the outer membrane serves as general entry gate into mitochondria (Fig. 1). The TOM complex consists of receptors, which recognize precursors on the cytosolic surface of mitochondria and membrane-embedded channels to transfer them across the outer membrane. Proteins of the outer membrane can be integrated straight into the outer membrane either via the TOM complex or, in case of beta-barrel proteins, with the help of the sorting and assembly machinery (SAM) complex. Cysteine-rich IMS proteins are imported in a reaction that is driven by their oxidation by the essential IMS oxidoreductase Mia40. Proteins of the mitochondrial matrix typically carry an N-terminal mitochondrial targeting sequence (MTS) that interacts with the TOM and TIM23 (translocase of the inner membrane) complex. The positively charged MTS is attracted by the membrane potential that is negatively charged on the matrix side. Moreover, the ATP-driven import motor of the TIM23 complex is required for matrix proteins to pass the inner membrane (IM). After complete translocation, the MTS is proteolytically removed by the matrix processing peptidase (MPP). Inner membrane proteins can either be stalled at the TIM23 complex and laterally released into the inner membrane, or integrated by the membrane insertase Oxa1 from the matrix side. Some inner membrane proteins like carrier proteins depend on a specific insertion complex called TIM22 complex.

1.2 Import of Mitoribosomal Proteins

In yeast, all mitoribosomal proteins (MRPs) except for Var1 (uS3m) are nuclear encoded, synthesized in the cytosol, and imported into the mitochondrial matrix. In contrast to all other mitochondrial matrix proteins, many MRPs do not have a conventional N-terminal MTS [41]. Structural data as well as computational predictions indicate that up to 25% of yeast MRPs lack such sequences [42]. In some cases, the N-terminal regions of the proteins still share the properties of MTSs but are not cleaved by

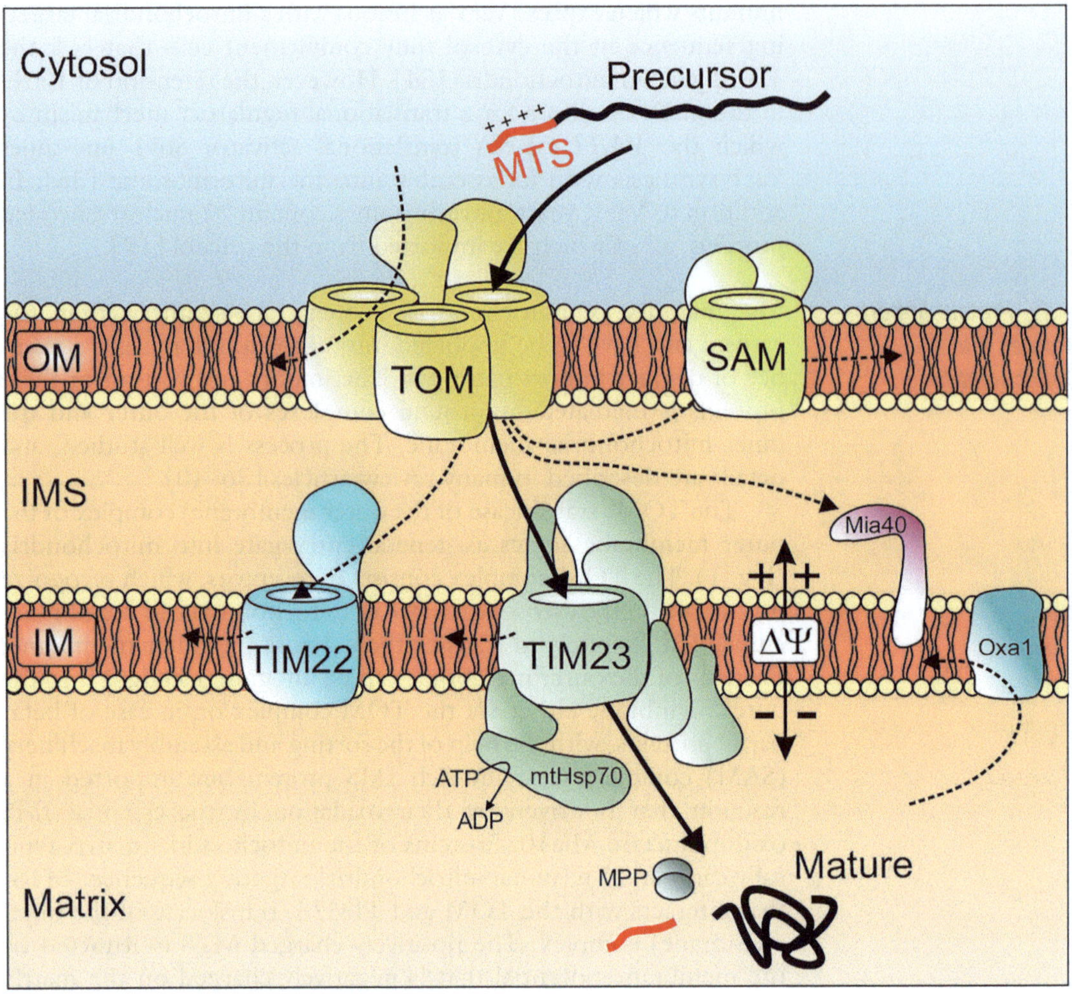

Fig. 1 Overview of protein import pathways into mitochondria. Proteins destined to the mitochondrial matrix are synthesized in the cytosol as precursor proteins. An N-terminal MTS facilitates their binding to outer membrane receptors and leads the way through channels of the TOM and TIM23 complex into the matrix where it is removed by MPP. The membrane potential and ATP hydrolysis of the import motor (mtHsp70) drives this import reaction. β-Barrel proteins are inserted into the outer membrane (OM) by the SAM complex. Other outer membrane proteins are laterally inserted by the TOM complex into the outer membrane. Mia40 promotes the import of cysteine-rich proteins into the IMS. Carrier proteins are integrated into the inner membrane (IM) by the TIM22 complex. Other inner membrane proteins can leave the TIM23 complex laterally or insert from the matrix with the help of the Oxa1 insertase

MPP. However, a number of proteins lack any targeting information in their N-termini, and internal regions are critical for their import. Only in very few cases, the import process was studied experimentally [42]. For example, Mrp10 (mS37) is a protein of the small subunit that carries an unconventional proline-rich MTS; this sequence is recognized by the TOM translocase but fails to be handed on to the TIM23 directly [43]. Instead, Mrp10 explores the IMS where it is oxidized by Mia40 before it further translocates into the matrix. Thus far, Mrp10 is the only known Mia40 substrate that reaches the matrix. Mrpl32 (bL32m) is another unconventional protein as its atypical presequence is not recognized by MPP and instead removed the m-AAA protease, thus by a degradation protease that serves in this case as a processing enzyme [44, 45]. Mrp17 (bS6m) is even more awkward as its N-terminus is not required for targeting and multiple features along its sequence facilitate its import into the matrix [42]. There are still many MRPs which lack obvious MTSs and for which the import modes await to be elucidated. The methods described in this chapter might provide the necessary strategies to study the biogenesis of these proteins.

2 Materials

2.1 Isolation of Mitochondria from Yeast Cells

1. Stock solution of a lab strain of *Saccharomyces cerevisiae*, such as W303 [46], YPH499 [47], or BY4742 [48].
2. For 1 L media: 20 g bacto peptone, 10 g yeast extract, adjust pH to 5.5 using HCl. Add 2% of the respective carbon source.
3. MP1 buffer: 100 mM Tris–HCl (pH not adjusted), 10 mM dithiothreitol (DTT). Add DTT freshly prior to use.
4. MP2 buffer: 1.2 M sorbitol, 20 mM potassium phosphate buffer, pH 7.4, 3 mg zymolyase (20T) per g wet weight of cells. Add zymolyase freshly prior to use.
5. PMSF: Freshly prepare prior to use a 200 mM solution of phenylmethylsulfonyl fluoride (PMSF) in ethanol (>99.8%, p.a.).
6. Homogenization buffer: 10 mM Tris–HCl, pH 7.4, 0.6 M sorbitol, 1 mM ethylenediaminetetraacetic acid (EDTA), 0.2% fatty acid-free bovine serum albumin (BSA), 1 mM PMSF. Add BSA and PMSF freshly prior to use.
7. SH buffer: 0.6 M sorbitol, 20 mM HEPES/KOH, pH 7.2.

2.2 Import of Mitoribosomal Proteins into Isolated Mitochondria

1. 2× import buffer: 1 M sorbitol, 160 mM KCl, 20 mM magnesium acetate, 4 mM potassium phosphate buffer, pH 7.2, 100 mM HEPES/KOH, pH 7.2. Store in aliquots at −20 °C (*see* **Notes 1** and **2**).
2. 0.2 M ATP: Dissolve in water and adjust the pH with KOH to 7.0. Make single use aliquots of 10 μL and store at −20 °C.
3. 0.2 M NADH: Dissolve in water. Make single use aliquots of 10 μL and store at −20 °C.
4. Isolated yeast mitochondria at concentration of 10 mg/mL, keep on single use aliquots at −80 °C.

5. 10 mg/mL Proteinase K (PK): Dissolve in water and store in single use aliquots of 50–100 μL at −20 °C.
6. ^{35}S-methionine: 11 mCi/mL. EasyTag Express Protein Labeling Mix (^{35}S) (Perkin Elmer).
7. ^{35}S-labeled mitochondrial precursor protein prepared in an *in vitro* translation reaction based on rabbit reticulocyte lysate. To this end, coupled transcription-translation methods can be used (e.g., the TnT Quick Coupled Transcription/Translation system from Promega). Alternatively, the radiolabeled proteins can be produces by separate *in vitro* transcription and translation reactions [49]. Store aliquots at −80 °C (*see* **Note 3**).
8. SH/KCl buffer: 0.6 M sorbitol, 20 mM HEPES/KOH, pH 7.2, 150 mM KCl. Store in 50 mL aliquots at −20 °C. Thawed aliquots can be kept at 4 °C.
9. VAO: 55.6 μg/mL valinomycin, 440 μg/mL antimycin A, 850 μg/mL oligomycin in ethanol (>99.8%, p.a.). Store at −20 °C.
10. Trypsin: Prepare a 10 mg/mL stock solution in water. Store aliquots of a 200 μg/mL concentration at −20 °C (*see* **Note 4**).
11. 50 mg/mL Soybean Trypsin inhibitor (TI): Dissolve in water. Store at −20 °C.
12. 20 mM Carbonyl cyanide-m-chlorophenylhydrazone (CCCP): Dissolve in DMSO. Store at −20 °C.
13. Apyrase: Prepare a 391 U/mL stock solution in water. Store in 100 μL aliquots at −20 °C.
14. Laemmli^{+}: 2% (w/v) sodium dodecylsulfate (SDS), 50 mM DTT, 10% (v/v) glycerol, 0.02% (w/v) bromophenol blue, 60 mM Tris–HCl, pH 6.8.

2.3 Immunoprecipitation of MRPs and Sample Preparation for Mass Spectrometry

1. Lysis buffer: 10 mM Tris–HCl, pH 7.5, 150 mM sodium chloride (NaCl), 0.5 mM EDTA, 10 mM magnesium chloride ($MgCl_2$), 0.5% Triton X-100, 1 mM PMSF. Prepare freshly prior to use (*see* **Note 5**).
2. Dilution buffer: 10 mM Tris–HCl, pH 7.5, 150 mM NaCl, 0.5 Mm EDTA; 1× cOmplete protease inhibitor cocktail, 1× PhosSTOP phosphatase inhibitor cocktail (Roche). Add protease cocktails freshly prior to use.
3. Wash buffer 1: 150 mM NaCl, 50 mM Tris–HCl, pH 7.5, 5% glycerol, 0.05% Triton X-100.
4. Wash buffer 2: 150 mM NaCl, 50 mM Tris–HCl, pH 7.5, 5% glycerol.
5. Elution buffer 1: 2 M urea, 50 mM Tris–HCl, pH 7.5, 1 mM DTT, 5 ng/μL trypsin. Prepare buffer freshly prior to use.

6. Trypsin in mass spectrometry (MS) grade: prepare prior to use a 15 ng/μL solution. Store at −80 °C.
7. Elution buffer 2: 2 M urea, 50 mM Tris–HCl, pH 7.5, 5 mM chloroacetamide (CAA). Prepare buffer freshly prior to use.
8. 10% trifluoroacetic acid (TFA): 10% TFA in MS grade H_2O.
9. Buffer A: 0.1% formic acid in MS grade H_2O.
10. Buffer B: 80% acetonitrile, 0.1% formic acid.
11. Buffer A*: 0.1% formic acid, 0.1% trifluoroacetic acid (TFA).
12. For GFP immunoprecipitation, GFP-Trap_A-beads from ChromoTek were used; ANTI-FLAG M2 Affinity Gel from Sigma-Aldrich were used for immunoprecipitation of FLAG-tagged proteins.

3 Methods

3.1 Isolation of Mitochondria from S. cerevisiae

Isolated mitochondria of the budding yeast *S. cerevisiae* are commonly used for several *in vitro* assays. The standard isolation procedure was first described by the Schatz laboratory [1]. The big advantages of using *S. cerevisiae* as donor organism is that the mitochondria can be isolated relatively easy and remain energized even upon a long storage period at −80 °C. The first main step of the procedure is to remove the cell wall enzymatically by zymolyase. The resulting spheroplasts are than opened mechanically by using a glass homogenizer. Afterward, mitochondria can be purified from the previously obtained cellular lysate by differential centrifugation. The finally collected fraction contains biochemically highly active mitochondria but also other cellular membranes such as microsomes. To get even purer mitochondria, further purification steps are mandatory [50, 51].

1. Grow yeast cells at 30 °C in a shaker to an OD_{600} of 0.8–1.5 in the desired culture medium (*see* **Notes 6–8**).
2. Harvest the cells by centrifugation at 2800 × *g* for 5 min at room temperature, discard the supernatant.
3. Resuspend the cells with distilled water and centrifuge again as in **step 2**.
4. Weigh the wet weight of the cell pellet.
5. Resuspend the cells in 2 mL of MP1 buffer per g wet weight.
6. Incubate the cells for 10 min shaking at 30 °C.
7. Centrifuge at 2100 × *g* for 5 min at room temperature, discard the supernatant.
8. Wash the cells in 50 mL of 1.2 M sorbitol and centrifuge as in **step** 7.

9. Resuspend the cells in 6.7 mL MP2 buffer per g wet weight.
10. Incubate for 30–60 min at 30 °C in a shaker (*see* **Note 9**).
11. Perform all the following steps on ice or 4 °C (*see* **Note 10**).
12. Harvest the spheroplasts by centrifugation at 1900 × *g* for 5 min at 4 °C, discard the supernatant.
13. Resuspend the spheroplasts in 6.7 mL ice-cold homogenization buffer per g wet weight.
14. Rupture the spheroplasts with a cooled glass homogenizer with a glass pistil (Sartorius) by douncing 10–12 times (*see* **Note 11**).
15. Wash the glass homogenizer and pistil with 6.7 mL ice-cold homogenization buffer and add it to the previously homogenized sample.
16. To remove cell debris, nuclei and intact cells centrifuge at 1900 × *g* for 5 min at 4 °C. Pour the supernatant to a new cooled bottle and discard the pellet.
17. Repeat **step 16** until no pellet is collected (*see* **Note 12**).
18. Collect the mitochondria by centrifugation of the previously collected supernatant at 17,700 × *g* for 12 min at 4 °C, discard the supernatant.
19. Resuspend mitochondria in 0.5–1 mL of SH buffer, depending on the size of the mitochondrial pellet (*see* **Note 13**).
20. Determine the protein concentration via Bradford assay [52] and adjust it to 10 mg/mL with SH buffer.
21. Make aliquots of 50 μL, freeze them in liquid nitrogen and store them at −80 °C (*see* **Note 13**).

3.2 Import of Mitoribosomal Proteins into Isolated Mitochondria

The *in vitro* import assay of mitochondrial precursor proteins into isolated mitochondria is a well-established method (Fig. 2). It is used to investigate the import capability and properties of a specific precursor protein. Therefore, radiolabeled precursor proteins synthesized in reticulocyte lysate [53] are used. The *in vitro* translated precursor protein is maintained in an import-competent state by chaperones that are part of the reticulocyte lysate. To investigate the variety of targeting signals and import pathways of MRPs, this *in vitro* assay can be carried out relatively easy and fast. Different reagents such as trypsin, CCCP, or apyrase can be used to dissect mechanistic properties of the specific import reaction of a given protein (*see* **Notes 27–29**).

1. Mix in the following order: 50 μL 2× import buffer, 42 μL H_2O, 1 μL ATP, 1 μL NADH, 5 μL mitochondria (equivalent to 50 μg protein) (*see* **Notes 13–15**).
2. Incubate the sample for 10 min at 30 °C to energize mitochondria (*see* **Note 16**).

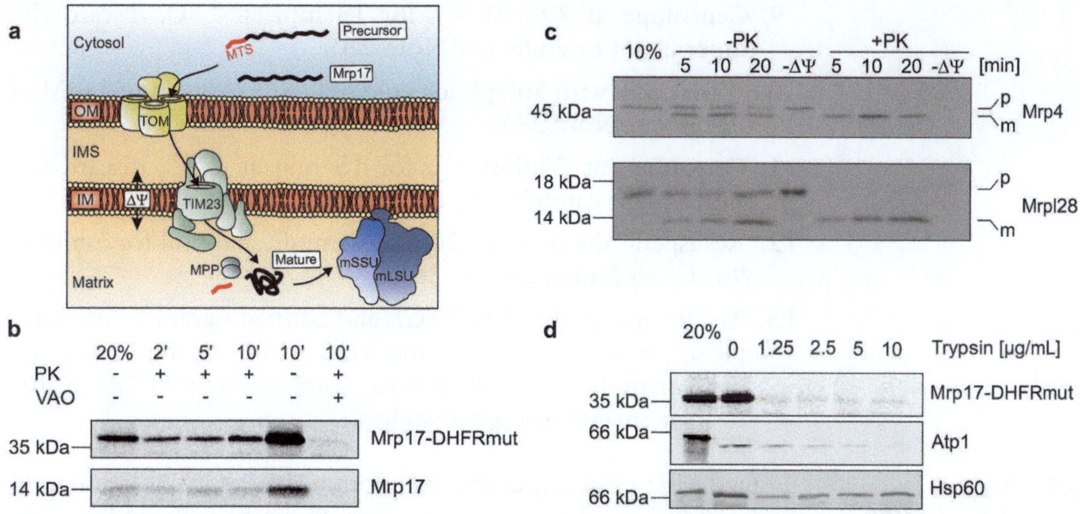

Fig. 2 *In vitro* experiments to study the import of mitoribosomal proteins. (**a**) Schematic representation of the import and assembly of MRPs. Many MRPs lack N-terminal MTSs. Mrp17 (bS6m) is shown here as a representative of this protein class. (**b**) *In vitro* import experiment with radiolabeled Mrp17-DHFRmut and Mrp17. The precursor form was synthesized in reticulocyte lysate (lane 1). The precursors were imported into isolated mitochondria for the indicated time points. Proteinase K (PK) was added to degrade nonimported precursors (lanes 2–4). Lane 5 shows imported precursor as well as precursors that are associated with mitochondria. As a negative control, mitochondria were treated with VAO that dissipates the membrane potential (lane 6). The mitoribosomal protein Mrp17 (bS6m) contains only four methionine residues. Adding DHFRmut C-terminally to the protein increases the signal in autoradiography but has no effect on the import rate. For both radiolabeled proteins, the same exposure time was used. Please note the considerably stronger signal in the DHFRmut fusion version. (**c**) Time course experiment with the radiolabeled mitoribosomal protein Mrp4 (uS2m) and Mrpl28 (mL40). In contrast to Mrp17 (bS6m), these two proteins carry cleavable MTSs. Without adding protease, the precursor (p) as well as the mature (m) form can be detected by autoradiography (lanes 2–4). PK treatment of mitochondria after the import reaction removes the precursor protein and only the imported mature form remains intact due to its inaccessibility to the protease (lanes 6–8). Pretreatment of mitochondria with VAO uncouples the membrane potential ($\Delta\Psi$); therefore, no import is possible. This sample can be used as a negative control (lanes 5 and 9). (**d**) Import experiment with trypsin pretreatment. Mitochondria were incubated with the indicated concentrations. After reisolation, mitochondria were incubated with radiolabeled precursors for 5 min. PK was added to remove nonimported proteins. The *in vitro* import of Mrp17-DHFRmut is more sensitive to the elimination of outer mitochondrial membrane proteins by trypsinization than the model import substrate Atp1 or Hsp60

3. Add 1 μL of radiolabeled precursor protein (*see* **Notes 17** and **18**).
4. Incubate the sample for 10 min at 30 °C (*see* **Notes 19** and **20**).
5. Add 900 μL of ice-cold SH buffer and transfer sample tube on ice to stop import reaction.
6. Add 10 μL of PK (*see* **Note 21**).
7. Incubate the sample for 30 min on ice.
8. Add 10 μL of PMSF. Mix rapidly to avoid precipitation of PMSF (*see* **Note 22**).

9. Centrifuge at 30,000 × *g* for 15 min at 4 °C, discard the supernatant carefully (*see* **Note 23**).
10. Wash pellet with 500 μL ice-cold SH/KCl containing 2 mM of PMSF (*see* **Note 24**).
11. Centrifuge at 30,000 × *g* for 15 min at 4 °C, discard the supernatant carefully (*see* **Note 23**).
12. Resuspend the pellet in 20 μL Laemmli$^+$ and boil for 5 min at 96 °C (*see* **Notes 25** and **26**).
13. Analyze sample by SDS-PAGE and autoradiography. An example is shown in Fig. 2. Imported matrix proteins that are proteolytically matured migrate faster on an SDS-PAGE due to their smaller molecular weight.

3.3 Immunoprecipitation of MRPs and Sample Preparation for Mass Spectrometry

Tagged MRPs can be purified by immunoprecipitation. Upon mild lysis conditions, also interaction partners can be isolated by coimmunoprecipitation. These can be identified by Western blotting or, with much better sensitivity and reliability, by mass spectrometry. Along with other MRPs from the small and large subunit also assembly factors as well as proteins involved in targeting and mitochondrial import can be identified (Fig. 3). Small C-terminal tags like the FLAG-tag allow specific purification and do not interfere with correct targeting and import [54]. Bigger tags like GFP might interfere or even block import. However, published datasets from genome-wide screens for the localization of GFP-tagged proteins, such as Yeast RGB, are available which allow it to quickly check which GFP-tagged proteins maintain a mitochondrial localization [55]. The high affinity of GFP-binding nanobodies allow very specific isolation of GFP-tagged proteins by affinity chromatography [56].

Here, we describe the immunoprecipitation of the fusion protein MTS-Var1-GFP which consists of the mitochondrial presequence of Cox4 (residues 1–25), the sequence of Var1 and GFP. This nuclear-encoded version of Var1 is imported into mitochondria and assembles with the mitoribosomal small subunit (mSSU). The procedure can be easily employed for the purification of any given GFP-tagged MRP from yeast cells.

1. Start with 500 μg isolated mitochondria from cells expressing the tagged protein (*see* **Note 30**).
2. Resuspend mitochondria in 200 μL lysis buffer and place the tubes on ice for 30 min while extensively pipetting every 10 min to lyse mitochondria mechanically (*see* **Note 31**).
3. Centrifuge for 30 s at 5000 × *g* at 4 °C and transfer lysate to a new tube (*see* **Note 32**).
4. Centrifuge cell lysates at 20,000 × *g* for 10 min at 4 °C.
5. Transfer lysates to a precooled tube and discard the pellet.

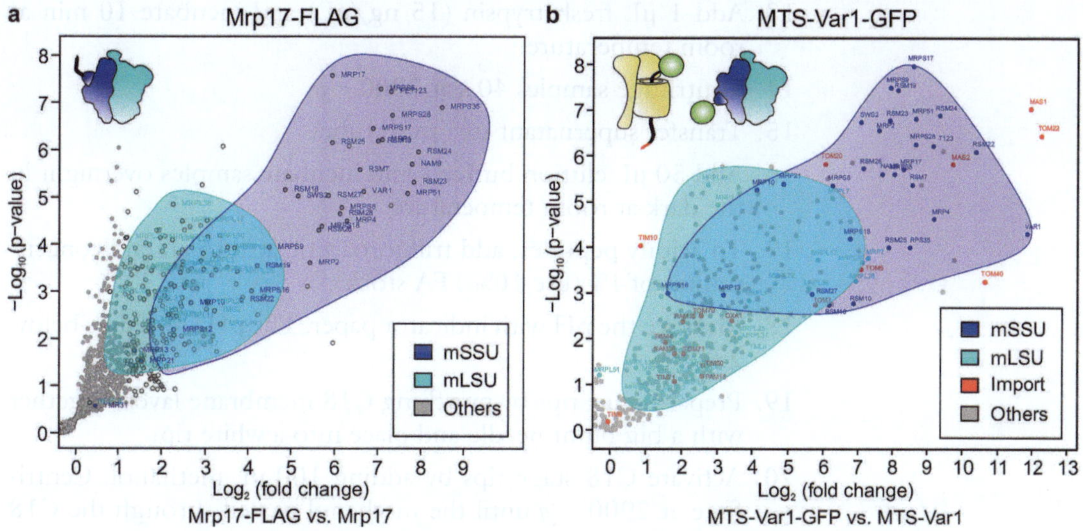

Fig. 3 Mass spectrometric analysis of the immunoprecipitation of Mrp17-FLAG and MTS-Var1-GFP. (**a**) Immunoprecipitation of the mSSU protein Mrp17-FLAG from whole-cell extracts. Cells were lysed by using a mild detergent (0.05% Triton X-100) and by glass bead lysis. The extracts were incubated on a tumbler for 2 h with anti-FLAG beads. Proteins coisolated with Mrp17-FLAG were analyzed by mass spectrometry basically as described before [62]. As a control yeast, cells expressing Mrp17-DHFR were used. Proteins of the mSSU (dark blue) show stronger enrichment than the proteins of the mitoribosomal large subunit mLSU (cyan). (**b**) Immunoprecipitation of mSSU protein MTS-Var1-GFP from isolated mitochondria. Mitochondria were lysed mechanically by pipetting and the usage of a mild detergent (0.5% Triton X-100). The extract was cleared by centrifugation and incubated with GFP-Traps_A for 1 h. Bound protein was analyzed by mass spectrometry. The enriched proteins show the same clustering of mSSU and mLSU proteins as identified in the Mrp17-FLAG experiment. Additionally, proteins of the import machinery were coisolated with MTS-Var1-GFP revealing the different stages of its import into mitochondria. Note, that proteins of the TOM can be copurified with the aggregation-prone MTS-Var1-GFP protein but not with the rapidly imported Mrp17-FLAG protein, revealing insights into the import process of both proteins under in vivo conditions

6. Add 300 μL precooled dilution buffer.
7. Add lysate to 25 μL of activated GFP-Trap_A beads (*see* **Note 33**).
8. Incubate for 1 h tumbling end-over-end at 4 °C to bind the tagged protein to the beads (*see* **Note 34**).
9. Centrifuge sample 2 min at 2000 × *g* (*see* **Note 32**). Discard the flow through.
10. Wash the beads three times with 800 μL wash buffer 1 to get rid of unspecific binding partners.
11. Wash the beads twice with 500 μL wash buffer 2 (*see* **Notes 32** and **35**). After adding wash buffer 2 for the first time, transfer the beads to a new tube to get rid of the detergent of the lysis buffer.
12. Add 50 μL elution buffer 1 and incubate the samples for 1 h at room temperature (*see* **Note 36**).

13. Add 1 μL fresh trypsin (15 ng/μL) and incubate 10 min at room temperature.
14. Centrifuge samples 40 s at 2000 × *g*.
15. Transfer supernatant to a fresh tube.
16. Add 50 μL elution buffer 2 and incubate samples overnight in the dark at room temperature.
17. To acidify peptides, add trifluoroacetic acid to an end concentration of 1% (use 10% TFA stock).
18. Measure the pH with indicator paper. The pH should be below 2.0.
19. Prepare stage tips by punching C18 membrane layers together with a big blunt needle and place into a white tip.
20. Activate C18 stage tips by adding 100 μL methanol. Centrifuge at 2900 × *g* until the methanol passed through the C18 filter in a microcentrifuge.
21. Wash twice with 100 μL buffer A. Centrifuge at 2900 × *g* until the buffer A passed through the C18 filter in a microcentrifuge.
22. Load acidified peptides onto stage tips. Centrifuge at 2900 × *g* until the samples passed through the C18 filter in a microcentrifuge.
23. Wash with 100 μL buffer A. Centrifuge in a microcentrifuge (*see* **Note 37**).
24. Elute peptides with 40 μL buffer B into 200 μL tubes.
25. To evaporate samples, centrifuge in a speedvac for 20 min at room temperature (*see* **Note 38**).
26. Add 1 μL of buffer A* to an end volume of 4 μL. The samples are now ready for mass spectrometry measurements.

4 Notes

1. BSA can be added to stabilize isolated mitochondria. For incubation, times longer than 30 min and temperatures of 30 °C or more buffers with BSA can be beneficial. Please note that only fatty acid-free BSA should be used as BSA of lower quality can result in the lysis of membranes.
2. BSA can be used to stabilize precursor proteins, though for many precursor proteins the presence of BSA has no positive or negative effect. If the import reaction is followed by crosslinking or TCA precipitation, BSA should be omitted.
3. The prepared radiolabeled lysate can be refrozen after usage at −80 °C. Quick freezing in liquid nitrogen is important. It is important to mention that this cannot be done with every

precursor protein. Lysates of precursors that fold tightly (e.g., DHFR) or are aggregation-prone can be used only once. The chaperones of the *in vitro* translation system keep the precursors import-competent.

4. Trypsin should be used directly after thawing and should be refrozen immediately to prevent self-autolysis of trypsin.
5. GFP-Trap_A beads require addition of 0.5 mM EDTA which, however, promotes disassociation of large and small mitoribosomal subunits. Addition of $MgCl_2$ and KCl can help to preserve monosomes. Furthermore, yeast strains lacking specific proteases (e.g., Pep4) or nucleases (e.g., Nuc1) can be useful to increase mitoribosome stability in extracts [57].
6. Yeast cells can be grown in media with different carbon sources. The choice of the carbon source has a strong influence on the mitochondrial volume in yeast cells, and hence on the yield of mitochondria that can isolated from these cultures. Carbon sources that force cells to respire such as glycerol, ethanol or lactate should be used. Strains that are not respiration-competent can be grown in media containing galactose or raffinose. Glucose should be avoided since it represses the expression of many mitochondrial proteins, and the yield of mitochondrial fractions from glucose-grown cultures therefore is very low. For selection purposes, selective media can be used.
7. To compare mitochondria from different strains, it is important that all strains were grown in the same media. Mitochondria of cells grown in YP media should not be compared to mitochondria of cells grown in synthetic media.
8. Typically, at least 2 L of yeast culture should be used for the isolation of mitochondria. It is recommended to use precultures of 15–25 mL which are diluted over 2–3 days into larger culture volumes, and to inoculate the final medium from these in overnight incubation.
9. To check for successful digestion of the cell wall, take twice 50 μL of the suspension and dilute one aliquot in 1.4 mL of H_2O and the second in 1.4 mL of 0.6 sorbitol. The OD_{600} of the H_20 sample should be 10–20% of that of the sorbitol sample.
10. The opening of the cell wall and later rupturing of the spheroplasts can lead to the release of hydrolytic enzymes from the yeast vacuole; therefore, it is important to carry out the procedure fast and on ice.
11. The douncing has to be carried out gently and slowly. If this step was done to harsh mitochondrial, membranes can be ruptured. It is possible that proteins of the outer membrane and the intermembrane space of mitochondria get lost.

12. It is crucial that no pellet is collected together with the supernatant. If so, further centrifugation steps might be necessary. The number of centrifugation steps might therefore vary and depend on the sample.
13. To avoid rupturing of the mitochondrial membranes, pipette tips should be clipped with a scissor.
14. Mitochondria should be thawed on ice and always kept on ice. Long incubation even on ice should be avoided since mitochondria will lose their import competence.
15. As negative control, a reaction mix containing VAO can be prepared. VAO will dissipate the membrane potential and, therefore, no import of matrix proteins will take place. VAO sample contains 50 μL 2× import buffer, 42 μL H_2O, 2 μL VAO, and 5 μL mitochondria (50 μg).
16. Incubation for 5 min at 25 °C might be sufficient. The temperature should be adjusted to the settings in **step 4**.
17. Typically, 1% (v/v) of the radiolabeled lysate is used per import reaction. If the content of methionine residues is low or the precursor protein has poor import efficiency, up to 20% reticulocyte lysate can be used.
18. We observed that precursor proteins with less than 4–6 methionine residues will give weak signals in autoradiography. Therefore, dihydrofolate reductase (DHFR) or DHFRmut can be fused C-terminally to the precursor protein. In DHFRmut, three point mutations prevent protein folding [58], which prevents stalling of import intermediates in the import machinery. The DHFR or DHFRmut domain will add seven additional methionine residues that lead to stronger radioactive signals (Fig. 2).
19. The incubation time depends on the precursor protein that is used. Some are imported rapidly within 2 min; other precursors need up to 30 min. Import kinetics can be done to examine the import efficiency of a precursor protein (Fig. 2) Therefore, a larger reaction mix might be used from which aliquots are taken after different incubation times. These aliquots (100 μL) can be quickly mixed with 900 μL ice-cold SH buffer to stop protein import.
20. The import rate can be controlled by the incubation temperature. Lower temperatures (25 or even 16 °C) will reduce the import rate whereas the overall efficiency stays more or less constant. Higher temperatures up to about 37 °C increase the speed of the import reaction but can also impair the import competence of proteins.

21. Proteinase K (PK) is used to degrade nonimported precursor proteins. Thereby, only imported proteins can be detected in the following autoradiography. It is important to wash PK off from the mitochondria and to block residual amounts with PMSF during sample preparation. Otherwise, after addition of detergents, all proteins are degraded.
22. PMSF stock solution should be kept at room temperature rather than on ice to prevent its crystallization and precipitation. In ethanol, PMSF remains stable for several hours and then loses activity. Alternatively, PMSF can be dissolved in DMSO where it is longer active; however, since DMSO can have negative effects on the integrity of membranes, ethanol is the preferred solvent in import experiments.
23. We observed that the pellet is often soft and fluffy after centrifugation and can be accidently removed. We recommend to leave around 100 μL of the supernatant in the tube and centrifuge again at 2900 × *g* for few seconds in a microcentrifuge. This will lead to a more solid pellet from which the supernatant can be easily removed.
24. Add the PMSF to the SH/KCl shortly before usage since it is rapidly inactivated in aqueous solutions.
25. It is possible that PK is not fully inactivated by PMSF treatment. Proteins of the sample will be degraded during SDS-PAGE leading to a specific loss of proteins larger than 45–60 kDa. To ensure that there is no residual active PK left, boil the samples immediately after adding Laemmli^{+}. Only for very hydrophobic proteins, avoid this boiling step since they may aggregate so that they will not enter the SDS gel.
26. Prepare additionally a control which contains 10% or 20% of the radiolabeled protein used per import reaction. Load this onto the gel in order to compare the intensities in the import reactions.
27. Trypsin can be used to degrade outer membrane proteins such as receptors. To pretreat mitochondria, add trypsin with a final concentration of 10 μg/mL to the import reaction mix on ice. After incubation for 10 min, add trypsin inhibitor (TI) with a final concentration of 300 μg/mL. To reisolate mitochondria, centrifuge at 16,000 × *g* for 10 min at 4 °C, discard the supernatant. Resuspend the pellet in an import reaction mix containing 100 μg/mL TI and continue with **step 2** of the import protocol. TI does not interfere with PK digestion. PMSF inactivates both PK and trypsin.
28. CCCP uncouples the membrane potential. To determine the dependency of a precursor on the membrane potential, add CCCP with concentrations from 0 to 80 μM into the import reaction mix. To prevent regeneration of the membrane

potential, additional oligomycin (final concentration 20 μM) can be added into the reaction mix. If necessary, CCCP can be inactivated by addition of DTT [59, 60].

29. ATP is necessary for the import of proteins into the matrix as it drives the Hsp70-dependent import motor. To dissipate ATP to AMP, 40 U/mL apyrase can be added into the import reaction mix.

30. Immunoprecipitation from lysed isolated mitochondria rather than from whole-cell extracts reduces the number of interaction partners arising from postlysis artifacts. However, since samples can differ considerably after cellular fractionation procedures, all mitochondrial isolations should be done under the same preparation conditions, preferentially in parallel, and many (at least four) biological replicates should be used. It is therefore recommended to make unbiased proteomic analyses from whole-cell extracts. Alternatively, cells from isotope-labeled cultures can be mixed before mitochondria are isolated and analyze the samples using established stable isotope labeling by amino acids in cell culture (SILAC) procedures [61].

31. Alternatively, samples can be lysed using glass beads. The sample gets resuspended in lysis buffer and transferred into screw-cap-tubes with glass beads (0.5 mm) inside. Lyse sample using a FastPrep-24 5G homogenizer (MP Biomedicals) with 3 cycles of 30 s, speed 8.0 m/s, 120 s breaks.

32. At this step, a sample can be collected and stored at −20 °C for later analysis via immunoblotting.

33. To activate the beads, wash them three times with the dilution buffer. Check the user manual as details depend on the type of beads used.

34. Incubation times can differ depending on the protein of interest. Shorter incubation reduces protein degradation; longer incubation times might yield in more bound protein.

35. If the target protein is unstable or rapidly degraded, wash only once with washing buffer 2.

36. To elute samples for immunoblotting instead of proceeding with sample preparation for mass spectrometry, add 100 μL of sample buffer (Laemmli$^+$). Boil GFP-Trap_A beads for 10 min at 96 °C to dissolve immunocomplexes from GFP-Trap_A beads. Pellet beads by centrifugation for 2 min at 2000 × *g* and use the supernatant for SDS-PAGE.

37. At this point, the tips can be stored in the fridge for some weeks until further use.

38. Time depends on the evaporator and temperature settings used.

Acknowledgments

We thank Markus Räschle and Lavanya Deenadayalu for help in the context of the experimental data shown in Figs. 2 and 3. The authors were supported by grants of the Deutsche Forschungsgemeinschaft (HE2803/9-2) and the Forschungsinitiative Rheinland-Pfalz BioComp.

References

1. Daum G, Böhni PC, Schatz G (1982) Import of proteins into mitochondria: cytochrome b_2 and cytochrome c peroxidase are located in the intermembrane space of yeast mitochondria. J Biol Chem 257:13028–13033
2. Rout S, Oeljeklaus S, Makki A et al (2021) Determinism and contingencies shaped the evolution of mitochondrial protein import. Proc Natl Acad Sci U S A 118. https://doi.org/10.1073/pnas.2017774118
3. Murcha MW, Wang Y, Narsai R et al (2014) The plant mitochondrial protein import apparatus – the differences make it interesting. Biochim Biophys Acta 1840:1233–1245. https://doi.org/10.1016/j.bbagen.2013.09.026
4. Hewitt V, Alcock F, Lithgow T (2011) Minor modifications and major adaptations: the evolution of molecular machines driving mitochondrial protein import. Biochim Biophys Acta 1808:947–954. https://doi.org/10.1016/j.bbamem.2010.07.019
5. Callegari S, Cruz-Zaragoza LD, Rehling P (2020) From TOM to the TIM23 complex – handing over of a precursor. Biol Chem 401:709–721. https://doi.org/10.1515/hsz-2020-0101
6. Finger Y, Riemer J (2020) Protein import by the mitochondrial disulfide relay in higher eukaryotes. Biol Chem 401:749–763. https://doi.org/10.1515/hsz-2020-0108
7. Imachi H, Nobu MK, Nakahara N et al (2020) Isolation of an archaeon at the prokaryote-eukaryote interface. Nature 577:519–525. https://doi.org/10.1038/s41586-019-1916-6
8. Ku C, Nelson-Sathi S, Roettger M et al (2015) Endosymbiotic origin and differential loss of eukaryotic genes. Nature 524:427–432. https://doi.org/10.1038/nature14963
9. Burger G, Gray MW, Forget L et al (2013) Strikingly bacteria-like and gene-rich mitochondrial genomes throughout jakobid protists. Genome Biol Evol 5:418–438. https://doi.org/10.1093/gbe/evt008
10. Lang BF, Burger G, O'Kelly CJ et al (1997) An ancestral mitochondrial DNA resembling a eubacterial genome in miniature. Nature 387:493–497
11. Ott M, Amunts A, Brown A (2016) Organization and regulation of mitochondrial protein synthesis. Annu Rev Biochem. https://doi.org/10.1146/annurev-biochem-060815-014334
12. Smits P, Smeitink JA, van den Heuvel LP et al (2007) Reconstructing the evolution of the mitochondrial ribosomal proteome. Nucleic Acids Res 35:4686–4703
13. Desmond E, Brochier-Armanet C, Forterre P et al (2011) On the last common ancestor and early evolution of eukaryotes: reconstructing the history of mitochondrial ribosomes. Res Microbiol 162:53–70. https://doi.org/10.1016/j.resmic.2010.10.004
14. van der Sluis EO, Bauerschmitt H, Becker T et al (2015) Parallel structural evolution of mitochondrial ribosomes and OXPHOS complexes. Genome Biol Evol 7:1235–1251. https://doi.org/10.1093/gbe/evv061
15. Tobiasson V, Amunts A (2020) Ciliate mitoribosome illuminates evolutionary steps of mitochondrial translation. elife 9. https://doi.org/10.7554/eLife.59264
16. Petrov AS, Wood EC, Bernier CR et al (2019) Structural patching fosters divergence of mitochondrial ribosomes. Mol Biol Evol 36:207–219. https://doi.org/10.1093/molbev/msy221
17. Amunts A, Brown A, Toots J et al (2015) Ribosome. The structure of the human mitochondrial ribosome. Science 348:95–98. https://doi.org/10.1126/science.aaa1193
18. Beckmann R, Herrmann JM (2015) Structural biology. Mitoribosome oddities. Science 348:288–289. https://doi.org/10.1126/science.aab1054
19. Desai N, Brown A, Amunts A et al (2017) The structure of the yeast mitochondrial ribosome. Science 355:528–531. https://doi.org/10.1126/science.aal2415

20. Englmeier R, Pfeffer S, Forster F (2017) Structure of the human mitochondrial ribosome studied in situ by cryoelectron tomography. Structure 25:1574–1581 e1572. https://doi.org/10.1016/j.str.2017.07.011
21. Greber BJ, Ban N (2016) Structure and function of the mitochondrial ribosome. Annu Rev Biochem. https://doi.org/10.1146/annurev-biochem-060815-014343
22. Itoh Y, Andrell J, Choi A et al (2021) Mechanism of membrane-tethered mitochondrial protein synthesis. Science 371:846–849. https://doi.org/10.1126/science.abe0763
23. Waltz F, Soufari H, Bochler A et al (2020) Cryo-EM structure of the RNA-rich plant mitochondrial ribosome. Nat Plants 6:377–383. https://doi.org/10.1038/s41477-020-0631-5
24. Ramrath DJF, Niemann M, Leibundgut M et al (2018) Evolutionary shift toward protein-based architecture in trypanosomal mitochondrial ribosomes. Science 362. https://doi.org/10.1126/science.aau7735
25. Zeng R, Smith E, Barrientos A (2018) Yeast mitoribosome large subunit assembly proceeds by hierarchical incorporation of protein clusters and modules on the inner membrane. Cell Metab 27:645–656 e647. https://doi.org/10.1016/j.cmet.2018.01.012
26. Tu YT, Barrientos A (2015) The human mitochondrial DEAD-box protein DDX28 resides in RNA granules and functions in mitoribosome assembly. Cell Rep. https://doi.org/10.1016/j.celrep.2015.01.033
27. Maiti P, Lavdovskaia E, Barrientos A et al (2021) Role of GTPases in driving mitoribosome assembly. Trends Cell Biol 31:284–297. https://doi.org/10.1016/j.tcb.2020.12.008
28. De Silva D, Tu YT, Amunts A et al (2015) Mitochondrial ribosome assembly in health and disease. Cell Cycle 14:2226–2250. https://doi.org/10.1080/15384101.2015.1053672
29. Singh AP, Salvatori R, Aftab W et al (2020) Molecular connectivity of mitochondrial gene expression and OXPHOS biogenesis. Mol Cell 79:1051–1065 e1010. https://doi.org/10.1016/j.molcel.2020.07.024
30. Münch C, Harper JW (2016) Mitochondrial unfolded protein response controls matrix pre-RNA processing and translation. Nature 534:710–713. https://doi.org/10.1038/nature18302
31. Mick DU, Dennerlein S, Wiese H et al (2012) MITRAC links mitochondrial protein translocation to respiratory-chain assembly and translational regulation. Cell 151:1528–1541. https://doi.org/10.1016/j.cell.2012.11.053
32. Richter-Dennerlein R, Oeljeklaus S, Lorenzi I et al (2016) Mitochondrial protein synthesis adapts to influx of nuclear-encoded protein. Cell 167:471–483 e410. https://doi.org/10.1016/j.cell.2016.09.003
33. Suhm T, Kaimal JM, Dawitz H et al (2018) Mitochondrial translation efficiency controls cytoplasmic protein homeostasis. Cell Metab 27:1309–1322 e1306. https://doi.org/10.1016/j.cmet.2018.04.011
34. Sanchirico M, Tzellas A, Fox TD et al (1995) Relocation of the unusual *VAR1* gene from the mitochondrion to the nucleus. Biochem Cell Biol 73:987–995
35. Seshadri SR, Banarjee C, Barros MH et al (2020) The translational activator Sov1 coordinates mitochondrial gene expression with mitoribosome biogenesis. Nucleic Acids Res 48:6759–6774. https://doi.org/10.1093/nar/gkaa424
36. Chacinska A, Koehler CM, Milenkovic D et al (2009) Importing mitochondrial proteins: machineries and mechanisms. Cell 138:628–644
37. Wiedemann N, Pfanner N (2017) Mitochondrial machineries for protein import and assembly. Annu Rev Biochem 86:685–714. https://doi.org/10.1146/annurev-biochem-060815-014352
38. Endo T, Yamano K (2009) Multiple pathways for mitochondrial protein traffic. Biol Chem 390:723–730
39. Gupta A, Becker T (2021) Mechanisms and pathways of mitochondrial outer membrane protein biogenesis. Biochim Biophys Acta Bioenerg 1862:148323. https://doi.org/10.1016/j.bbabio.2020.148323
40. Mokranjac D (2020) How to get to the other side of the mitochondrial inner membrane - the protein import motor. Biol Chem 401:723–736. https://doi.org/10.1515/hsz-2020-0106
41. Woellhaf MW, Hansen KG, Garth C et al (2014) Import of ribosomal proteins into yeast mitochondria. Biochem Cell Biol:1–10. https://doi.org/10.1139/bcb-2014-0029
42. Bykov YS, Flohr T, Boos F et al (2021) Widespread use of unconventional targeting signals in mitochondrial ribosome proteins. EMBO J: e109519. https://doi.org/10.15252/embj.2021109519
43. Longen S, Woellhaf MW, Petrungaro C et al (2014) The disulfide relay of the intermembrane space oxidizes the ribosomal subunit mrp10 on its transit into the mitochondrial matrix. Dev Cell 28:30–42. https://doi.org/10.1016/j.devcel.2013.11.007

44. Nolden M, Ehses S, Koppen M et al (2005) The m-AAA protease defective in hereditary spastic paraplegia controls ribosome assembly in mitochondria. Cell 123:277–289
45. Bonn F, Tatsuta T, Petrungaro C et al (2011) Presequence-dependent folding ensures MrpL32 processing by the m-AAA protease in mitochondria. EMBO J 30:2545–2556. https://doi.org/10.1038/emboj.2011.169
46. Thomas BJ, Rothstein R (1989) Elevated recombination rates in transcriptionally active DNA. Cell 56:619–630
47. Sikorski RS, Hieter P (1989) A system of shuttle vectors and host strains designed for efficient manipulation of DNA in *Saccharomyces cerevisiae*. Genetics 122:19–27
48. Brachmann CB, Davies A, Cost GJ et al (1998) Designer deletion strains derived from *Saccharomyces cerevisiae* S288C: a useful set of strains and plasmids for PCR-mediated gene disruption and other applications. Yeast 14:115–132
49. Mokranjac D, Neupert W (2007) Protein import into isolated mitochondria. In: Leister D, Herrmann JM (eds) Mitochondria. Practical protocols, vol 372. Humana Press, Totowa, pp 277–286
50. Meisinger C, Sommer T, Pfanner N (2000) Purification of *Saccharomcyes cerevisiae* mitochondria devoid of microsomal and cytosolic contaminations. Anal Biochem 287:339–342. https://doi.org/10.1006/abio.2000.4868
51. Sickmann A, Reinders J, Wagner Y et al (2003) The proteome of *Saccharomyces cerevisiae* mitochondria. Proc Natl Acad Sci U S A 100: 13207–13212
52. Bradford MM (1976) A rapid and sensitive method for quantitation of microgram quantities of protein utilising the principle of protein-dye binding. Anal Biochem 72:248–254
53. Pelham HRB, Jackson RJ (1976) An efficient mRNA-dependent translation system from reticulocyte lysates. Eur J Biochem 67:247–256
54. Couvillion MT, Soto IC, Shipkovenska G et al (2016) Synchronized mitochondrial and cytosolic translation programs. Nature 533:499–503. https://doi.org/10.1038/nature18015
55. Dubreuil B, Sass E, Nadav Y et al (2019) YeastRGB: comparing the abundance and localization of yeast proteins across cells and libraries. Nucleic Acids Res 47:D1245–D1249. https://doi.org/10.1093/nar/gky941
56. Braun MB, Traenkle B, Koch PA et al (2016) Peptides in headlock--a novel high-affinity and versatile peptide-binding nanobody for proteomics and microscopy. Sci Rep 6:19211. https://doi.org/10.1038/srep19211
57. Kehrein K, Schilling R, Moller-Hergt BV et al (2015) Organization of mitochondrial gene expression in two distinct ribosome-containing assemblies. Cell Rep 10:843–853. https://doi.org/10.1016/j.celrep.2015.01.012
58. Vestweber D, Schatz G (1988) Point mutations destabilizing a precursor protein enhance its post-translational import into mitochondria. EMBO J 7:1147–1151
59. Hines V, Schatz G (1993) Precursor binding to yeast mitochondria. A general role for the outer membrane protein Mas70p. J Biol Chem 268:449–454
60. Backes S, Hess S, Boos F et al (2018) Tom70 enhances mitochondrial preprotein import efficiency by binding to internal targeting sequences. J Cell Biol 217:1369–1382. https://doi.org/10.1083/jcb.201708044
61. Morgenstern M, Stiller SB, Lubbert P et al (2017) Definition of a high-confidence mitochondrial proteome at quantitative scale. Cell Rep 19:2836–2852. https://doi.org/10.1016/j.celrep.2017.06.014
62. Zöller E, Laborenz J, Kramer L et al (2020) The intermembrane space protein Mix23 is a novel stress-induced mitochondrial import factor. J Biol Chem 295:14686–14697. https://doi.org/10.1074/jbc.RA120.014247

Chapter 11

Systematic Analysis of Assembly Intermediates in Yeast to Decipher the Mitoribosome Assembly Pathway

Samuel Del'Olio and Antoni Barrientos

Abstract

Studies of yeast mitoribosome assembly have been historically hampered by the difficulty of generating mitoribosome protein-coding gene deletion strains with a stable mitochondrial genome. The identification of mitochondrial DNA-stabilizing approaches allows for the generation of a complete set of yeast deletion strains covering all mitoribosome proteins and known assembly factors. These strains can be used to analyze the integrity and assembly state of mitoribosomes by determining the sedimentation profile of these structures by sucrose gradient centrifugation of mitochondrial extracts, coupled to mass spectrometry analysis of mitoribosome composition. Subsequent hierarchical cluster analysis of mitoribosome subassemblies accumulated in mutant strains reveals details regarding the order of protein association during the mitoribosome biogenetic process. These strains also allow the expression of truncated protein variants to probe the role of mitochondrion-specific protein extensions, the relevance of protein cofactors, or the importance of RNA-protein interactions in functional sites of the mitoribosome. In this chapter, we will detail the methodology involved in these studies.

Key words Yeast mitoribosome gene deletion strain, Sucrose gradient, Mitochondrial ribosome, Mitoribosome profile, Gradient fractionation, Immunoblotting, Mass spectrometry, Mitoribosome assembly intermediate, Clustering analysis

1 Introduction

Ribosome biogenesis involves the processing and modification of ribosomal RNAs (rRNAs) coordinated with the incorporation of ribosomal proteins. For the ribosomes present in mitochondria (mitoribosomes), their assembly is complicated by the dual genetic origin, nuclear and mitochondrial, of their rRNA and protein (MRP) components [1]. Furthermore, mitoribosome composition and structure vary significantly among organisms [2], leading to species-specific assembly constraints. The biogenetic process requires numerous nucleus-encoded transacting factors, some of which are conserved in bacterial systems, in mitochondria across species, or species-specific, acting in all steps of the process.

Antoni Barrientos and Flavia Fontanesi (eds.), *The Mitoribosome: Methods and Protocols*,
Methods in Molecular Biology, vol. 2661, https://doi.org/10.1007/978-1-0716-3171-3_11,

Several approaches have been used to address aspects of the mitoribosome biogenesis pathway in yeast and other organisms. Searching for mitochondrial RNA-binding proteins in mammalian cells, or proteomics analyses of the mitoribosome interactome have allowed for the identification of potential mitoribosome assembly factors [3–7]. Also, studies in patients with mitochondrial translation efficiency disorders have allowed identifying new assembly factors and the relevance of MRPs for mitoribosome assembly across tissues [1, 8–10]. Other studies have leveraged the better characterized bacterial ribosome assembly pathway [11] and screens in yeast [12, 13] to identify potential mitoribosome assembly factors that were subsequently characterized in human cultured cells or mouse models [14–16]. Furthermore, structures of the mitochondrial ribosome from human HEK293 cells [17–23] and the human parasite *Trypanosoma brucei* [24–27] in native states of assembly have revealed insights into the timing of rRNA folding and protein incorporation during the final steps of ribosomal maturation.

Two approaches have so far been used to disclose the overall assembly pathway of the mitochondrial ribosome. First, a study performed in human HeLa cells analyzed the order of MRP assembly using stable pulse-chase isotope labeling in cell culture (SILAC) and mass spectrometry analysis of the human 55S mitochondrial monosome [28]. The approach is based on the model that the kinetics of incorporation of different MRPs into 55S mitoribosomes indicates their relative assembly order [28]. This study provided a useful albeit low-resolution draft of the assembly pathway by defining sets of early, intermediate, and late assembly proteins [28]. Second, a study performed by our group on the yeast *Saccharomyces cerevisiae* characterized a collection of strains systematically deleted for one of the mitochondrial large 54S subunit (mtLSU) proteins or known assembly factors [29]. The *S. cerevisiae* mitoribosome is a 74S ribonucleoprotein particle whose mtSSU is formed by a 15S rRNA and 34 MRPs [30], and its mtLSU by a 21S rRNA and 46 MRPs [31]. The two rRNAs and one of the mtSSU proteins, Var1 (renamed as uS3m) are encoded in the mitochondrial genome (mtDNA), whereas all other proteins are encoded in the nuclear genome. In our study, the mtLSU subassemblies that accumulated in each deletion strain were obtained by sucrose gradient sedimentation fractionation, their composition deciphered by mass spectrometry, and their potential order in the assembly pathway established by clustering analysis. We are now applying this approach to the characterization of the yeast 37S small subunit (mtSSU) and will describe it in full detail in this chapter.

Studies of yeast mitoribosome assembly have been historically difficult due to the loss of mtDNA inherent to all strains defective in mitochondrial translation. The approach presented here takes advantage of the identification of multicopy suppressors of the mtDNA instability, which affords the possibility of generating mitoribosome deletion strains carrying mtDNA [32]. All the *S. cerevisiae* strains used were isogenic to the respiratory-robust wild-type strain W303I^0 carrying intronless mtDNA to distinguish mitoribosome assembly defects due to altered processing of the intron-containing 21S rRNA gene. To prevent the mtDNA loss that occurs in yeast mitoribosome assembly mutants [29], all strains are engineered to overexpress a previously identified suppressor of this phenotype (the ribonucleotide reductase catalytic subunit *RNR1*) [32, 33]. Furthermore, since the mtDNA-encoded mtSSU protein Var1 will not be synthesized when mitochondrial translation is impaired, which can be a confounding variable, all the strains were transformed with a construct that successfully relocates a recoded version of the *VAR1* gene (*VAR1*^{U}) to the nucleus as described [34].

The systematic analysis of mitoribosome assembly intermediates in deletion strains described in this chapter is limited by the potential heterogeneity of the subassemblies analyzed in each strain. This approach does not distinguish among "on pathway" assembled intermediates, "dead-end" aberrant subassemblies, and potential artifact subassemblies resulting from fragmentation of larger assemblies during experimental manipulation. The global survey of mitoribosome mutant yeast strains involved in this approach provides a strong model for mitoribosome subunit assembly that will serve as a framework for further pathway refinement studies in wild-type cells.

2 Materials

Prepare all solutions using analytical grade reagents and ultrapure water obtained by deionizing distilled water to a resistivity of 18 MΩ-cm at 25 °C. Store solutions at room temperature (RT, 25 °C) unless otherwise indicated.

2.1 Generation of a Collection of Yeast Strains KO for Genes Coding for Mitoribosome Proteins and Assembly Factors

2.1.1 Amplification of Gene-Targeted KanMX4 Deletion Cassettes

1. Agarose.
2. 1× TBE: 100 mM Tris–HCl, 100 mM boric acid, 2 mM ethylenediaminetetraacetic acid (EDTA) pH 8.0.
3. DNA extraction kit.

2.1.2 Yeast Transformation

1. Yeast strain a/α W303I^0 + *VAR1^U* + *RNR1* (genotype: ***diploid**, ade2-1 his3-11,15 leu2-3,112 trp1-1 ura3-1, URA3::pRS316-VAR1^U, LEU2::YEplac181-RNR1*, ρ$^+$ I^0) [29].
2. Synthetic growth media: 0.67% (w/v) yeast nitrogenous base without amino acids (WO), 2% (w/v) glucose. Add 2% (w/v) agar for solid media. Sterilize by autoclaving.
3. AHW – 50× auxotrophic markers: 2 mg/mL adenine, 2.5 mg/mL histidine, 2.5 mg/mL tryptophan. Dissolve 1 g of adenine, 1.25 g histidine, and 1.25 g tryptophan in 400 mL of water and bring up to a final volume of 500 mL (*see* **Note 1**). Sterilize by filtration through a 0.2 μm filter and store at RT in either a sterile light-sensitive amber bottle or wrap the bottle in aluminum foil as tryptophan is photosensitive.
4. TEL solution: 10 mM Tris–HCl pH 7.5, 1 mM EDTA, 100 mM lithium acetate. Sterilize by filtration through a 0.2 μm filter and store at RT.
5. 10 mg/mL sheared and denatured salmon sperm carrier DNA.
6. 40% PEG-TEL solution: 40% (w/v) polyethylene glycol (PEG 3500 or 4000), 10 mM Tris–HCl pH 7.5, 1 mM EDTA, 100 mM lithium acetate. Sterilize by filtration through a 0.45 μm filter and store at RT.
7. YPD liquid growth media: 1% (w/v) yeast extract, 2% (w/v) peptone, 2% (w/v) glucose. Sterilize by autoclaving.
8. 50 mg/mL Geneticin (G418). Dissolve in water, sterilize by filtration through a 0.2 μm syringe tip filter, and store at 4 °C.

2.1.3 Genomic DNA Extraction and Validation of KanMX4 Cassette Integration

1. Haploid yeast KO strain in the W303I^0 + *VAR1^U* + *RNR1* background. For example, strain αW303-I^0 Δ*uL1* + *VAR1^U* + *RNR1* (*genotype MAT*α, *ade2-1 his3-11,15 leu2-3,112 trp1-1 ura3-1, URA3::pRS316-VAR1, LEU2::YEplac181-RNR1*, Δ*uL1::KanMX*, ρ$^+$ I^0) [29].
2. Solution A: 50 mM Tris–HCl pH 7.5, 10 mM EDTA pH 8.0, 0.3% (v/v) β-mercaptoethanol, 0.5 mg/mL 100 T Zymolyase.
3. 10% (w/v) sodium dodecyl sulfate (SDS).
4. 8 M ammonium acetate.
5. Isopropanol.
6. 80% (v/v) ethanol.

2.2 Sucrose Gradient Analysis to Establish the Mitoribosome Profile and Detect Assembly Intermediates

2.2.1 Growth of Yeast Cells

1. WO-Gal (low-glucose) synthetic growth media: 0.67% (w/v) yeast nitrogenous base without amino acids, 2% (w/v) galactose, 0.5% (w/v) glucose. Dissolve 6.71 g of yeast nitrogenous base without amino acids, 20 g of galactose, and 5 g of glucose in 900 mL of water. Adjust pH to 5.8–6.0 with potassium hydroxide and bring up to a final volume of 1 L with water. For solid media, additionally, dissolve 20 g/L (2% w/v) of agar. Sterilize by autoclaving (*see* **Note 2**). For WO-Gal-AHW, add 20 mL of 50× AHW per 1 L of sterile WO-Gal to achieve 1× working strength.

2.2.2 Isolation of Mitochondria

1. Tris-DTT buffer: 100 mM Tris–HCl pH 8.8, 10 mM dithiothreitol (DTT).
2. 1.2 M Sorbitol: Dissolve 874.416 g of sorbitol in 3 L of water. Bring to a final volume of 4 L. To remove traces of metals and other ionic contaminants, deionize sorbitol overnight at RT with AG 501-X8 resin from Bio-Rad (Hercules, CA). Filter through a 0.2 μm filter and store at 4 °C.
3. Digestion buffer: 1.2 M sorbitol, 20 mM potassium phosphate pH 7.4 (*see* **Note 3**), 0.25 mg/mL 100 T zymolyase.
4. Homogenization buffer: 10 mM Tris–HCl pH 7.5, 1 mM EDTA pH 8.0, 0.2% (w/v) bovine serum albumin (BSA), 1 mM phenylmethylsulfonyl fluoride (PMSF), 0.6 M sorbitol. Prepare fresh before use.
5. SH buffer: 0.6 M sorbitol, 20 mM HEPES pH 7.4. Sterilize by filtration through a 0.2 μm filter and store at 4 °C.
6. Floor centrifuge capable of reaching 14,000 × *g* at 4 °C.
7. 50 mL glass dounce with a loose Teflon pestle.
8. 10 mL glass dounce with a tight Teflon pestle.

2.2.3 Preparation of a 10–30% Linear Sucrose Gradient

1. 10% sucrose gradient buffer: 10 mM Tris pH 7.4, 100 mM NH_4Cl, 10 mM $MgCl_2$, 1 mM PMSF, 0.05% (w/v) digitonin (*see* **Note 4**), 1× yeast protease inhibitor cocktail, 0.1 U/μL RNase inhibitor, 10% (w/v) sucrose (*see* **Note 5**).
2. 30% sucrose gradient buffer: 10 mM Tris pH 7.4, 100 mM NH_4Cl, 10 mM $MgCl_2$, 1 mM PMSF, 0.05% (w/v) digitonin, 1× yeast protease inhibitor cocktail, 0.1 U/μL RNase inhibitor, 30% (w/v) sucrose.
3. Polypropylene ultracentrifuge tubes (13 × 51 mm).
4. Biocomp Gradient Master with marker block, layering cannula, MagnaBase tube holder, 4 mm short tube caps and leveling tool.

Table 1
Recipe to prepare solution for extraction of mitoribosomes

Extraction buffer			
Reagent	**[Stock]**	**[Final]**	**1× Volume (500 μL)**
Tris pH 7.4	1 M	10 mM	5 μL
NH_4Cl	1 M	100 mM	50 μL
$MgCl_2$	1 M	10 mM	5 μL
PMSF	100 mM	1 mM	5 μL
DTT	100 mM	0.25 mM	1.25 μL
Digitonin (high purity)	10%	0.6%	30 μL
Yeast protease inhibitor cocktail	100×	1×	5 μL
RNase inhibitor	40 U/μL	0.1 U/μL	1.25 μL
H_2O	–	–	397.5 μL

This solution can be made at the same time as preparing the sucrose gradient buffers. Ensure the digitonin used is high purity and is prepared the same day as the experiment

2.2.4 Mitoribosome Extraction and Sedimentation Analysis

1. Extraction buffer: 10 mM Tris–HCl pH 7.4, 100 mM NH_4Cl, 10 mM $MgCl_2$, 1 mM PMSF, 0.25 mM DTT, 0.6% (w/v) high-purity digitonin (*see* **Note 4**), 1× yeast protease inhibitor cocktail, 0.1 U/μL RNase inhibitor (Table 1).
2. Ultracentrifuge capable of reaching 200,000 × *g*.
3. Swinging-bucket rotor for ultracentrifuge (i.e., SW55ti).
4. 4× LB (4× Laemmli sample buffer): 200 mM Tris–HCl pH 6.8, 4% (w/v) SDS, 40% (v/v) glycerol, 4% (v/v) β-mercaptoethanol, 0.05% (w/v) bromophenol blue.

2.2.5 SDS-PAGE and Immunoblot Analysis of Sedimented Mitoribosomal Particles

1. 12% SDS-PAGE polyacrylamide gel: Gels are made in-house from a 12% resolving gel (12%/0.32% (w/v) acrylamide/bis-acrylamide, 375 mM Tris–HCl pH 8.8, 0.1% (w/v) SDS, 0.03% (w/v) ammonium persulfate (APS), 0.08% (w/v) N,N, N′,N′-Tetramethyl ethylenediamine (TEMED)) and a 6% stacking gel (6%/0.16% (w/v) acrylamide/bis-acrylamide, 125 mM Tris–HCl pH 6.8, 0.1% (w/v) SDS, 0.05% (w/v) APS, 0.15% (w/v) TEMED). The recipe for preparing one gel can be found in Tables 2 and 3, scale up as necessary.
2. Pre-stained protein ladder (available commercially).
3. Ponceau protein staining solution: 0.2% (w/v) Ponceau S in 3% (w/v) Trichloroacetic acid (TCA).
4. Nitrocellulose membrane (0.45 μm).

Table 2
Recipe for 12% acrylamide resolving gel

Resolving gel – 12% acrylamide			
Reagent	**[Stock]**	**[Final]**	**1× Volume (1 gel)**
H_2O	–	–	10.35 mL
Acrylamide/Bis-acrylamide (37.5:1)	30%/0.8%	12%/0.32%	9 mL
Tris–HCl pH 8.8	3 M	375 mM	2.8125 mL
SDS	10%	0.1%	225 μL
APS	10%	0.03%	75 μL
TEMED	~99%	0.08%	18.75 μL

Table 3
Recipe for 6% acrylamide stacking gel

Stacking gel – 6% acrylamide			
Reagent	**[Stock]**	**[Final]**	**1× Volume (1 gel)**
H_2O	–	–	5.28 mL
Acrylamide/Bis-acrylamide (37.5:1)	30%/0.8%	6%/0.16%	1.6 mL
Tris–HCl pH 6.8	1 M	125 mM	1 mL
SDS	10%	0.1%	80 μL
APS	10%	0.05%	40 μL
TEMED	~99%	0.15%	12 μL

5. Wash buffer: 10 mM Tris–HCl pH 8.0, 1 mM EDTA pH 8.0, 150 mM NaCl, 0.1% (v/v) Triton X-100.
6. Blocking buffer: 5% (w/v) nonfat dry milk in wash buffer.
7. Primary antibodies against mitoribosome small and large subunit proteins and relevant assembly factors (a list can be found in Ref. [29]).
8. Horseradish-peroxidase (HRP) conjugated secondary antibodies specific for the host species of primary antibodies.
9. Enhanced chemiluminescence (ECL) solution: ECL solution is made in-house by combining equal volumes of solution 1 (100 mM Tris–HCl pH 8.5, 2.5 mM luminol, 400 μM p-coumaric acid) and solution 2 (100 mM Tris–HCl pH 8.5, 0.036% (v/v) H_2O_2) (*see* **Note 6**). Store solutions 1 and 2 protected from light and at 4 °C, where they are stable for

Table 4
Recipe to prepare chemiluminescence solution 1

ECL solution 1			
Reagent	**[Stock]**	**[Final]**	**1× Volume (50 mL)**
H_2O	–	–	44.28 mL
Tris–HCl pH 8.5	1 M	100 mM	5 mL
Luminol	250 mM	2.5 mM	500 µL
p-Coumaric acid	90 mM	400 µM	220 µL

Table 5
Recipe to prepare chemiluminescence solution 2

ECL solution 2			
Reagent	**[Stock]**	**[Final]**	**1× Volume (50 mL)**
H_2O	–	–	44.94 mL
Tris–HCl pH 8.5	1 M	100 mM	5 mL
H_2O_2	30%	0.036%	60 µL

1–2 months. The recipe for preparing these solutions can be found in Tables 4 and 5.

2.3 Mass Spectrometry Analysis of Assembly Intermediates Compositions and Clustering Analysis

2.3.1 Methanol/Chloroform Protein Precipitation of Sucrose Gradient Fractions Containing Assembly Intermediates and Mitoribosomal Subunits

1. Methanol.
2. Chloroform.
3. High-speed vacuum concentrator.

2.3.2 Clustering Analysis of Mass Spectrometry Data to Assign Hierarchical Assembly Clusters

1. RStudio software.
2. CSV file of proteomics results.

3 Methods

3.1 Generation of a Collection of Yeast Strains KO for Genes Coding for Mitoribosome Proteins and Assembly Factors

To generate a library of yeast mitoribosome knockout (KO) strains, we obtained a collection of deletion strains in which genes were replaced with a KanMX4 cassette conferring resistance to Geneticin. The strains are in a BY4741/4742 background and are available as part of the yeast knockout collection from Horizon Discovery Biosciences (Cambridge, United Kingdom). For this method, the KanMX4 cassette in each strain flanked by ~500 bp homologous to the MRP to be deleted was transformed in a diploid a/α W303I^0 + *VAR1*^{U} + *RNR1* [29]. Following sporulation and tetrads dissection, haploid deletion strains were obtained (*see* Fig. 1a) (*see* **Note 7**). Some strains, however, are not available as part of the collection and are created by replacing the entire ORF with a KanMX4 cassette or a *HIS5* cassette.

3.1.1 Amplification of Gene-Targeted KanMX4 Deletion Cassettes

1. Design primers to amplify the KanMX4 cassette with an additional ~500 bp flanking the resistance marker that is homologous to the 5′- and 3′-UTRs of the corresponding mitoribosomal protein or assembly factor (*see* Fig. 1a). If the desired yeast null mutant is unavailable from the deletion strains in the Yeast Knockout Collection from Horizon Discovery Biosciences, *see* **Note 8**.
2. Extract genomic DNA following **steps 3–13**.
3. Culture yeast KO strain overnight in 10 mL of WO liquid media with 1× AHW.
4. Pellet 1 mL of confluent cells by centrifuging at 1500 × *g* for 5 min at RT.
5. Wash cells by resuspending the pellet in 500 μL of sterile water and centrifuge again at 1500 × *g* for 5 min at RT.
6. Resuspend cells in 150 μL of solution A and incubate at 37 °C for 1 h without shaking to digest the cell wall.
7. Add 20 μL of 10% (w/v) SDS and vortex to mix.
8. Add 100 μL of 8 M ammonium acetate and vortex to mix.
9. Incubate at −20 °C for 15 min.
10. Centrifuge at 10,000 × *g* for 10 min at 4 °C.
11. Transfer 180 μL to a new microcentrifuge tube and add 120 μL of isopropanol. Vortex to mix and incubate for 5 min at RT to precipitate DNA.
12. Pellet the DNA by centrifuging at 10,000 × *g* for 15 min at 4 °C. Dispose of the supernatant and wash the pellet with 300 μL of 80% ethanol.
13. Centrifuge at 10,000 × *g* for 1 min at 4 °C. Dispose of supernatant and air-dry the pellet at RT. Resuspend the pellet in 30 μL of water.

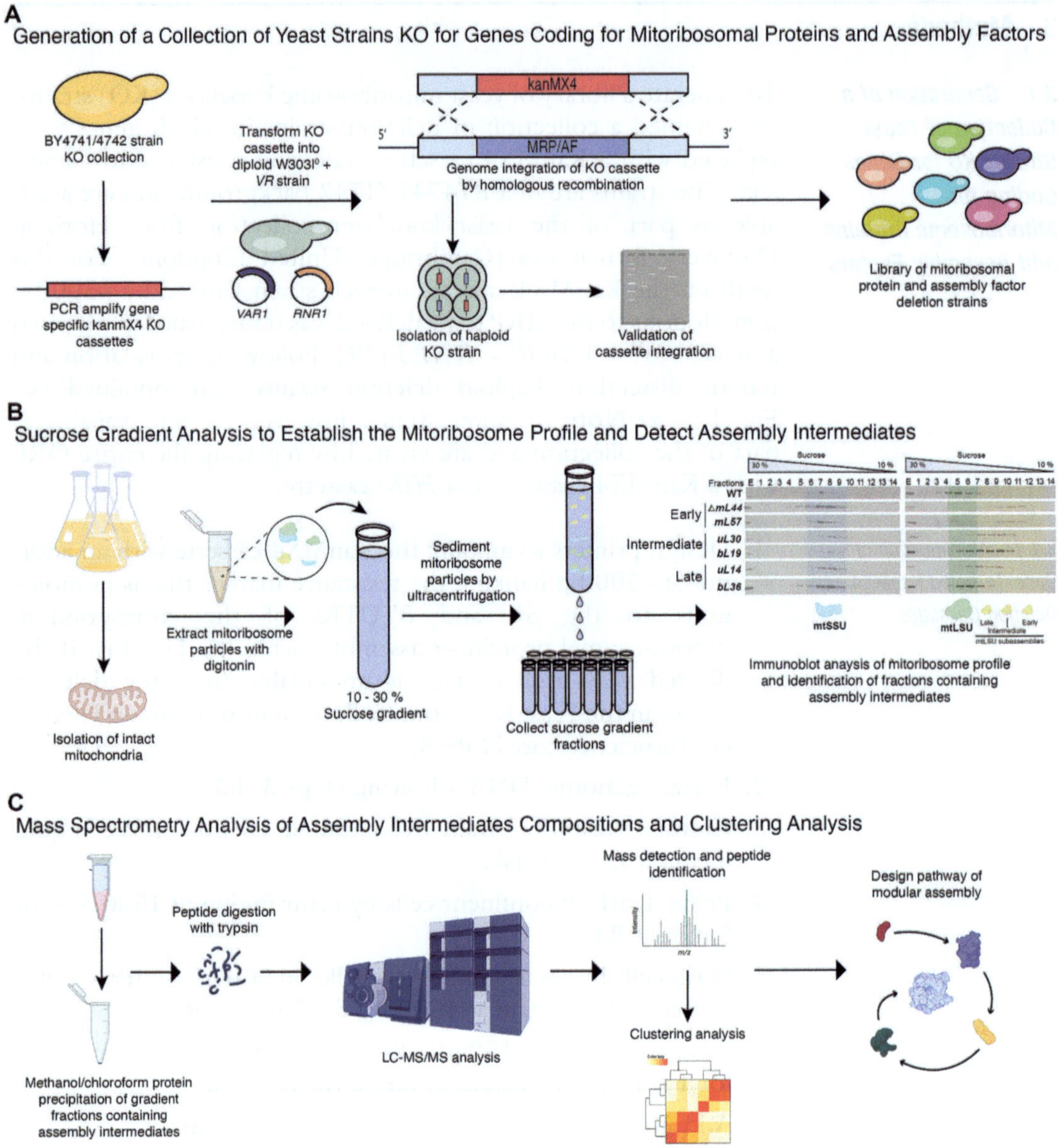

Fig. 1 Graphical outline of methods. Schematic detailing the procedures for (**a**) generating a library of yeast deletion strains, (**b**) separating mitoribosome particles by sucrose gradient sedimentation. (Figure modified from Ref. [29]) and (**c**) identifying assembly intermediates by tandem mass spectrometry. (The figure was prepared with Adobe Illustrator and BioRender)

14. Using genomic DNA extracted from the yeast KO strain of interest, amplify by PCR the gene-specific KO cassette. Isolate the correct product by separation on a 1% (w/v) agarose gel in 1× TBE followed by DNA extraction with a commercially available kit.
15. Use the purified KO cassette for transformation into the diploid a/α W303I^0 + *VAR1*^{U} + *RNR1* strain.

3.1.2 Yeast Transformation

1. Inoculate a starter culture with the a/α W303I^0 + *VAR1^U* + *RNR1* strain in 10 mL of WO liquid media supplemented with the required auxotrophic markers (AHW) at a 1× working concentration and grow overnight at 30 °C, shaking at 200 × *g*.
2. Dilute starter culture in 10 mL of fresh WO liquid media supplemented with 1× AHW solution to an optical density at 600 nm (OD_{600}) = 0.1.
3. Continue to grow cells for 4–6 h until the culture reaches an OD_{600} = 0.6–1.0.
4. Collect 1 mL of culture in a sterile microcentrifuge tube and pellet cells by centrifuging at 1500 × *g* for 5 min at RT.
5. Wash cells by resuspending the cell pellet in 1 mL of sterile TEL and centrifuging again at 1500 × *g* for 5 min at RT to pellet the cells.
6. Resuspend cells in 100 μL of sterile TEL. In a separate sterile tube, premix 1–10 μg of the purified KO cassette from **step 2** in Subheading 3.1.1 with 10 μL of 10 mg/mL denatured salmon sperm carrier DNA (*see* **Note 9**). Add mixture to the cells and incubate for 30 min at RT without shaking.
7. Add 700 μL of sterile PEG-TEL and gently mix by pipetting. Incubate for 30 min at RT without shaking.
8. Heat shock at 42 °C for 10–15 min and transfer cells to ice (4 ° C) for 2 min.
9. Centrifuge at 1500 × *g* for 5 min at RT to pellet cells and resuspend in 1 mL of YPD liquid media. Outgrow cells for 30–60 min at 30 °C to allow recovery and increase transformation efficiency.
10. Pellet cells by centrifuging at 1500 × *g* for 5 min at RT. Wash cell pellet by resuspending in 1 mL of sterile water and centrifuge again at 1500 × *g* for 5 min at RT.
11. Resuspend the cell pellet in 100 μL of sterile water and spread on a WO solid media plate containing the selective resistance marker geneticin at final 500 μg/mL and the auxotrophic requirements (AHW) (*see* **Note 10**).
12. Incubate plates at 30 °C; colonies should appear in 4–5 days. Pick colonies and test them by genotyping.

3.1.3 Genomic DNA Extraction and Validation of KanMX4 Cassette Integration

1. Extract genomic DNA as explained in Subheading 3.1.1 (**steps 2–13**).
2. Validate correct *KanMX4* cassette integration by PCR using primers that amplify regions spanning the KanMX4 cassette and the 5′- and 3′-UTRs to ensure deletion in the correct genetic locus (*see* Fig. 1a).
3. Assess that the deletions strains retain mtDNA (*see* **Note 7**).

3.2 Sucrose Gradient Analysis to Establish the Mitoribosome Profile and Detect Assembly Intermediates

The overall approach is depicted in Fig. 1b.

3.2.1 Culture of Yeast Cells

S. cerevisiae can utilize a variety of different carbon sources to transduce energy and fuel growth. Fermentable carbon sources, such as glucose and galactose, drive energy production through glycolysis-based metabolism. Alternatively, cells grown with non-fermentable carbon sources like glycerol and ethanol adapt to an aerobic-based metabolism, generating ATP through cellular respiration. Yeast cells are highly sensitive to changes in metabolism and regulate mitochondria quantity, morphology, and proteome depending on the available carbon source [35–37]. When available, yeast preferentially utilize glucose to generate energy via glycolysis, regardless of the available oxygen [35, 38]. Consequently, under high glucose conditions, the expression of genes encoding enzymes of the TCA cycle and many respiratory complex proteins is down-regulated, resulting in impaired OXPHOS complex assembly and activity [39–41]. This phenomenon is known as "glucose repression" and inhibits the use of other carbon sources and respiration-based metabolism via signaling, transcriptional, and posttranslational regulation [35, 38, 41]. Many studies have proposed that mitoribosome activity is tightly regulated to coordinate the synthesis of mtDNA-encoded proteins in response to OXPHOS complex assembly demands [42–44]. Interestingly, a recent study suggests that mitochondrial translation capacity is maintained under high glucose conditions and may be uncoupled from the repressive effects on OXPHOS assembly and activity [45]. Galactose however is a non-repressive fermentable carbon source. For this reason, cells are grown in WO-Gal to isolate mitochondria in this protocol (*see* **Note 11**).

1. Yeast strains should be spread on WO-Gal-AHW solid media agar plates from a rho+ glycerol stock stored at −80 °C. Incubate plate at 30 °C and use within 2–3 days to inoculate cultures.
2. For each yeast strain, inoculate four separate flasks at a starting OD_{600} = 0.01 in 100 mL of WO-Gal-AHW medium (4× 100 mL). Grow starter cultures for 24–48 h at 30 °C with shaking, taking periodic OD_{600} measurements to estimate doubling time (*see* **Note 12**).
3. Per flask, dilute starter culture into 1 L of WO-Gal-AHW (4× 1 L per strain) and grow at 30 °C with shaking until OD_{600} = 1.5–2.0 (*see* **Note 13**). Be careful not to exceed 20 generations of growth (*see* **Note 14**). Proceed directly to mitochondrial isolation.

3.2.2 Isolation of Mitochondria

Isolation of mitochondria as described in this protocol yields a highly enriched mitochondrial preparation with intact-membrane organelles. A high-quality mitochondrial isolate is essential for meaningful downstream proteomics analysis. This method is slightly modified from Horn et al. [46] to increase the removal of common non-mitochondrial co-contaminants such as the ER and Golgi. In brief, this protocol is based on the homogenization of spheroplasts followed by the isolation of mitochondria by differential centrifugation. Mitochondria prepared using this technique have intact membranes and can be used for a variety of downstream *in organello* analyses such as oxygen rate consumption, protein synthesis and transport, and transcription.

1. Harvest cells in polypropylene centrifuge bottles by centrifuging at 900 × *g* for 5 min. Wash cells once with sterile RT water.
2. Resuspend cells in Tris-DTT buffer at a ratio of 1 mL per 2 g of cells and incubate at 30 °C for 10 min with gentle shaking (*see* **Notes 15** and **16**).
3. Pellet the cells by centrifugation at 2200 × *g* for 6 min and wash cells once with 1.2 M sorbitol using 150 mL per 10 g of cells (*see* **Note 17**).
4. Resuspend cells in digestion buffer using 1 mL for every 0.15 g of cells (*see* **Note 18**). Incubate at 30 °C for 30 min with gentle shaking. At the end of the incubation period immediately place bottles containing spheroplasts on ice to stop the digestion reaction (*see* **Note 19**). All subsequent steps must be done at 4 °C.
5. Pellet spheroplasts by centrifugation at 1250 × *g* for 5 min at 4 ° C. Wash once with ice-cold 1.2 M sorbitol.
6. Resuspend washed spheroplasts with ice-cold homogenization buffer using 1 mL for every 0.15 g of cells (*see* **Note 20**). Transfer to a prechilled 50 mL glass dounce tissue homogenizer fitted to a loose Teflon pestle and homogenize with 6–8 strokes (*see* **Note 21**).
7. Centrifuge homogenate at 1700 × *g* for 6 min at 4 °C. Repeat this step on the supernatant in a new centrifuge bottle.
8. Centrifuge the supernatant at 14,000 × *g* for 6 min at 4 °C to pellet mitochondria.
9. Resuspend pelleted mitochondria in 10 mL of SH buffer (*see* **Note 22**). Centrifuge at 2200 × *g* for 5 min at 4 °C.
10. Centrifuge the supernatant containing mitochondria at 14,000 × *g* for 6 min at 4 °C. Wash mitochondrial pellet with 10 mL of ice-cold SH buffer to remove additional contaminants (*see* **Note 23**). Resuspend mitochondria in 200–600 μL of cold SH buffer, depending on pellet size (*see* **Note 24**).

11. Determine protein concentration using the desired method and adjust to 10 mg/mL. Prepare 2–4 mg aliquots that will be used for mitoribosome extraction and sucrose gradient sedimentation in the subsequent steps. Use the remainder of the mitochondria to prepare samples for immunoblot analysis (*see* **Note 25**). Flash freeze samples in liquid nitrogen and store at −80 °C until use (*see* **Note 26**).

3.2.3 Preparation of a 10–30% Linear Sucrose Gradient

Density gradients are a valuable separation technique that can be used to isolate particles in a mixture of biomolecules or further purify organelles. The approach described here uses rate zonal separation in which particles sediment based on mass and size (not only density) and all particles will eventually accumulate at the bottom of the tube if centrifuged long enough. We describe here the preparation of a continuous 10–30% sucrose linear gradient prepared by using a titled tube rotation approach, the Biocomp Instruments Gradient Master. These gradients have been shown to successfully separate mitoribosome subpopulations representing assembly intermediates in addition to fully assembled subunits and monosomes from a mitochondrial lysate [29, 47].

1. Prepare both 10% and 30% sucrose gradient buffers according to the recipe in Table 6 (*see* **Notes 27** and **28**). Place solutions on ice to cool.
2. Begin preparing the gradient by placing the 10 mL ultracentrifuge tube in the marker block. Mark the upper edge with a fine

Table 6
Recipe for 10% and 30% sucrose solutions to prepare sucrose gradients

Gradient buffer Reagent	[Stock]	[Final]	10% sucrose 2× Volume (5 mL)	30% sucrose 2× Volume (5 mL)
Tris pH 7.4	1 M	10 mM	50 μL	50 μL
NH_4Cl	1 M	100 mM	500 μL	500 μL
$MgCl_2$	1 M	10 mM	50 μL	50 μL
PMSF	100 mM	1 mM	50 μL	50 μL
Digitonin (high purity)	10%	0.05%	25 μL	25 μL
Yeast protease inhibitor cocktail	100×	1×	50 μL	50 μL
RNase inhibitor	40 U/μL	0.1 U/μL	12.5 μL	12.5 μL
Sucrose	50%	10%/30%	1000 μL	3000 μL
H_2O	–	–	3262.5 μL	1262.5 μL

The recipe indicates the volumes required for two sample. Scale up as necessary tip marker (*see* **Note 29**). This will indicate the "stopping point" when filling with sucrose gradient solution. Place marked ultracentrifuge tube in the MagnaBase tube holder.

3. Fill the ultracentrifuge tube with the 10% sucrose solution until the volume is slightly above the marked stopping point (~2 mm) (*see* **Note 30**).
4. Using the provided layering cannula attached to a syringe, insert the cannula tip to the bottom of the tube and slowly fill with the 30% sucrose solution. As the solution is added from the bottom of the tube, the lighter 10% sucrose layer will be pushed toward the top of the tube. A clear interface between the two layers should be visible. Fill solution until just above the marked stopping point. Once added, swiftly and carefully remove the cannula.
5. Place 4 mm tube cap on ultracentrifuge tube (*see* **Note 31**).
6. Use a leveling tool to ensure that the magnetic plate on the gradient master is perfectly level.
7. Carefully place the tube holder with sucrose gradients onto the center of the gradient master plate (*see* **Note 32**). Ensure that tubes are balanced.
8. Use gradient master to make a 10–30% linear sucrose gradient: Using the gradient master manual interface, select 'Grad' → 'List' → 'SW55' → '10–30%' → 'Run' (*see* **Note 33**).
9. Gently remove gradients from the tube holder and keep them at 4 °C until used in the next step (*see* **Note 34**).

3.2.4 Mitoribosome Extraction and Sedimentation Analysis

To ensure mitoribosomes and any possible assembly complexes remain intact, mitoribosomes are extracted with the mild nonionic detergent digitonin. A crucial aspect in maintaining the association between the small and large subunits in the mitoribosome is the ratio of monovalent to divalent cations in both the extraction and sucrose gradient buffers [43, 48]. A 10:1 ratio of monovalent to divalent ions using 100 mM NH_4Cl and 10 mM $MgCl_2$ is described here. Additionally, the use of an RNase inhibitor to prevent the degradation of rRNA and the bound mRNA may help to stabilize the monosome. Take care to avoid the use of EDTA and other ion chelating agents as this will disrupt the interaction between subunits. An example of the dissociation caused by EDTA can be observed in Fig. 1b. Sample fractionation in this protocol is performed with a straight needle; however, a side hole needle or piston fractionator fitted with a trumpet needle will provide increased resolution.

1. Remove 2–4 mg aliquot of isolated mitochondria from −80 °C and thaw on ice (*see* **Note 35**). Use immediately upon thawing.

2. Centrifuge at 10,000 × *g* for 10 min at 4 °C to pellet mitochondria.
3. Remove the supernatant and resuspend the mitochondrial pellet in 400 μL of ice-cold freshly prepared extraction buffer. Incubate on ice for 10 min.
4. Centrifuge at 24,000 × *g* for 15 min at 4 °C. Collect supernatant containing mitoribosomal and soluble mitochondrial proteins as mitochondrial extract (*see* **Note 36**).
5. Carefully pipette the remaining supernatant (~360 μL) onto the top of the prepared 10–30% sucrose gradient (*see* **Note 37**). Add slowly to ensure that the sample does not mix with the upper sucrose layer.
6. Gently place sucrose gradient into the swinging-bucket rotor. Cap tightly and ensure that the rotor is balanced. Ultracentrifuge at 200,000 × *g* for 3 h: 10 min at 4 °C (*see* **Note 38**). Once finished, very carefully remove the gradient from the rotor. Avoid any disturbances at this point as this will disrupt the sedimentation profile of protein complexes in the gradient. Keep gradient cold until proceeding to fractionation.
7. Fractionate sucrose gradient by piercing the bottom of the ultracentrifuge tube with a standard straight beveled needle. Collect 14 equal volume fractions dropwise in microcentrifuge tubes (*see* **Note 39**). Immediately place all fractions on ice.
8. Immediately prepare fraction samples for immunoblotting by combining 75 μL of each fraction with 25 μL of 4× LB (*see* **Note 40**). Store the remainder of each fraction at −80 °C until used for protein precipitation and subsequent mass spectrometry analysis.

3.2.5 SDS-PAGE and Immunoblot Analysis of Sedimented Mitoribosomal Particles

To identify assembly intermediates for downstream proteomics, the mitoribosome sedimentation profile and distribution of subassembly particles can be analyzed by immunoblot. Comparison of assembly intermediates with the fully assembled subunit in the WT may suggest if the protein deleted acts in an early or late stage of the assembly process. For example, if an early assembly protein is deleted, intermediates will accumulate in the lightest fractions of the gradient and be significantly smaller than the WT particle. Alternatively, assembly intermediates in late assembly mutants will be larger and have a sedimentation profile like WT. Figure 1b shows the distribution of subassemblies that accumulate in the early, intermediate, and late stages of mtLSU assembly. Use as many different tested antibodies possible to ensure visualization of intermediates by immunoblot. Alternatively, although not described here, gradient fractions could be screened by northern blotting or RT-qPCR for the presence of mitochondrial rRNA. Fractions identified to harbor intermediates will be further analyzed by mass spectrometry.

1. In a 12% SDS-PAGE gel made with a 16-lane comb, load from left to right a pre-stained protein ladder, 10 μL of the "total extract" from **step 4** in Subheading 3.2.4, and 30 μL each of the fractions 1 through 14 from **step 8** in Subheading 3.2.4.
2. Begin electrophoresis at 100 V and allow samples to migrate through the stacking portion of the gel (~30 min) (*see* **Note 41**). Once the Laemmli sample buffer dye front enters the resolving gel, increase the voltage to 140 V and run until the dye front reaches the bottom of the gel (~4 h).
3. Transfer proteins to a nitrocellulose membrane by semidry transfer at 250 mA for 5 h. Alternatively, a wet transfer can be used.
4. After transfer, wash the membrane with water for ~15 min. To check for protein degradation and assess the distribution of protein across fractions, incubate the membrane with ponceau staining solution with mild shaking for ~15 min (*see* **Note 42**).
5. Using the protein ladder as a reference guide, cut the membrane in strips such that each piece corresponds with the molecular weight of the mtSSU or mtLSU protein to be probed against by the predetermined antibody.
6. Rinse membranes in wash buffer with mild shaking for 5–10 min or until the ponceau stain has been washed out.
7. Block membranes by incubating in blocking buffer for 30 min with mild shaking. Alternatively, membranes can be blocked overnight at 4 °C.
8. Incubate membranes with respective primary antibodies for 1 h at RT with mild shaking (*see* **Notes 43–45**).
9. Rinse membranes by incubating in wash buffer with mild shaking for 10 min. Perform three washes in total.
10. Incubate membranes in secondary antibody for 1 h at RT with mild shaking (*see* **Note 46**).
11. Rinse membranes again by performing three washes in wash buffer with mild shaking for 10 min each.
12. Develop membranes by incubating them in ECL solution and subsequently exposing them to autoradiography film. Identify fractions corresponding to any assembly intermediates in addition to fractions representing the mtSSU, mtLSU, and monosome. These fractions will be used for protein precipitation and quantitative mass spectrometry.

3.3 Mass Spectrometry Analysis of Assembly Intermediates Compositions and Clustering Analysis

The overall approach is depicted in Fig. 1c.

3.3.1 Methanol/Chloroform Protein Precipitation of Sucrose Gradient Fractions Containing Assembly Intermediates and Mitoribosomal Subunits

An essential step in ensuring accurate identification of assembly intermediates and collection of high-quality data is to prepare samples for mass spectrometry analysis by precipitating protein from sucrose gradient fractions. Methanol/chloroform extraction allows for the elimination of lipids, detergents, nucleic acids, salts, and other contaminants that may otherwise interfere with ionization or data interpretation (*see* Fig. 1c).

1. Per 100 μL of sample, add 400 μL of methanol and vortex well to mix (*see* **Note 47**).
2. Add 100 μL of chloroform and vortex well to mix (*see* **Note 48**).
3. Add 300 μL of water and vortex well to mix (*see* **Note 49**). The sample should become cloudy as phase separation begins to occur.
4. Centrifuge at 14,000 × *g* for 2 min.
5. Carefully remove the aqueous (top) phase with a pipette and discard. Proteins reside at the thin interphase between the organic and aqueous layers and are easily disturbed.
6. Add 400 μL of methanol and vortex well to mix (*see* **Note 50**).
7. Centrifuge at 14,000 × *g* for 3 min.
8. Remove as much solvent as possible by pipetting without disturbing the protein pellet.
9. Dry the pellet using a high-speed vacuum concentrator (*see* **Note 51**).
10. Process samples by LC-MS/MS (liquid chromatography-tandem mass spectrometry) analysis.

3.3.2 Quantitative Mass Spectrometry Analysis of Assembly Intermediates and Data Clustering Analysis to Determine Hierarchical Mitoribosome Protein Assembly

Analyzing samples by nano LC-MS/MS allows for highly sensitive quantification and accurate detection of proteins in a mixture. This method enables proper identification and classification of assembly intermediates as well as any previously unknown assembly factors, which may help in defining a de novo assembly pathway. Liquid chromatography (LC) coupled to mass spectrometry (MS) allows for the physical separation of components in a mixture prior to mass analysis. Tandem mass spectrometry (MS/MS) involves using two mass analyzers and further fragmentation and separation of precursor ions to provide more sensitive detection of proteins and other biomolecules. In short, the process of LC-MS/MS begins with the enzymatic cleavage of proteins into small peptides, typically by trypsin digestion. Peptides are then separated by liquid chromatography and eluted based on their affinity for the stationary phase and the solvent used. Following this, peptides are passed to the mass spectrometer and ionized by electrospray or another equivalent ionization method. The mass analyzer then separates the precursor ions based on their mass-to-charge ratio (m/z). These "parent" ions are then further fragmented by collision-induced dissociation (CID). The ion fragments are separated by the second mass analyzer before finally being measured by the detector. The detector will produce a mass spectrum for each ion. An increasingly popular and cost-efficient method for MS-based quantitation and peptide identification is the use of isobaric stable isotope labeling. For example, tandem mass tag (TMT) mass spectrometry yields robust and reproducible protein quantitation while enabling sample multiplexing up to 16 plex. Using data analysis platforms such as Proteome Discoverer or MaxQuant, the fragmentation spectra are referenced against a protein sequence database by search algorithms including MASCOT, SEQUEST, X!TANDEM, Andromeda, or any other commonly used database search engine. All bioinformatics and statistical analysis should be done using the TMT reporter intensity. However, in this chapter, we present the use of values from the MASCOT algorithm and ion search engine. MASCOT is a powerful tool that provides probabilistic scoring and integrates peptide mass fingerprints, protein sequence queries, and tandem MS ion searches. Based on this, a protein score or "MASCOT score" is distilled in which the value represents the combined score of each ion fragment spectra matching peptide sequences within the same protein. This value, however, is a probability, and for increased sensitivity, TMT reporter intensity should be used in place of MASCOT scores. LC-MS/MS analysis should be performed in at least biological triplicate. Differential regulation across all groups is tested by ANOVA, and Tukey's HSD test can be used to make comparisons between pairs. MASCOT scores can be used for downstream hierarchical clustering analysis to determine assembly modules as described below.

1. In preparation for clustering analysis and to make comparisons between strains, MASCOT scores for each protein should be normalized to WT and $\log_2$ transformed $\left(\text{Log}_2 \frac{\text{mutant}_{\text{MASCOT}}}{\text{WT}_{\text{MASCOT}}}\right)$. Log transformation reduces heteroscedasticity and results in a more normal distribution of the data which in turn increases the strength of statistical tests.
2. To facilitate cluster analysis, arrange data in a CSV file such that rows represent observations (mutant strains) and columns indicate variables (MASCOT score for each protein).
3. Import the dataset into RStudio and load the packages "gplots" and "RColorBrewer" (*see* **Note 52**).

4. Use the heatmap.2 function to generate a heatmap of the dataset. The function will automatically perform hierarchical clustering for both rows and columns and reorder the corresponding dendrograms based on the mean. This will yield vertical clusters of proteins that maintain similar stabilities across mutant strains. Vertical clustering of these stability profiles yields distinct mitoribosome structural modules (*see* Fig. 2). One such example from the yeast mtLSU is the central protuberance, in which proteins composing this structural module are detected at similar levels across all mutant strains (*see* Fig. 2). Horizontal clustering generates assembly clusters representing mutant strains with similar mitoribosome protein subassembly profiles (*see* Fig. 2). The similarity of the mitoribosome protein composition for mutant strains in each assembly cluster indicates that the deleted proteins act at the same or similar stage of assembly.
5. Assembly clusters can then be classified as either early, intermediate, or late incorporation modules based on the relative stability and detection of all other mitoribosomal proteins. For example, mutant strains in the earliest assembled cluster will have subassemblies in which most proteins are decreased or undetectable, as the early assembly proteins serve as a scaffold for subsequent clusters. On the contrary, mutant strains of a late assembly cluster will have subassemblies in which most proteins are present at WT levels. In the case of the yeast mtLSU, the first assembled cluster, depicted in red, correlates with the most underrepresentation of mitoribosomal proteins and has extensive contact with the ribosomal RNA (*see* Fig. 2), whereas subassemblies of proteins in the last assembled cluster, shown in orange, are characterized by either WT levels or an increase in most mitoribosomal proteins (*see* Fig. 2). Constituents of late assembly clusters typically map to the surface of the mitoribosome (*see* Fig. 2).
6. Additional insight from phenotypic experiments such as ribosomal RNA levels, cellular respiration, and protein synthesis capacity combined with structural comparison to bacterial counterparts allows for the construction of a more granular pathway and a hierarchical ranking of clusters.

4 Notes

1. The final concentrations in 500 mL are: 2 mg/mL adenine, 2.5 mg/mL histidine, and 2.5 mg/mL tryptophan. Pure adenine is poorly soluble in water due to its purine ring structure. To overcome this, first bring water to a boil before dissolving adenine. The compound should dissolve easily and be stable at RT for 1–2 months. Once slightly cooled, dissolve the histidine and tryptophan and bring up to 500 mL with water before filtering.

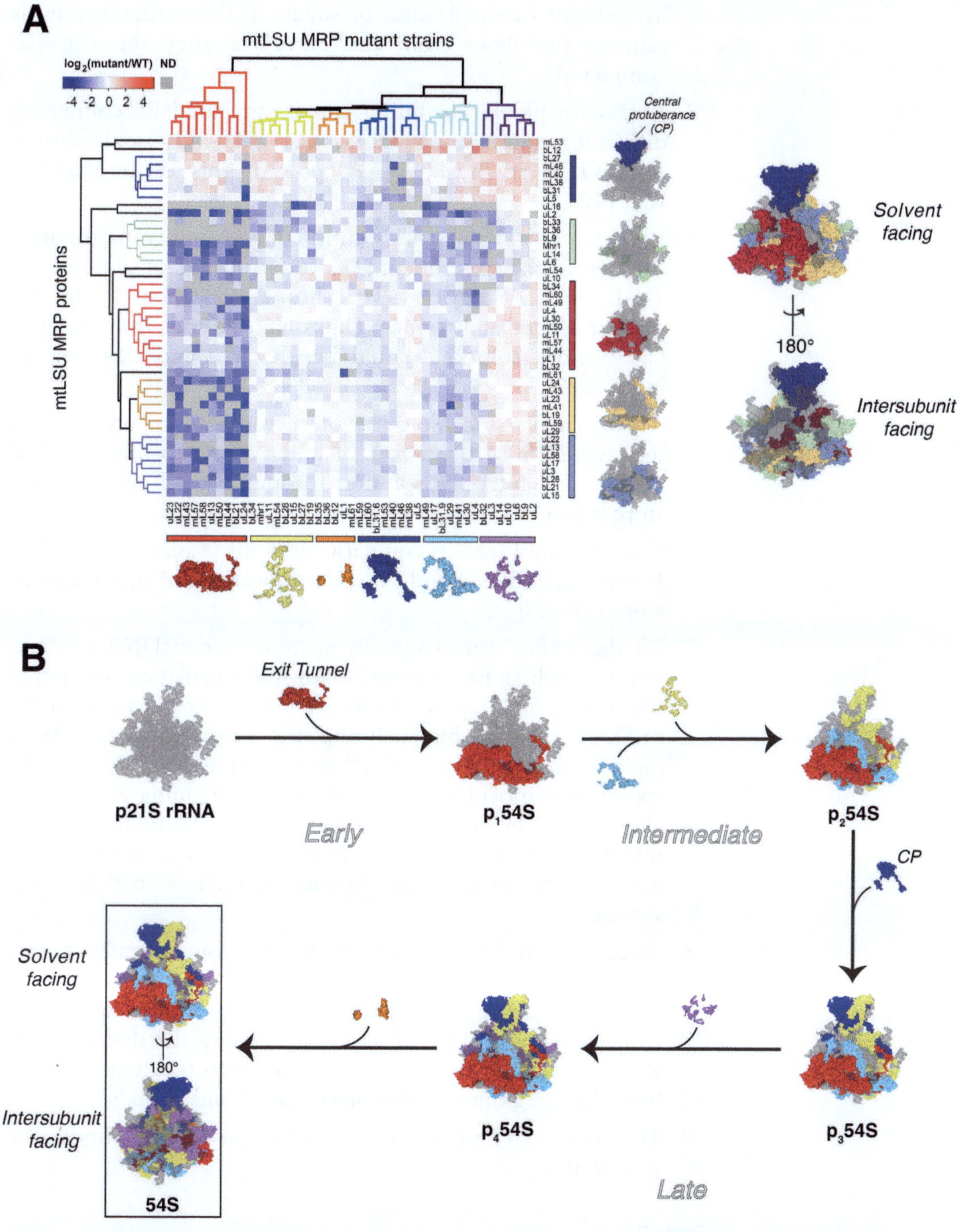

Fig. 2 Clustering analysis and elucidation of hierarchical mitoribosome large subunit assembly pathway. (**a**) Heatmap of relative protein stability (vertical axis) in each yeast deletion strain (horizontal axis) as determined by mass spectrometry of subassembly containing sucrose gradient fractions. Structural modules are clustered on the vertical axis and their structures are mapped to the mtLSU with the 21S rRNA shown in grey. Assembly modules are clustered on the horizontal axis with their representative cryo-EM structures shown below. Each cluster is color-coded to match the corresponding dendrogram. (Figure modified from Ref. [29]). (**b**) Assembly pathway mapping clusters defined in panel A to the mtLSU. Clusters are denoted either early (red), intermediate (cyan, yellow, blue), or late (purple, orange). The 21S rRNA is shown in grey. All structures are prepared using PDB: 3J9M with proteins shown as surface

2. To prevent caramelization of sugars in the media, promptly remove the flasks from the autoclave after the run has completed.
3. Potassium phosphate buffer can be prepared by combining different volumes of monobasic KH_2PO_4 and dibasic K_2HPO_4 to reach pH 7.4. A guide on this can be found at Ref. [49].
4. Prepare a 10% (w/v) stock solution of high-purity digitonin on the day of the experiment. Dissolve 100 mg of digitonin in 1 mL water. To fully dissolve it, heat at 96 °C for 10 min, and place it immediately on ice. Digitonin dilutions from the 10% (w/v) stock solution should be prepared and used on the same day.
5. A 50% (w/v) stock solution of sucrose can be prepared by dissolving 250 g of sucrose in water. Bring to 500 mL final volume and filter the solution through a 0.45 μm filter and store it at 4 °C.
6. Both luminol and p-coumaric acid are easily dissolved in DMSO at 250 mM and 90 mM, respectively. Aliquots can be stored at −20 °C.
7. All the strains were routinely analyzed for mtDNA content. During each strain creation, following sporulation and tetrad dissection, all 4 spores from 20 tetrads were examined for mtDNA content by analyzing the respiratory growth of diploids from crosses with ρ^0 tester strains. Additionally, the spores were inoculated in liquid media containing glucose and allowed to divide for 5–20 generations. Subsequently, the mtDNA content in at least 100 colonies from each culture was estimated by analyzing diploids from crosses with ρ^0 tester strains.
8. To generate deletion strains not available commercially, create a de novo KO cassette by amplifying the KanMX4 cassette from one of the mutants in the collection using primers in Table 7, conjugated to ~80 bp of sequence flanking the desired gene. Repeat this step with a second set of primers to extend flanking homology regions to ~160 bp on the 5′ and 3′ ends.
9. The total volume of the premixed solution should be 20–30 μL.

Table 7
Primer pair used to generate library of yeast KO strains

Description	Sequence (5′ – 3′)
Amplification of kanMX4 cassette	F: CGTACGCTGCAGGTCGAC R: ATCGATGAATTCGAGCTCG

10. For plates, AHW is spread fresh onto the plate using a sterile cell spreader. Assuming an average volume of 25 mL per plate, evenly spread 500 μL of the 50× stock to obtain a 1× working concentration.
11. Minimal media is used to allow for auxotrophic selection. In this case, AHW is supplemented in the media, omitting uracil and leucine to ensure plasmids carrying the *VAR1*^{U} and *RNR1* genes are selected, to suppress the loss of mtDNA.
12. Doubling times should be calculated for each strain. Calculating the doubling time will allow an estimation of how many generations (doublings) the cell population will undergo over a given time. This will be essential for achieving the desired OD in the 1 L flasks when harvesting cells for mitochondrial isolation. Doubling time can be expressed as:

$$t_d(\text{doubling time}) = \frac{t\ (\text{time passed})}{g\ (\text{number of generations})}$$

To calculate the doubling time of each strain, we must first determine the number of generations or doubling events that occur between two OD_{600} measurements. We can use an expression of the exponential growth function $N = N_0e^{kt}$ in terms of doubling. Where:

$N = OD_{600}$ measured at the time interval; N_0 = initial OD_{600} measured at time zero; t = time interval between OD_{600} measurements; k (doubling rate constant) $= \frac{\ln(2)}{t_d}$.

Simplifying this equation, we get: $N = N_0 2^{\frac{t}{t_d}}$

We can now calculate generations passed by substituting $\frac{t}{t_d}$ for g, yielding: $N = N_0 2^g$

After solving for the number of generations passed (g), we can easily calculate the doubling time using: $g = \frac{t}{t_d}$

Example: We inoculate a culture with a starting $OD_{600} = 0.01$. After 24 h, the OD_{600} measured is 0.30. In this case: $N = 0.30$, $N_0 = 0.01$, $t = 24$ h. Solve for g: $N = N_0 2^g$; $0.30 = (0.01)2^g$; $30 = 2^g$; $\ln(30) = g\ln(2)$;

$\frac{\ln(30)}{\ln(2)} = g; g = 4.91$ generations.

Solve for t_d: $t_d = \frac{t}{g}$; $t_d = \frac{24}{4.91}$; $t_d = 4.89$ h

Do this calculation for all four flasks of each strain and take the average to improve accuracy.
13. After calculating average doubling times for each strain, we can use these values to determine the optimal cell density for inoculating 1 L cultures to reach $OD_{600} = 1.5–2.0$ within a predefined timeframe. Typically, we grow the cultures overnight for 18–22 h, and begin harvesting cells at 9 am the next day.

Example: A 100 mL starter culture with a doubling time of 4.89 h is measured to have an $OD_{600} = 2.62$ at 1 pm. Calculate the optimal OD_{600} to start 1 L overnight cultures such that they reach the desired OD_{600} of 1.5–2.0 for mitochondrial isolation the next day at 9 am. In this case: $N = 1.75$; $t = 20$ h; $t_d = 4.89$ h

We can already determine the number of doubling events (g) that will occur during a time (t) based on the doubling time (t_d). Solve for g: $t_d = \frac{t}{g}; 4.89 = \frac{20}{g}; g = 4.09$ generations.

Now that we know the number of generations, we can use the exponential growth function again to calculate the initial OD_{600} necessary to reach an OD_{600} of 1.75 over 4.09 doublings.

Solve for N_0: $N = N_0 2^g$; $1.75 = N_0 2^{4.09}$; $1.75 = N_0(17.03)$; $N_0 = 0.10$. Inoculate 1 L flasks at $OD_{600} = 0.10$

14. To ensure the retention of plasmids carrying genes that suppress the loss of mtDNA, do not exceed 20 generations of growth in culture. Keep track of generations passed for each strain using the equation described above.
15. This solution should be prepared immediately before use, adding DTT from a 1 M stock stored at −20 °C.
16. Throughout this protocol, cells can be manually resuspended using either a glass stir rod or a small plastic spatula unless stated otherwise.
17. Use RT sorbitol for this step. Cold sorbitol may reduce the efficiency of digestion in the next step.
18. Digestion buffer should be prepared fresh, dissolving the zymolyase just before use. Warm the prepared digestion buffer to 30 °C before resuspending cells to allow for efficient digestion.
19. If optimization would be required, spheroplast conversion efficiency can be assessed by osmotic sensitivity. For this purpose, resuspend a 50 μL aliquot of zymolyase-treated cells in 500 μL of 1.2 M sorbitol and another 50 μL aliquot in water. In sorbitol, spheroplasts will remain intact but in water should lyse immediately. Spheroplast conversion efficiency can be estimated by measuring OD_{600}. If at least 90% of cells are converted to spheroplasts, the OD_{600} of the sorbitol suspension should be ~10 fold higher.
20. Prepare homogenization buffer during the previous centrifugation steps and chill on ice. Fatty acid-free BSA should be dissolved just before use to protect against protein aggregation and mitochondria membrane uncoupling. A 100 mM stock solution of PMSF in pure ethanol should be prepared on the day of the experiment and diluted to 1 mM in the homogenization buffer before use.

21. Ensure that each stroke results in proper homogenization. The pestle should completely reach the bottom of the dounce, creating a vacuum that shears spheroplasts as they pass into the space between the pestle and the bottom of the homogenizer. Do not overfill the dounce as this will lead to inefficient homogenization. Sample volume should not exceed 75% of the total capacity of the dounce (e.g., for a 50 mL homogenizer, do not fill more than 35 mL). If the sample volume is too large, homogenization can be done with several rounds, keeping the already homogenized aliquots on ice in a prechilled glass beaker.
22. First, suspend the mitochondria pellet in the 10 mL of SH buffer by gently scraping it with a small plastic spatula. Transfer the mixture to a prechilled 10 mL glass dounce with a tight Teflon pestle. Do not create a vacuum. Instead, perform 3–5 gentle strokes by hand or until there are no visible clumps. After resuspension, transfer to 50 mL polypropylene centrifuge tubes.
23. Mitochondria isolation frequently involves the co-purification of contaminants from other organelles. Most notable are ER protein contaminants. After this high-speed centrifugation, a white ring or "fluffy halo layer" will be visible surrounding the iron-hued mitochondria in the center of the pellet. Using a small plastic spatula, physically remove the white outer ring of contaminants. Resuspend the enriched mitochondria pellet in 10 mL of ice-cold SH buffer and centrifuge again.
24. Use a P1000 tip with approximately 1 cm cut off to resuspend the final mitochondrial pellet gently.
25. We typically prepare immunoblot samples at 4 mg/mL in a 100 μL total volume. (400 μg mitochondrial isolate + 25 μL 4× LB + 3 μL 100 mM PMSF + 1 μL 100× yeast protease inhibitor cocktail + up to 100 μL with water). Samples are run on a 12% SDS-PAGE gel using 40 μg (10 μL) per sample and transferred to a nitrocellulose membrane. The quality of the isolated mitochondria can be determined by staining with ponceau and probing for the mitochondrial transporter Porin. Furthermore, these samples can be used to determine steady-state levels of other mitoribosomal proteins by immunoblot. Evaluating the stability of other mitoribosomal proteins will begin to elucidate the hierarchical order of protein assembly.
26. Analyze the samples as soon as possible since even at −80 °C, proteins can undergo some progressive proteolytic degradation.

27. Carefully mix these solutions and allow them to sit on ice for 5–10 min or until all bubbles have vanished, as air bubbles will disrupt the preparation of the gradient.
28. The extraction buffer used in the next section can be prepared at this time and kept on ice until used.
29. The marker block will have a lower notch and a slightly raised upper edge, be sure to use the upper edge for marking.
30. This step can be done using a P1000 pipette. Be sure not to generate any bubbles.
31. The tube caps have a small hole on one side. Place the cap at roughly a 45° angle with one side flush against the lip of the tube and the other side with the hole lifted. Close cap by downward motion, forcing any excess air and buffer to be expelled through this hole. Ensure that there are no bubbles between the sucrose solution and the bottom of the cap. If bubbles persist, add additional 10% sucrose solution on top of the gradient before placing the cap again. Excess sucrose that escapes through the hole can be removed by pipetting.
32. The magnetic force when placing the tube holder onto the plate can be very strong and may disrupt the gradient. To mitigate this, slide the tube holder onto the plate from the side, so that the bottom of the tube holder is parallel with the plate. Slide the tube holder until it is centered on the gradient master plate.
33. After initiating the run command, the gradient master will use tilted tube rotation with a predetermined time, angle, and speed to create a consistent and reproducible linear gradient. This will take approximately 2 min to complete.
34. Keep the cap on the gradient up until the moment of loading the sample.
35. If using freshly isolated mitochondria, proceed directly to **step 2**.
36. Save 40 μL of this as the "total extract" to serve as input control for immunoblot analyses. Combine this 40 μL of the extract with 25 μL of 4× LB and 35 μL of water to prepare the sample for immunoblot.
37. The volume of the extract will likely exceed the remaining space available in the ultracentrifuge tube. In this case, remove 100–200 μL from the top of the gradient before adding the mitochondrial extract. Be sure to only insert the pipette tip just below the surface of the gradient solution and pipette gently to not disturb the formation of the gradient.
38. Use slow acceleration and braking settings to avoid disturbing the gradient.

39. The total volume of the gradient is approximately 5 mL. Collecting 14 equal volume fractions yields an individual fraction volume of 357 μL. When fractionating, this volume is estimated by collecting ~6 drops per fraction, assuming consistent drop size. A visual reference for 357 μL can be used to determine when enough drops have been collected for each fraction. Using a microcentrifuge tube filled with 357 μL water, draw a horizontal line at the meniscus. This will serve as a reference for the desired volume in each fraction during collection. A simple alternative approach for fraction collection is by careful pipetting from the top of the gradient, although it can potentially be less accurate.
40. Samples can either be immediately run on a 12% SDS-PAGE gel or stored at −80 °C until use.
41. Before running the gel, ensure that there are no bubbles at the bottom of the gel, as this will cause uneven electrophoresis of samples. Remove bubbles by using a syringe fitted with a 45° needle.
42. If bands are not visible after a 15 min incubation with ponceau, increase the wash with water to 1 h.
43. Antibody dilutions should be optimized beforehand within a range of 1:200–1:2000. An ideal starting point however is 1: 500.
44. Primary antibodies should be prepared in blocking buffer and can be reused exhaustively until intensity diminishes. To increase the longevity of the solution, add sodium azide at a final concentration of 0.02% (w/v) and store the prepared antibody at −20 °C. However, consider that azide can inhibit the antigen-antibody interaction.
45. To increase the signal, primary antibodies can alternatively be incubated overnight at 4 °C.
46. Secondary antibody dilution should be within the range 1: 5000–1:10,000 and prepared in blocking buffer.
47. At this point, each gradient fraction should be ~280 μL. In this case, add 1.12 mL of methanol to each sample.
48. For the gradient fractions described in this protocol, add 280 μL of chloroform per sample.
49. Add 840 μL of water if continuing extraction from the sucrose gradient fraction described here.
50. Add 1.12 mL of methanol for proceeding with the extraction of 280 μL sucrose gradient fractions.
51. Do not overdry the pellet, as this will make its resuspension difficult.
52. The "RColorBrewer" package is optional for data analysis and only serves to allow color customization of the heatmap.

Acknowledgment

This research was supported by the National Institutes of Health grant R35-GM118141 (AB).

References

1. Ferrari A, Del'Olio S, Barrientos A (2021) The diseased mitoribosome. FEBS Lett 595(8): 1025–1061. https://doi.org/10.1002/1873-3468.14024
2. Petrov AS, Wood EC, Bernier CR et al (2019) Structural patching fosters divergence of mitochondrial ribosomes. Mol Biol Evol 36(2): 207–219
3. Jourdain AA, Koppen M, Wydro M et al (2013) GRSF1 regulates RNA processing in mitochondrial RNA granules. Cell Metab 17(3):399–410
4. Antonicka H, Choquet K, Lin ZY et al (2017) A pseudouridine synthase module is essential for mitochondrial protein synthesis and cell viability. EMBO Rep 18(1):28–38
5. Tu YT, Barrientos A (2015) The human mitochondrial DEAD-box protein DDX28 resides in RNA granules and functions in mitoribosome assembly. Cell Rep 10(6):854–864
6. Maiti P, Kim HJ, Tu YT et al (2018) Human GTPBP10 is required for mitoribosome maturation. Nucleic Acids Res 13. https://doi.org/10.1093/nar/gky938
7. Antonicka H, Sasarman F, Nishimura T et al (2013) The mitochondrial RNA-binding protein GRSF1 localizes to RNA granules and is required for posttranscriptional mitochondrial gene expression. Cell Metab 17(3):386–398
8. Gopisetty G, Thangarajan R (2016) Mammalian mitochondrial ribosomal small subunit (MRPS) genes: a putative role in human disease. Gene 589(1):27–35
9. Dalla Rosa I, Durigon R, Pearce SF et al (2014) MPV17L2 is required for ribosome assembly in mitochondria. Nucleic Acids Res 42(13): 8500–8515
10. Reyes A, Favia P, Vidoni S et al (2020) RCC1L (WBSCR16) isoforms coordinate mitochondrial ribosome assembly through their interaction with GTPases. PLoS Genet 16(7): e1008923
11. Shajani Z, Sykes MT, Williamson JR (2011) Assembly of bacterial ribosomes. Annu Rev Biochem 80:501–526
12. Barrientos A, Korr D, Barwell KJ et al (2003) *MTG1* codes for a conserved protein required for mitochondrial translation. Mol Biol Cell 14(6):2292–2302
13. Paul MF, Alushin GM, Barros MH et al (2012) The putative GTPase encoded by *MTG3* functions in a novel pathway for regulating assembly of the small subunit of yeast mitochondrial ribosomes. J Biol Chem 287(29): 24346–24355
14. Rackham O, Busch JD, Matic S et al (2016) Hierarchical RNA processing is required for mitochondrial ribosome assembly. Cell Rep 16(7):1874–1890
15. Dennerlein S, Rozanska A, Wydro M et al (2010) Human ERAL1 is a mitochondrial RNA chaperone involved in the assembly of the 28S small mitochondrial ribosomal subunit. Biochem J 430(3):551–558
16. Metodiev MD, Spahr H, Loguercio Polosa P et al (2014) NSUN4 is a dual function mitochondrial protein required for both methylation of 12S rRNA and coordination of mitoribosomal assembly. PLoS Genet 10(2): e1004110
17. Brown A, Rathore S, Kimanius D et al (2017) Structures of the human mitochondrial ribosome in native states of assembly. Nat Struct Mol Biol 24(10):866–869
18. Cipullo M, Gesé GV, Khawaja A et al (2021) Structural basis for late maturation steps of the human mitoribosomal large subunit. Nat Commun 12(1):3673. https://doi.org/10.1038/s41467-021-23617-8
19. Hillen HS, Lavdovskaia E, Nadler F et al (2021) Structural basis of GTPase-mediated mitochondrial ribosome biogenesis and recycling. Nat Commun 12(1):3672. https://doi.org/10.1038/s41467-021-23702-y
20. Lenarčič T, Jaskolowski M, Leibundgut M et al (2021) Stepwise maturation of the peptidyl transferase region of human mitoribosomes. Nat Commun 12(1):3671. https://doi.org/10.1038/s41467-021-23811-8
21. Itoh Y, Khawaja A, Laptev I et al (2022) Mechanism of mitoribosomal small subunit biogenesis and preinitiation. Nature 606:603. https://doi.org/10.1038/s41586-022-04795-x
22. Cheng J, Berninghausen O, Beckmann R (2021) A distinct assembly pathway of the human 39S late pre-mitoribosome. Nat Commun 12(1):4544. https://doi.org/10.1038/s41467-021-24818-x
23. Chandrasekaran V, Desai N, Burton NO et al (2021) Visualizing formation of the active site

in the mitochondrial ribosome. elife 10. https://doi.org/10.7554/eLife.68806

24. Saurer M, Ramrath DJF, Niemann M et al (2019) Mitoribosomal small subunit biogenesis in trypanosomes involves an extensive assembly machinery. Science 365(6458): 1144–1149
25. Lenarčič T, Niemann M, Ramrath DJF et al (2022) Mitoribosomal small subunit maturation involves formation of initiation-like complexes. Proc Natl Acad Sci U S A 119(3). https://doi.org/10.1073/pnas.2114710118
26. Jaskolowski M, Ramrath DJF, Bieri P et al (2020) Structural insights into the mechanism of mitoribosomal large subunit biogenesis. Mol Cell 79(4):629–644
27. Tobiasson V, Gahura O, Aibara S et al (2021) Interconnected assembly factors regulate the biogenesis of mitoribosomal large subunit. EMBO J 40(6):e106292. https://doi.org/10.15252/embj.2020106292
28. Bogenhagen DF, Ostermeyer-Fay AG, Haley JD et al (2018) Kinetics and mechanism of mammalian mitochondrial ribosome assembly. Cell Rep 22(7):1935–1944
29. Zeng R, Smith E, Barrientos A (2018) Yeast mitoribosome large subunit assembly proceeds by hierarchical incorporation of protein clusters and modules on the inner membrane. Cell Metab 27(3):645–656
30. Desai N, Brown A, Amunts A et al (2017) The structure of the yeast mitochondrial ribosome. Science 355(6324):528–531
31. Amunts A, Brown A, Bai X et al (2014) Structure of the yeast mitochondrial large ribosomal subunit. Science 343:1485–1489
32. De Silva D, Fontanesi F, Barrientos A (2013) The DEAD-Box protein Mrh4 functions in the assembly of the mitochondrial large ribosomal subunit. Cell Metab 18:712–725
33. van Dyck L, Neupert W, Langer T (1998) The ATP-dependent PIM1 protease is required for the expression of intron-containing genes in mitochondria. Genes Dev 12(10):1515–1524
34. Sanchirico M, Tzellas A, Fox TD et al (1995) Relocation of the unusual *VAR1* gene from the mitochondrion to the nucleus. Biochem Cell Biol 73(11–12):987–995
35. Gancedo JM (1998) Yeast carbon catabolite repression. Microbiol Mol Biol Rev 62(2): 334–361. https://doi.org/10.1128/MMBR.62.2.334-361.1998
36. Visser W, van Spronsen EA, Nanninga N et al (1995) Effects of growth conditions on mitochondrial morphology in Saccharomyces cerevisiae. Antonie Van Leeuwenhoek 67(3): 243–253. https://doi.org/10.1007/BF00873688
37. Morgenstern M, Stiller SB, Lubbert P et al (2017) Definition of a high-confidence mitochondrial proteome at quantitative scale. Cell Rep 19(13):2836–2852. https://doi.org/10.1016/j.celrep.2017.06.014
38. Kim JH, Roy A, Jouandot D 2nd et al (2013) The glucose signalg network in yeast. Biochim Biophys Acta 1830(11):5204–5210. https://doi.org/10.1016/j.bbagen.2013.07.025
39. Zaman S, Lippman SI, Zhao X et al (2008) How *Saccharomyces* responds to nutrients. Annu Rev Genet 42:27–81
40. Ulery TL, Jang SH, Jaehning JA (1994) Glucose repression of yeast mitochondrial transcription: kinetics of derepression and role of nuclear genes. Mol Cell Biol 14(2):1160–1170
41. Kayikci O, Nielsen J (2015) Glucose repression in Saccharomyces cerevisiae. FEMS Yeast Res 15(6). https://doi.org/10.1093/femsyr/fov068
42. Richter-Dennerlein R, Oeljeklaus S, Lorenzi I et al (2016) Mitochondrial protein synthesis adapts to influx of nuclear-encoded protein. Cell 167(2):471–483
43. Couvillion MT, Soto IC, Shipkovenska G et al (2016) Synchronized mitochondrial and cytosolic translation programs. Nature 533(7604): 499–503
44. Tang JX, Thompson K, Taylor RW et al (2020) Mitochondrial OXPHOS biogenesis: co-regulation of protein synthesis, import, and assembly pathways. Int J Mol Sci 21(11). https://doi.org/10.3390/ijms21113820
45. Anderson JM, Box JM, Stuart RA (2022) The mitospecific domain of Mrp7 (bL27) supports mitochondrial translation during fermentation and is required for effective adaptation to respiration. Mol Biol Cell 33(1):ar7. https://doi.org/10.1091/mbc.E21-07-0370
46. Horn D, Fontanesi F, Barrientos A (2008) Exploring protein-protein interactions involving newly synthesized mitochondrial DNA-encoded proteins. Methods Mol Biol 457:125–139
47. Choi A, Barrientos A (2021) Sucrose gradient sedimentation analysis of mitochondrial ribosomes. Methods Mol Biol 2192:211–226. https://doi.org/10.1007/978-1-0716-0834-0_16
48. Vignais PV, Stevens BJ, Huet J et al (1972) Mitoribosomes from *Candida utilis*. Morphological, physical, and chemical characterization of the monomer form and of its subunits. J Cell Biol 54(3):468–492
49. Gomori G. Preparation of buffers for use in enzyme studies. In: Press A. Methods in enzymology. Elsevier; 1955. p. 138–146

Chapter 12

Metabolic Labeling of Mitochondrial Translation Products in Whole Cells and Isolated Organelles

Priyanka Maiti and Flavia Fontanesi

Abstract

Mitochondria retain their own genome and translational apparatus that is highly specialized in the synthesis of a handful of proteins, essential components of the oxidative phosphorylation system. During evolution, the players and mechanisms involved in mitochondrial translation have acquired some unique features, which we have only partially disclosed. The study of the mitochondrial translation process has been historically hampered by the lack of an in vitro translational system and has largely relied on the analysis of the incorporation rate of radiolabeled amino acids into mitochondrial proteins *in cellulo* or *in organello*. In this chapter, we describe methods to monitor mitochondrial translation by labeling newly synthesized mitochondrial polypeptides with [S^{35}]-methionine in either yeast or mammalian whole cells or isolated mitochondria.

Key words Mitochondrial translation, Protein synthesis, Yeast, Human cells, [S^{35}]-methionine, Newly synthesized polypeptides, Pulse-chase labeling

1 Introduction

In all eukaryotes, only a handful of proteins are encoded in the mitochondrial genome (mtDNA) and synthesized by mitochondrial ribosomes (mitoribosomes). In most cases, mitochondrion-encoded polypeptides are highly hydrophobic transmembrane subunits of the oxidative phosphorylation system (OXPHOS) complexes. Human mtDNA encodes for 13 OXPHOS subunits: ND1, ND2, ND3, ND4, ND4L, ND5 and ND6 from Complex I, CYTB from Complex III, COX1, COX2, and COX3 from Complex IV, and ATP6 and ATP8 from Complex V. The baker's yeast *Saccharomyces cerevisiae* lacks a canonical mitochondrial Complex I and its mtDNA encodes for seven OXPHOS subunits: CytB from Complex III; Cox1, Cox2, and Cox3 from Complex IV; and Atp6, Atp8, and Atp9 from Complex V. Additionally, the yeast mitochondrial genome encodes for the Var1 (or uS3m) protein, a structural

Antoni Barrientos and Flavia Fontanesi (eds.), *The Mitoribosome: Methods and Protocols*,
Methods in Molecular Biology, vol. 2661, https://doi.org/10.1007/978-1-0716-3171-3_12,

component of the mitoribosome small subunit. Mitochondrion-encoded proteins play key catalytic roles, and their expression is essential for the process of oxidative phosphorylation.

During the evolution from the mitochondrion prokaryotic ancestor, the mitoribosomes, mitochondrial translation factors, and the process of mitochondrial translation itself have acquired unique features, which reflect the characteristics of the translated proteins, and differ significantly from their bacterial and cytosolic counterparts. Moreover, differences exist between lower and higher eukaryotes regarding the mitoribosome structure and ancillary factors required for its biogenesis and function. For example, yeast mitochondria contain mRNA-specific translational activators that bind to the mRNA 5′-untranslated region (5′-UTR) and promote translation initiation [1, 2], through molecular mechanism/s that remain to be fully elucidated. On the contrary, in addition to lacking a 5′-methylated cap and TATA box, mammalian mRNAs also lack long 5′-UTRs, and the mechanism of mRNA delivery to the mitoribosome has only begun to emerge [3].

Recent developments in cryogenic electron microscopy (cryo-EM) techniques and data analysis have greatly advanced our understanding of mitoribosome composition, biogenesis, and mechanisms of translation [4]. However, the approaches available to monitor the synthesis of mtDNA-encoded proteins remain limited. Due to the lack of an *in vitro* mitochondrial translation system, mitochondrial protein synthesis has been traditionally analyzed in whole cells or *in organello* by metabolic labeling of newly synthesized polypeptides. These methods are based on the incorporation of a labeled amino acid, normally [S^{35}]-methionine, in mtDNA-encoded proteins, either in isolated intact mitochondria or in cells poisoned with a cytosolic translation inhibitor. Exposure to [S^{35}]-methionine for short periods of time (pulse) allows for evaluating the rate of translation of a specific mitochondrial protein or the overall mitochondrion-encoded proteome. Following the pulse phase, the pool of labeled mitochondrial polypeptides can be chased over time with unlabeled methionine to determine their stability, incorporation into the mitochondrial inner membrane, and/or assembly into the OXPHOS complexes.

Here, we describe pulse-chase protocols to label newly synthesized mitochondrial proteins in whole yeast (Subheading 3.1) or mammalian (Subheading 3.3) cells in culture. Additionally, we outline methods for the analysis of yeast and mammalian mitochondrial translation *in organello* (Subheadings 3.2 and 3.4), including the isolation of intact mitochondria and the labeling of newly synthesized polypeptides. Lastly, we include a protocol to resolve and detect the labeled proteins by denaturing electrophoresis and autoradiography (Subheading 3.5).

2 Materials

Prepare all solutions using distilled ultrapure H_2O. Prepare and store all reagents at room temperature unless otherwise indicated. Always store, handle, and dispose any radioactive material using the appropriate protective equipment and following your institution's guidelines.

2.1 Labeling of Mitochondrion-Encoded Proteins in Yeast Cells

1. *Saccharomyces cerevisiae* yeast strains of interest. We commonly use the respiratory-competent W303-1A (*mat a, ade2-1, his3-11,15, leu2-3-112, trp1-1, ura3-1, can1-100,* ρ^+) strain [5] and its derivatives.
2. Synthetic galactose-containing liquid media (YN-Gal): 2% (w/v) galactose, 6.7 g/L yeast nitrogen base (YN), 30 mg/L adenine, 30 mg/L uracil, 50 mg/L leucine, 50 mg/L tryptophan, 50 mg/L histidine (*see* **Note 1**). Dissolve 100 g of galactose in H_2O to a final volume of 500 mL and sterilize by filtration to make a 20% galactose stock. Dissolve 13.4 g of YN in H_2O to a final volume of 1 L and autoclave to make a 2× YN stock solution. For 1 L of media mix 500 mL 2× YN solution, 100 mL 20% galactose stock solution, 15 mL sterile 2 g/L adenine solution, 15 mL sterile 2 g/L uracil solution, 5 mL sterile 10 g/L leucine solution, 5 mL sterile 10 g/L tryptophan solution, 5 mL sterile 10 g/L histidine solution.
3. Labeling buffer: 2% (w/v) galactose, 40 mM K phosphate buffer pH 6. To prepare 1 M K phosphate buffer pH 6 stock, mix 13.2 mL of 1 M K_2HPO_4, 86.8 mL of 1 M KH_2PO_4, and confirm that the pH = 6.0 (if the pH needs to be adjusted, use phosphoric acid or KOH). For 50 mL solution, mix 2 mL 1 M phosphate buffer pH 6 stock, 5 mL 20% galactose stock, and 43 mL H_2O.
4. 10 mg/mL cycloheximide solution freshly prepared (*see* **Note 2**).
5. 10 mCi/mL [^{35}S]-methionine (specific activity 1175 Ci/mmol). Aliquot and store at −80 °C.
6. 200 mM methionine. Dissolve 2.98 g of methionine in 100 mL of H_2O.
7. 4 mg/mL puromycin in H_2O.
8. Rodel's mix: 1.85 M NaOH, 7.4% (v/v) β-mercaptoethanol, 10 mM phenylmethylsulfonyl fluoride (PMSF). For 10 mL of solution, mix 3.7 mL 5 M NaOH stock solution, 0.74 β-mercaptoethanol, 1 mL 0.1 M PMSF solution dissolved in 100% EtOH and 4.56 mL H_2O. Prepare fresh before use and protect from light.
9. 50% (w/v) Trichloroacetic acid (TCA). Dissolve 50 g of TCA in 100 mL of H_2O.

10. 0.5 M Tris base (no pH adjusted). Dissolve 30.29 g of Tris base in 500 mL of H_2O.
11. 1× Laemmli buffer: 2% (w/v) sodium dodecyl sulfate (SDS), 10% (v/v) glycerol, 60 mM Tris–HCl, pH 6.8, 2.5% (v/v) β-mercaptoethanol, 0.02% (w/v) bromophenol blue. Aliquot and store at −20 °C.

2.2 Labeling of Newly Synthesized Proteins in Isolated Yeast Mitochondria

Avoid the use of detergent to wash glassware and plasticware used for mitochondria isolation since remaining traces of detergent could disrupt the mitochondrial membranes' integrity.

2.2.1 Mitochondria Isolation

1. Complete galactose-containing liquid media (YP-Gal): 2% (w/v) galactose, 1% (w/v) yeast extract, 2% (w/v) peptone. Comparable results can be obtained by growing the cells in YN-Gal media (*see* **step 2** in Subheading 2.1) if the strains require plasmid selection.
2. Reducing buffer: 100 mM Tris–HCl pH 8.8, 10 mM DTT. For 10 mL of solution, mix 548 μL 1.825 M Tris–HCl pH 8.8, 1 mL 0.1 M DDT, and 8.472 mL H_2O. Prepare fresh before use.
3. Cell wall digestion buffer: 1.2 M sorbitol, 20 mM potassium phosphate buffer pH 7.4, 0.25 mg/mL zymolyase-100T. For 10 mL of solution, mix 60 mL 2 M sorbitol, 2 mL 1 M potassium phosphate buffer pH 7.4, and 38 mL H_2O. Dissolve 25 mg of zymolyase-100T in the solution. Prepare fresh before use.
4. 1.2 M deionized sorbitol. Dissolve 218.6 g of sorbitol in H_2O to a final volume of 1 L. Add 25 g of AG501-X8 mixed bed resin (Biorad) and stir for 2 h at room temperature to remove all ions present in the solution. This is important to remove trace metal contaminants that can damage mitochondria. Sterilize by filtration and store at 4 °C.
5. Homogenization buffer: 0.6 M deionized sorbitol, 10 mM Tris–HCl pH 7.5, 1 mM EDTA pH 8, 0.2 mM PMSF. For 100 mL of solution, mix 50 mL 1.2 M deionized sorbitol, 1 mL 1 M Tris–HCl pH 7.5, 0.2 mL 0.5 M EDTA pH 8, and 48.8 mL H_2O. Sterilize by filtration and store at 4 °C. Just before use, add 200 μL freshly prepared 0.1 M PMSF solution dissolved in ethanol.
6. Iso-osmotic mitochondrial (SEH) buffer: 0.6 M deionized sorbitol, 1 mM EDTA pH 8, 20 mM HEPES pH 7.4. For 100 mL of solution, mix 50 mL 1.2 M deionized sorbitol, 2 mL 1 M HEPES pH 7.4, 0.2 mL 0.5 M EDTA pH 8, and 47.8 mL H_2O. Sterilize by filtration and store at 4 °C.
7. 50 and 10 mL glass/Teflon homogenizers.

2.2.2 In Organello Translation

1. 2 mg/mL amino acid mix: dissolve 20 mg each of the amino acids alanine, arginine, aspartic acid, asparagine, glutamic acid, glutamine, glycine, histidine, isoleucine, leucine, lysine, phenylalanine, proline, serine, threonine, tryptophan, and valine in 10 mL H_2O. Sterilize by filtration, aliquot, and store at 4 °C.
2. Translation buffer: 0.6 M sorbitol, 150 mM KCl, 15 mM potassium phosphate buffer pH 7.4, 20 mM HEPES pH 7.4, 12.66 mM $MgSO_4$, 4 mM ATP, 0.5 mM GTP, 5 mM phosphoenolpyruvate, 5 mM alpha-ketoglutarate, 24.26 μg/mL amino acid mix, 66.67 μM cysteine, 12.13 μg/mL tyrosine, 10 μg/mL pyruvate kinase. Prepare fresh before use (Table 1).
3. 10 mCi/mL [^{35}S]-methionine (specific activity 1175 Ci/mmol).
4. 200 mM methionine solution.
5. Wash buffer: 0.6 M sorbitol, 1 mM EDTA pH 8, 5 mM methionine. Sterilize by filtration and store at 4 °C.
6. 1× Laemmli buffer: 2% (w/v) sodium dodecyl sulfate (SDS), 10% (v/v) glycerol, 60 mM Tris–HCl, pH 6.8, 2.5% (v/v) β-mercaptoethanol, 0.02% (w/v) bromophenol blue. Aliquot and store at −20 °C.

2.3 Labeling of Mitochondrion-Encoded Proteins in Mammalian Cell Cultures

1. Embryonic kidney 293T (HEK293T) cells (*see* **Note 3**).
2. Complete growth medium: Dulbecco's Modified Eagle Medium (DMEM medium) containing 4.5 g/L glucose and supplemented with 10% fetal bovine serum (FBS). In addition, if working with respiratory deficient cells, supplement the media with 110 mg/L pyruvate, 100 μg/mL uridine, 3 mM formate, and 1× GlutaMAX (ThermoFisher) (*see* **Note 4**).
3. DMEM without methionine: Dulbecco's Modified Eagle Medium (DMEM medium) without methionine and cysteine, containing 4.5 g/L glucose and supplemented with 10% FBS and 6.3 mg/mL cysteine. In addition, if working with respiratory deficient cells, supplement the media with 110 mg/L pyruvate, 100 μg/mL uridine, 3 mM formate and 1× GlutaMAX (ThermoFisher).
4. 1× PBS without calcium and magnesium.
5. 0.2 mg/mL collagen (type I rat tail) solution dissolved in 1× PBS.
6. 1× Trypsin-EDTA.
7. 2 mg/mL emetine solution freshly prepared (*see* **Note 2**).
8. 2 mg/mL anisomycin solution freshly prepared (*see* **Note 2**).
9. 10 mCi/mL [^{35}S]-methionine (specific activity 1175 Ci/mmol).

Table 1
Composition of *in organello* translation buffers

Component	Stock concentration	Volume	Final concentration
***In organello* translation buffer for yeast mitochondria**			
Sorbitol	2 M	450 μL	0.6 M
KCl	1 M	225 μL	150 mM
K-phosphate buffer pH 7.4	1 M	22.5 μL	15 mM
HEPES pH 7.4	1 M	30 μL	20 mM
$MgSO_4$	1 M	19 μL	12.66 mM
ATP	200 mM	30 μL	4 mM
GTP	50 mM	15 μL	0.5 mM
Amino acid mix	2 mg/mL each	18.2 μL	24.26 μg/mL each
Cysteine	10 mM	10 μL	66.67 μM
Tyrosine	1 mg/mL	18.2 μL	12.13 μg/mL
Phosphoenolpyruvate (PEP)	250 mM	30 μL	5 mM
α-ketoglutarate	500 mM	15 μL	5 mM
Pyruvate kinase	10 mg/mL	1.5 μL	10 μg/mL
H_2O	–	615.6 μL (Total vol. 1.5 mL)	–
***In organello* translation buffer for mammalian mitochondria**			
Translation mix	2×	5 mL	1×
Amino acid mix	100× (6 mg/mL each)	0.1 mL	1× (60 μg/mL each)
Cysteine	6 mg/mL	0.1 mL	60 μg/mL
Tyrosine	3 mg/mL	0.2 mL	60 μg/mL
ATP	200 mM	0.25 mL	5 mM
GTP	50 mM	4 μL	20 μM
Creatine phosphate	1 M	60 μL	6 mM
Creatine kinase	10 mg/mL	60 μL	60 μg/mL
H_2O	–	4.226 mL (Total vol. 10 mL)	–

10. RIPA buffer: 50 mM Tris–HCl pH 8150 mM NaCl, 1% Nonidet P-40 (NP-40), 0.5% sodium deoxycholate, 0.1% sodium dodecyl sulfate, 2 mM EDTA pH 8. Store at 4 °C. Just before use, add 1 mM PMSF (in 1 mL RIPA buffer, add 10 μL 0.1 M PMSF stock solution dissolved in 100% EtOH) and 1× EDTA-free protease inhibitor cocktail.
11. 4× Laemmli buffer: 8% (w/v) sodium dodecyl sulfate (SDS), 40% (v/v) glycerol, 240 mM Tris–HCl, pH 6.8, 10% (v/v) β-mercaptoethanol, 0.08% (w/v) bromophenol blue. Aliquot and store at −20 °C.

2.4 Labeling of Newly Synthesized Proteins in Isolated Mammalian Mitochondria

2.4.1 Mitochondria Isolation

1. T-K-Mg buffer: 10 mM Tris–HCl pH 7.4, 10 mM KCl, 0.5 mM $MgCl_2$. Sterilize by filtration and store at 4 °C.
2. 1 M sucrose buffer: 1 M sucrose, 10 mM Tris–HCl pH 7.4. Sterilize by filtration and store at 4 °C.
3. ST buffer: 0.32 M sucrose, 10 mM Tris–HCl, pH 7.4. Sterilize by filtration and store at 4 °C.
4. Glass/Teflon homogenizer.
5. 50× EDTA-free protease inhibitor solution.

2.4.2 In Organello Translation

1. 2× Translation mix: 200 mM mannitol, 20 mM sodium succinate, 160 mM KCl, 10 mM $MgCl_2$, 2 mM sodium phosphate, 50 mM HEPES. Adjust the pH at 7.4, sterilize by filtration, and store at −20 °C.
2. 100× amino acid mix: 6 mg/mL each alanine, arginine, asparagine, aspartate, glutamine, glutamate, glycine, histidine, isoleucine, leucine, lysine, phenylalanine, proline, serine, threonine, tryptophan and valine. Sterilize by filtration, aliquot, and store at 4 °C.
3. Translation buffer: 1× translation mix, 1× amino acid mix, 60 μg/mL cysteine, 60 μg/mL tyrosine, 5 mM ATP, 20 μM GTP, 6 mM creatine phosphate, 60 μg/mL creatine kinase. Prepare fresh before use (Table 1).
4. 10 mCi/mL [^{35}S]-methionine (specific activity 1175 Ci/mmol).
5. 1× Laemmli buffer: 2% (w/v) sodium dodecyl sulfate (SDS), 10% (v/v) glycerol, 60 mM Tris–HCl, pH 6.8, 2.5% (v/v) β-mercaptoethanol, 0.02% (w/v) bromophenol blue. Aliquot and store at −20 °C.
6. 50× EDTA-free protease inhibitor.
7. 250 units/μL Benzonase nuclease.

2.5 Protein Separation and Detection by Electrophoresis, Electrotransfer, and Autoradiography

1. Apparatus for vertical electrophoresis (including glasses, spacers, combs, and casting system) and electrotransfer apparatus (*see* **Notes 5** and **6**).
2. 30% (w/v) acrylamide – 0.2% (w/v) BIS-acrylamide solution. Mix 150 g of acrylamide and 1 g of BIS-acrylamide in a beaker. Add 400 mL of warm water and stir on a magnetic plate until fully dissolved. Bring up to 500 mL volume with H_2O and filter through a paper filter to remove any remaining crystal. Handle acrylamide and BIS-acrylamide powder under a fume hood. Store at 4 °C.
3. 1.825 M Tris–HCl pH 8.8. Dissolve 110.5 g of Tris base in 400 mL of H_2O. Adjust the pH to 8.8 with HCl. Bring up to 500 mL volume with H_2O.
4. 0.6 M Tris–HCl pH 6.8. Dissolve 36.34 g of Tris base in 400 mL of H_2O. Adjust the pH to 6.8 with HCl. Bring up to 500 mL volume with H_2O.
5. 10% (w/v) SDS. Dissolve 5 g of SDS in 50 mL of H_2O. Handle SDS powder under a fume hood.
6. 10% (w/v) ammonium persulfate (APS). Dissolve 1 g of APS in 10 mL of H_2O. Store at 4 °C.
7. *N,N,N′,N′*-Tetramethylethylenediamine (TEMED). Handle TEMED under a fume hood. Store at 4 °C.
8. 1× Running buffer: 50 mM Tris–HCl pH 8.3, 384 mM glycine, 0.1% (w/v) SDS. To prepare a 5× stock solution, dissolve 60 g of Tris base, 288 g of glycine, and 10 g of SDS in H_2O to a final volume of 2 L. Prepare the 1× running buffer by diluting 200 mL of 5× buffer stock with 700 mL of H_2O. Precisely adjust the pH to 8.3 with HCl and bring up to 1 L volume with H_2O.
9. Pre-stained molecular weight marker (*see* **Note** 7).
10. 0.45 μm nitrocellulose membrane.
11. Transfer buffer: 192 mM glycine, 25 mM Tris base, 20% (v/v) methanol. Dissolve 57.7 g of glycine, and 12.1 g of Tris base in 3 L of H_2O. Add 800 mL of methanol and bring up to 4 L volume with H_2O.
12. Western blot rinse buffer (WRB): 10 mM Tris–HCl pH 8, 1 mM Ethylenediaminetetraacetic acid (EDTA) pH 8, 0.15 M NaCl, 0.1% (v/v) Triton X-100. Dissolve 60.57 g of Tris base in 400 mL of H_2O. Adjust the pH to 8 with HCl. Bring up to 500 mL volume with H_2O to make a 1 M Tris–HCl stock solution. Add 73.06 g of EDTA to 400 mL of H_2O. Stir on a magnetic plate and adjust the pH to 8 (EDTA will start dissolving as the pH increases) with NaOH (approximately 10 g of NaOH). Bring up to 500 mL volume with H_2O to make a 0.5 M EDTA stock solution. Dissolve 146.1 g of

NaCl in 500 mL of H_2O to make a 5 M stock solution. Add 450 mL of H_2O to a beaker and slowly add 50 mL of Triton X-100 while stirring the solution on a magnetic plate. Stir until fully homogenous to make a 10% Triton X-100 stock solution. To prepare 1 L of WRB mix 10 mL Tris–HCl pH 8 (1 M stock), 2 mL EDTA pH 8 (0.5 M stock), 30 mL NaCl (5 M stock), 10 mL Triton X-100 (10% stock), 1948 mL H_2O.

13. 5% (w/v) non-fat milk solution prepared in WRB. Dissolve 10 g of non-fat milk powder in 100 mL WRB. Store at 4 °C.

3 Methods

3.1 Labeling of Mitochondrion-Encoded Proteins in Yeast Cells

1. Inoculate the strain/s of interest in 10 mL of YN-Gal liquid media with the required amino acids and nucleobases supplementation (*see* **Note 8**). Grow the yeast cultures overnight at 30 °C with constant shaking at 200 rpm until reaching an optical density at 600 nm of wavelength (OD_{600}) equal to 2–4.
2. Transfer an aliquot of the precultures to fresh 10 mL YN-Gal liquid media to obtain a final OD_{600} of 0.4–0.6.
3. Grow the cultures for 3–4 h at 30 °C with constant shaking at 200 rpm until reaching a confluency of 0.8–1.2 OD_{600} (*see* **Note 9**).
4. Transfer to a 1.5 mL centrifuge tube a volume of culture equivalent to an OD_{600} of 0.6.
5. Pellet the cells by centrifugation at 2500 × *g* for 2 min and remove the supernatants.
6. Wash the cell pellets once by resuspending the cells in 0.5 mL of labeling buffer. Centrifuge at 2500 × *g* for 2 min and remove the supernatants.
7. Resuspend the cells in 0.5 mL of labeling buffer.
8. Add 10 μL of a freshly prepared 10 mg/mL cycloheximide solution and mix by inversion. Incubate at 30 °C for 2.5 min (*see* **Note 10**).
9. Add 1.5 μL [^{35}S]-methionine and mix by inversion (*see* **Note 11**). Incubate the cell suspensions at 30 °C for 5–10 min (*see* **Note 12**) to follow the rate of mitochondrial translation or pulse.
10. Optional: to follow the stability of the newly synthesized mitochondrial proteins (chase), to a separate set of samples, after the pulse add 0.1 mL of 200 mM cold methionine and 10 μL of a 4 mg/mL puromycin solution. Mix by inversion and incubate at 30 °C for 1–4 h (*see* **Note 13**).
11. Pellet the cells by centrifugation at 8000 × *g* for 30 s and remove the supernatants.

12. Add 75 μL of Rodel's mix and immediately vortex the samples to lyse the cells.
13. Add 0.5 mL of H_2O and 575 μL of 50% TCA solution. Mix by inversion and incubate on ice for 15 min.
14. Centrifuge the samples at 18,000 × *g* for 10 min at 4 °C to pellet the proteins. Carefully remove the supernatants without disturbing the pellets.
15. Add 0.5 mL of 0.5 M Tris base without resuspending the pellets. Centrifuge at 18,000 × *g* for 5 min at 4 °C. Carefully remove the supernatants without disturbing the pellets.
16. Add 0.5 mL of H_2O without resuspending the pellets. Centrifuge at 18,000 × *g* for 5 min at 4 °C. Carefully remove the supernatants without disturbing the pellets.
17. Resuspend the pellets in 20 μL of 1× Laemmli buffer (*see* **Note 14**). The samples are ready for electrophoretic analysis. Load the entire volume in a 17.3% SDS-polyacrylamide gel (*see* Subheading 3.5).

3.2 Labeling of Newly Synthesized Proteins in Isolated Yeast Mitochondria

Several methods are available for the isolation of mitochondria from yeast cells. The protocol presented here is based on the method by Herrmann et al. [6] with some modifications, and we have found it more suitable for the preparation of intact high-quality mitochondria competent for *in organello* translation.

3.2.1 Mitochondria Isolation

1. Inoculate the strain/s of interest in 50 mL of YP-Gal liquid media and grow the yeast cultures overnight at 30 °C with constant shaking at 200 rpm until reaching an optical density at 600 nm of wavelength (OD_{600}) equal to 2–4.
2. Transfer an aliquot of the precultures to fresh 2–6 flasks of 1 L YP-Gal liquid media and grow the cultures overnight at 30 °C with constant shaking at 200 rpm until reaching a confluency of 0.8–1.2 OD_{600}.
3. Harvest the cells by centrifugation at 900 g for 5 min at room temperature.
4. Combine the cells in a single preweighted centrifuge bottle and wash them with 200 mL H_2O.
5. Weight the cell pellet (*see* **Note 15**) and resuspend it in reducing buffer using 1 mL of buffer for 2 g of cells.
6. Incubate at 30 °C for 10 min with gentle shaking.
7. Add 150 mL 1.2 M deionized sorbitol and centrifuge at 2200 g for 10 min at room temperature.
8. Resuspend the cell pellets in cell wall digestion buffer using 1 mL of buffer for 0.15 g of cells.

9. Incubate at 30 °C for 30 min with gentle shaking. After approximately 80% of the cells have been converted to spheroplasts, transfer the sample on ice (*see* **Note 16**). From this point, all buffers should be chilled, and samples kept on ice.
10. Centrifuge at 2200 g for 10 min at 4 °C to pellet the spheroplasts.
11. Resuspend the pellet in homogenization buffer using 1 mL of buffer for 0.15 g of cells. Homogenize the sample in a glass/Teflon homogenizer using 10 stokes (*see* **Note 17**).
12. Centrifuge at 1500 g for 5 min at 4 °C to pellet unbroken cells, nuclei, and cell debris.
13. Transfer the supernatant to a clean centrifuge tube (*see* **Note 18**) and centrifuge at 8000 g for 10 min at 4 °C to pellet mitochondria.
14. Resuspend mitochondria in 20 mL SEH buffer using a loose glass/Teflon homogenizer (*see* **Note 19**).
15. Centrifuge at 1500 g for 5 min at 4 °C to pellet broken mitochondria and residual cellular debris.
16. Transfer the supernatant to a clean centrifuge tube and centrifuge at 8000 g for 10 min at 4 °C to pellet mitochondria.
17. Gently resuspend mitochondria in 0.2–0.5 mL SEH buffer (*see* **Note 20**).
18. Quantify mitochondrial proteins by your standard laboratory protein quantification method (Folin, Bradford, or Lowry reagents can be used).
19. Store mitochondria on ice until the next step (*see* Subheading 3.2.2). Unused mitochondria can be snap frozen in liquid N_2 and stored at −80 °C (*see* **Note 21**).

3.2.2 In Organello Translation

1. Pellet 50 μg of mitochondria from **step 19** of Subheading 3.2.1 by centrifugation at 8000 g for 10 min at 4 °C.
2. Eliminate the supernatant and resuspend the mitochondria in 50 μL translation buffer. Incubate at 30 °C for 5 min.
3. Add 1 μL [^{35}S]-methionine and mix by inversion. Incubate at 30 °C for 15–30 min to radiolabel newly synthesized mitochondrial proteins. This is the pulse phase.
4. Stop the labeling reaction by adding 15 μL 0.2 M cold methionine.
5. Centrifuge 18,000 g for 5 min at 4 °C.
6. Eliminate the supernatant and wash mitochondria once with 0.5 mL wash buffer. Centrifuge 18,000 g for 5 min at 4 °C.
7. Eliminate the supernatant and resuspend the pellet in 20 μL 1× Laemmli buffer. The samples are ready for electrophoretic analysis. Load the entire volume in a 17.3% SDS-polyacrylamide gel (*see* Subheading 3.5).

3.3 Labeling of Mitochondrion-Encoded Proteins in Mammalian Cell Cultures

1. Coat the wells of a 6-well plate with collagen by adding 1 mL 0.2 mg/mL collagen solution per well (*see* **Note 22**). Incubate the plate overnight at room temperature or at 37 °C for 3–4 h. Remove the collagen and wash once with 1 mL 1× PBS.
2. Seed 2.5–3 × 10^5 HEK293T cells per well (*see* **Note 23**) in DMEM complete growth media. Incubate overnight at 37 °C in a CO_2 incubator. The cells should reach 70–80% confluency (*see* **Note 24**).
3. Remove the media and wash twice with 1 mL 1× PBS (*see* **Note 25**).
4. Add 1 mL DMEM medium without methionine and incubate the cells for 20 min at 37 °C in a CO_2 incubator.
5. To block the cytosolic translation, add 50 μL/well of 2 mg/mL emetine for pulse-only experiments or 50 μL/well of 2 mg/mL anisomycin for pulse-chase experiments (*see* **Note 26**).
6. Incubate the cells with the selected inhibitor of cytoplasmic translation for 15 min at 37 °C in a CO_2 incubator.
7. Slowly add dropwise 8 μL [^{35}S]-methionine to each well and make sure that the solution is homogenously mixed.
8. Incubate the cells for 10–30 min (*see* **Note 27**) at 37 °C in a CO_2 incubator. This is the pulse phase.
9. Remove the media and wash twice with 1 mL 1× PBS. For pulse-only experiments, proceed to **step 10** (*see* **Note 28**). For pulse-chase experiments, add 1 mL of complete growth media and incubate the cells for 1–24 h (*see* **Note 29**) at 37 °C in a CO_2 incubator. This is the chase phase. Remove the media, wash with 1 mL 1× PBS and proceed to **step 10**.
10. Add 0.1 mL/well of trypsin and incubate for 2–3 min at 37 °C in a CO_2 incubator for the cells to completely detach from the plate.
11. Neutralize the trypsin by adding 1 mL of complete DMEM medium and transfer to a 1.5 mL microcentrifuge tube, place on ice.
12. Centrifuge at 400 g for 5 min at 4 °C.
13. Discard the supernatant and wash once with 1 mL cold 1× PBS. Centrifuge at 400 g for 5 min at 4 °C and discard the supernatant. Optional: at this point, the cell pellet can be stored at −80 °C until further use.
14. Resuspend the pellet in 50 μL RIPA buffer supplemented with protease inhibitors, vortex well, and incubate on ice for 30 min.
15. Centrifuge at 20,000 g for 5 min at 4 °C and transfer the supernatant to a new 1.5 mL microcentrifuge tube.

16. Quantify protein concentration by your standard laboratory protein quantification method (Bradford or Lowry reagents can be used).
17. Prepare samples for electrophoresis by diluting 100 μg of proteins in a final volume of 50 μL 1× Laemmli buffer. Load the entire sample in a 17.3% SDS-polyacrylamide gel (*see* Subheading 3.5).

3.4 Labeling of Newly Synthesized Proteins in Isolated Mammalian Mitochondria

To achieve a robust metabolic labeling of mitochondrial proteins, it is critical to use freshly isolated functional mitochondria with intact membranes. Isolation of mitochondria from exponentially growing cultured cells is based on the protocol by Enriquez and Attardi [7] with some modifications [8, 9].

3.4.1 Mitochondria Isolation

1. Grow HEK293T cells in complete growth media to obtain 10–15, 150 mm plates at 80–90% cell confluency.
2. To collect the cells, remove the growth media, wash once with 10 mL 1× PBS and add 1 mL/plate trypsin. Incubate at 30 °C for a few minutes or until the cells start detaching from the plate. Neutralize the trypsin with 10 mL of complete growth media and transfer the cell suspension in a 50 mL centrifuge tube.
3. Centrifuge at 500 g for 5 min. Eliminate the supernatant and resuspend the cell pellet in 10 mL cold 1× PBS. Combine cells in one preweighted tube and centrifuge at 500 g for 5 min. Perform an additional wash with 10 mL cold 1× PBS.
4. Weight the cell pellet and resuspend it in cold T-K-Mg Buffer using 1 mL of buffer for 0.15 g of cells.
5. Incubated on ice for 5 min to allow cell swelling.
6. Homogenize cells with Teflon/glass homogenizer until ~80% are broken (10 strokes, *see* **Note 17**).
7. Dilute the solution to a final concentration of 0.25 M sucrose by adding 1 M sucrose buffer (*see* **Note 30**).
8. Centrifuge the homogenate at 1500 g for 3 min at 4 °C to pellet unbroken cells, debris, and nuclei.
9. Collect the supernatant and centrifuge it at 8000 g for 10 min at 4 °C to pellet the mitochondria.
10. Discard the supernatant, gently resuspend the pellet in 10 mL of ST buffer (*see* **Note 31**).
11. Repeat the centrifugation at 1500 g for 3 min at 4 °C to ensure the complete elimination of heavy contaminants.
12. Collect the supernatant and centrifuge at 8000 g for 10 min at 4 °C.

13. Resuspend the isolated mitochondria in 300 μL of ST buffer and add 1× EDTA-free protease inhibitor (*see* **Note 32**).
14. Quantify mitochondrial proteins by your standard laboratory protein quantification method (Bradford or Lowry reagents can be used).
15. Store mitochondria on ice until the next step (*see* Subheading 3.4.2). Unused mitochondria can be snap frozen in liquid N_2 and stored at −80 °C for different uses.

3.4.2 In Organello Translation

The protocol for metabolic labeling of mitochondrial newly synthesized proteins in mitochondria isolated from cells in culture presented here is based on the method for mitochondrial translation analysis in mitochondria isolated from mouse tissues [10, 11] with some modifications.

1. Use 1 mg of freshly isolated mitochondria from Subheading 3.4.1.
2. Centrifuge mitochondria at 8000 g for 5 min at 4 °C.
3. Discard the supernatant and wash mitochondria in 1 mL of translation buffer.
4. Centrifuge at 8000 g for 5 min at 4 °C. Discard the supernatant and resuspend the mitochondrial pellet in 1 mL of translation buffer prewarmed at 37 °C (*see* **Notes 33** and **34**).
5. Add 8 μL of [^{35}S]-methionine.
6. Incubate at 37 °C for 30–60 min. This is the pulse phase.
7. Centrifuge samples at 8000 g for 5 min at 4 °C.
8. Discard the supernatant and resuspend the mitochondrial pellet in 100 μL of 1× Laemmli buffer supplemented with 1× EDTA-free protease inhibitor.
9. Add 1.5 μL of benzonase and incubate at 37 °C for 30 min. Benzonase helps to reduce sample viscosity due to the presence of nucleic acids.
10. Load 50 μL in a 17.3% SDS-polyacrylamide gel (*see* Subheading 3.5). Store the remaining samples at −80 °C (*see* **Note 35**).

3.5 Protein Separation and Detection by Electrophoresis, Electrotransfer, and Autoradiography

1. Assemble the SDS-PAGE casting system following the manufacturer's indication.
2. Prepare a 17.3% resolving gel solution by mixing the components listed in Table 2 and pouring it between the glasses. Cover the top of the resolving gel with a thin layer of isopropanol to obtain a flat gel surface after polymerization. Allow the gel to polymerize for approximately 45 min.

Table 2
Composition of SDS-polyacrylamide gels

Stock solution	Gel composition: 17.3% Resolving gel – medium size	17.3% Resolving gel – long size	5% Stacking gel
30% (w/v) acrylamide – 0.2% (w/v) BIS-acrylamide	12.6 mL	25.2 mL	1.083 mL
1.825 M Tris–HCl pH 8.8	4.52 mL	9.04 mL	–
0.6 M Tris–HCl pH 6.8	–	–	0.65 mL
10% (w/v) SDS	220 μL	440 μL	65 μL
10% (w/v) APS	110 μL	220 μL	32.5 μL
TEMED	11 μL	22 μL	6.5 μL
H_2O	4.539 mL	9.078 mL	4.663 mL
Final volume	22 mL	44 mL	6.5 mL

The reagents required for the preparation of SDS-polyacrylamide gels used for the separation of radiolabeled newly synthesized mitochondrial proteins under denaturing conditions are indicated. The dimensions of a medium-size gel correspond to 14.5 cm (W) × 16.5 cm (H) and a long size gel to 16.5 cm (W) × 21.5 cm (H)

3. Once the resolving gel is fully polymerized, discard the isopropanol overlay and wash out its traces with abundant H_2O. Carefully dry the gel surface with a piece of filter paper.
4. Prepare a 5% staking gel solution by mixing the components listed in Table 2 and pouring it over the resolving gel. Insert a proper comb and allow the staking gel to polymerize for approximately 30 min.
5. Once the gel is fully polymerized, remove the comb and insert the gel into the electrophoretic chamber. Fill the chamber with 1× running buffer. Wash the wells with a pipette tip to remove unpolymerized polyacrylamide residues.
6. Load in the gel the pre-stained molecular weight marker in the amount recommended by the manufacturer and the samples in the amounts indicated above (*see* **Note 36**).
7. Run the electrophoresis until the bromophenol blue dye reaches the bottom of the gel (*see* **Note 37**).
8. Recover the gel. Cut and dispose of the staking gel. Transfer the resolving gel in a tray containing transfer buffer. Soak in the same buffer one piece of nitrocellulose membrane and four pieces of filter paper. Assemble the transfer sandwich as follows: two pieces of filter paper, resolving gel, nitrocellulose membrane, and two pieces of filter paper. Exercise particular care in removing air bubbles in between any layer with the help of a glass rod. Place sandwich in the electrotransfer apparatus with

the gel facing the cathode and the membrane facing the anode. Perform the electrotransfer following manufacturer recommendation (*see* **Note 38**).

9. Recover the nitrocellulose membrane and air-dry it at room temperature under a chemical fume hood.
10. Wrap the membrane in thin cellophane (*see* **Note 39**), place it in a cassette holder and put an X-ray film on top of it. Expose the membrane for different amounts of time at room temperature before developing (*see* **Note 40**).
11. After obtaining the signal of the desired intensity, recover the nitrocellulose membrane. Soak the membrane in WRB to rehydrate it. Block the membrane with a 5% milk solution in WRB for 1 h at room temperature, replacing the milk with fresh solution every 20 min. For total protein normalization, perform a Western blot analysis (*see* Figs. 1 and 2) following your standard lab procedure (*see* **Note 41**). Commonly used antibodies against loading control proteins are reported in Table 3.

4 Notes

1. The amino acids and nucleobases used to supplement the YN-Gal described in this method are based on the auxotrophic requirements of the W303-1A strain. When working with strains carrying plasmids or constructed in different genetic backgrounds, amino acids and nucleobases media supplementation will need to be adjusted according to strain requirements.
2. Cycloheximide, emetine, and anisomycin do not dissolve easily in water. Prewarm the water in a thermoblock set at 60 °C before adding to the inhibitor powder. Vortex extensively until completely dissolved.
3. We describe here the use of HEK293T cells for the analysis of mitochondrial translation *in vivo* and *in organello*, but the protocol can be applied to other mammalian cells in culture, including HeLa, 143B, fibroblasts, and others [12].
4. An antibiotic-antimycotic solution can be added to the growth media to prevent bacterial and/or fungal contamination. However, care should be taken to avoid antibiotics that are mitoribosome inhibitors. For example, gentamicin, a commonly used antibiotic in mammalian cell culture, is a known mitochondrial translation inhibitor [13].
5. For the proper resolution of the radiolabeled mitochondrial proteins, large polyacrylamide gels must be used. A minimum

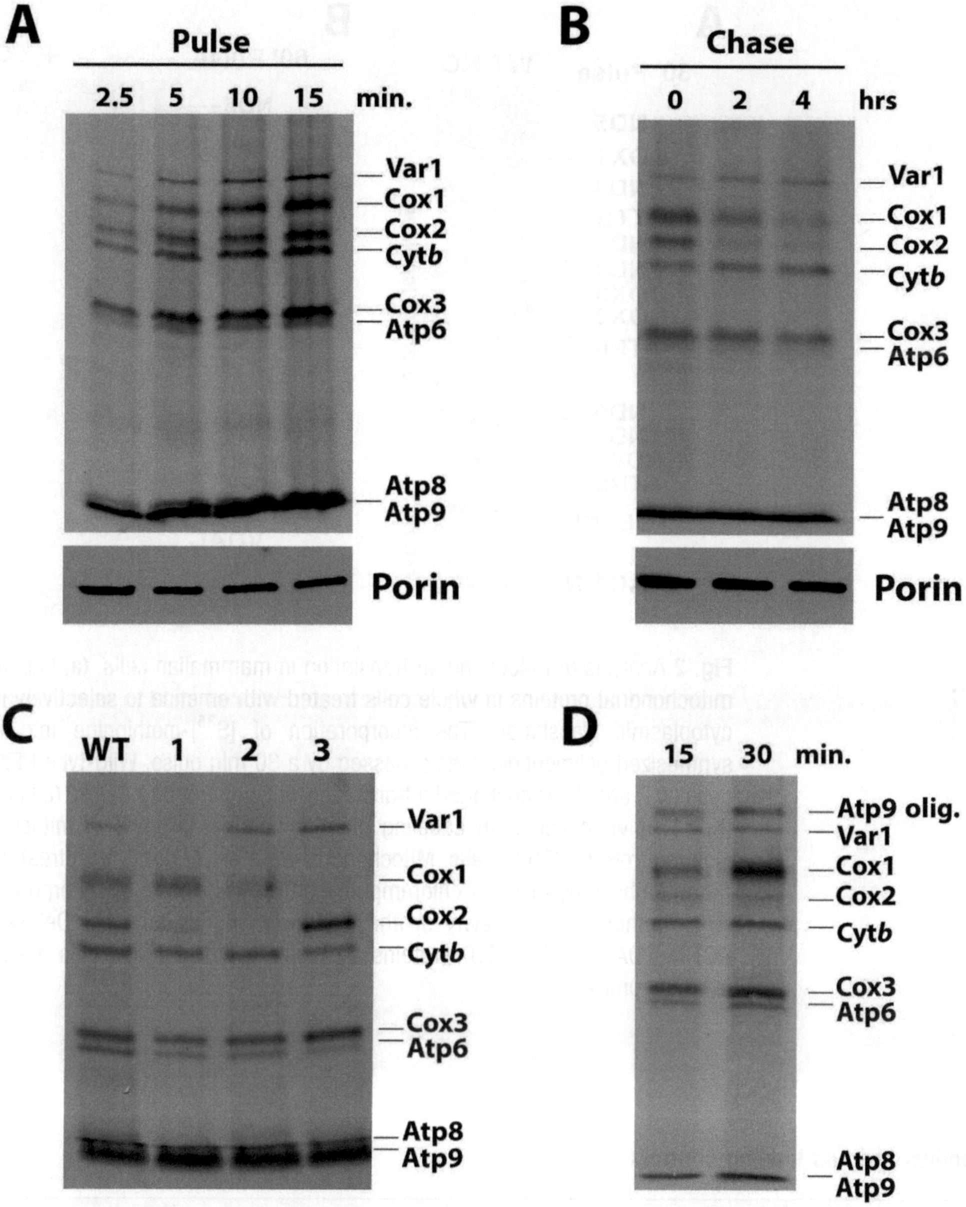

Fig. 1 Analysis of mitochondrial translation in yeast cells. (**a**) Labeling of mitochondrial proteins in whole wild-type cells treated with cycloheximide to selectively inhibit cytoplasmic translation. To determine the rate of mitochondrial translation, we performed a time course with 2.5, 5, 10, and 15 min pulses. (**b**) Stability of mitochondrion-encoded proteins analyzed by pulse-chase metabolic labeling. Whole wild-type cells were labeled for 10 min (pulse) with [S^{35}]-methionine in presence of cycloheximide followed by chases of 2 and 4 h. (**c**) Whole cell metabolic labeling of newly synthesized mitochondrial proteins in wild-type (WT) and mutant strains. 1: C-terminus truncation of the *VAR1* translational activator Sov1. The strain carries the allotopically expressed *VAR1* gene and is unable to synthesize the mitochondrion-encoded Var1 [15]. 2: Null mutant of the *COX2* translational activator Pet111. The strain is unable to synthesize COX2 and shows a downregulation of COX1 synthesis [16, 17]. 3: Null mutant of the *COX1* translational activator Pet309 [18]. (**d**) *In organello* mitochondrial translation. Mitochondria isolated from the wild-type strain W303-1A are incubated with [S^{35}]-methionine for 15 and 30 min pulses. Detection of Porin by Western blot analysis was used as a loading control

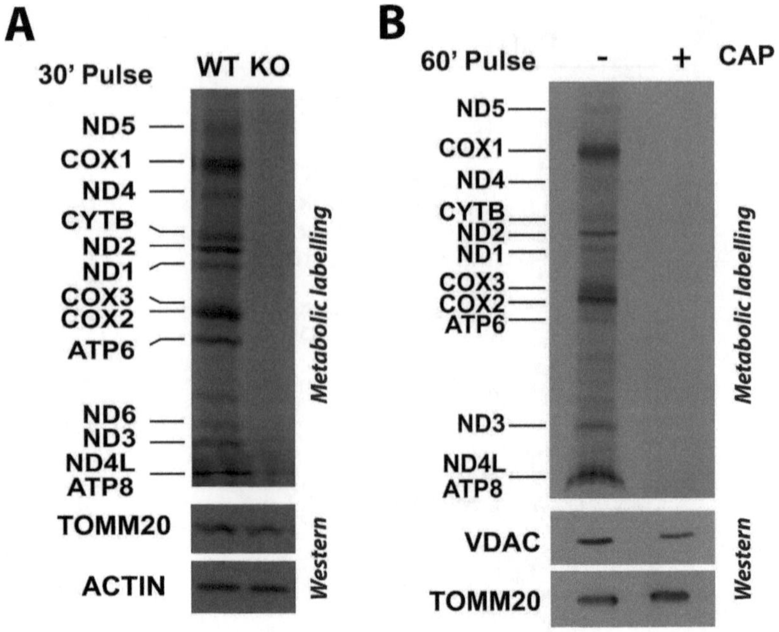

Fig. 2 Analysis of mitochondrial translation in mammalian cells. (**a**) Labeling of mitochondrial proteins in whole cells treated with emetine to selectively inhibit cytoplasmic translation. The incorporation of [S^{35}]-methionine into newly synthesized polypeptides was assessed by a 30 min pulse. Wild-type HEK293T cells (WT) and the mitochondrial translation-deficient mutant *GTPBP10* knockout (KO) [19] were used. (**b**) Labeling of mitochondrial proteins in mitochondria isolated from HEK293T cells. Mitochondria were (+) or not (−) pretreated with the mitoribosome inhibitor chloramphenicol (CAP), followed by incorporation of [S^{35}]-methionine into newly synthesized polypeptides for 1 h. Detection of ACTIN, VDAC, or TOMM20 proteins by Western blot analysis was used as a loading control

Table 3
Antibodies used as loading controls

Antibody	Source	Use
Porin	Abcam # ab110326	Yeast whole cell and *in organello* labeling
VDAC	Abcam # ab14734	Mammalian whole cell and *in organello* labeling
TOMM20	Santa Cruz # sc-11415	Mammalian whole cell and *in organello* labeling
Beta-ACTIN	Abcam # ab170325	Yeast and mammalian whole cell labeling
β-ACTIN	Proteintech # 60008-1-Ig	Mammalian whole cell labeling

glasses vertical length (H in Table 2) of 16 and 20 cm for yeast and mammalian proteins, respectively, should be used.

6. Both wet tank and semidry electrotransfer can be used. However, the semidry transfer has the advantage to not generate liters of potentially radioactively contaminated transfer buffer that require appropriate disposal.
7. The molecular weight marker should include proteins with a molecular weight range of 10–80 kDa, frequently referred to as the Low Range.
8. For consistent results, use a fresh yeast culture grown in a YN-Gal plate to inoculate the liquid media.
9. It is important to use yeast cells in the early/mid logarithmic growth phase ($OD_{600} = 0.6$–1.5). It is a good practice when using a new strain to determine its division time in YN-Gal media and adjust accordingly the preculture dilution.
10. When using non-thermosensitive strains, comparable results are obtained by performing this and the following steps at room temperature.
11. We purchase [^{35}S]-methionine from PerkinElmer (cat# NEG709A). The [^{35}S]-methionine volume indicated may need to be adjusted depending on the source of the radioisotope and its specific activity. Additionally, due to the radioactive decay, the intensity of the signal decreases with the aging of the [^{35}S]-methionine solution. We use [^{35}S]-methionine reagent up to 3 months old. However, a significant decrease in signal intensity is already observed with a month-old solution.
12. To properly determine the rate by which the different translation products are synthesized, it is essential to observe a linear increase in signal intensity over increasing pulse time. We recommend performing a time course including 2.5, 5, 10, 15, and 30 min pulses (*see* Fig. 1). In the case in which the total signal reaches a plateau already after a few minutes of labeling, it is possible to dilute 1:2 the [^{35}S]-methionine reagent with a cold 10 μM methionine solution to reduce the specific activity and slow down the incorporation of the [^{35}S]-methionine in the newly synthesized proteins [14].
13. To follow the kinetics of newly synthesized protein degradation, perform a time course with multiple chase times (1, 2, and 4 h) (*see* Fig. 1). Shorter incubation times may be needed for mutant strains or conditions characterized by high instability of newly synthesized proteins.
14. It is important to carefully remove the entire supernatant after the wash with H_2O. It is possible to aspirate with a needle connected to the vacuum the small drops of supernatant present on the tube walls. Residues of TCA not completely removed during the washes can lower the pH of the Laemmli buffer, which in consequence turns yellow. If this occurs, add

to the resuspended sample 1–3 μL of 0.5 M Tris base, 1 μL at the time, until the color of the solution turns dark blue.

15. You should expect to obtain 3–5 g of cell wet weight for liter of initial culture.
16. To test for cells conversion to spheroplasts, add 50 μL cell suspension to either 2 mL water or 2 mL 1.2 M sorbitol. Measure absorbance at 600 nm wavelength. The water sample should have OD_{600} 10–20% lower than the sorbitol sample. Optimization of zymolyase digestion time may be needed depending on the strain used and zymolyase source.
17. To ensure proper homogenization, the pestle should create a vacuum in the dounce that will break the plasma membrane by negative pressure. The number of strokes may need to be optimized.
18. The low-speed pellet is rather loose. It is possible to repeat the low-speed centrifugation step (1500 g, 5 min, 4 °C) with the supernatant to eliminate residual cellular debris.
19. Gently resuspend the mitochondria as they pass in the space between the dounce and the loose Teflon pestle. Avoid any suction that would break the mitochondria.
20. The volume of SEH buffer can be adjusted depending on mitochondrial pellet size.
21. It is preferable to use freshly isolated mitochondria for yeast *in organello* [^{35}S]-methionine labeling. However, mitochondria previously frozen once can be used for this assay, although with lower efficiency of [^{35}S]-methionine incorporation.
22. Collagen coating is useful to avoid the loss of cells, like HEK293T that poorly attach to plasticware during the [^{35}S]-methionine labeling experiment, but it is not necessary for cells that are strongly attached to plasticware.
23. The cell number may need to be adjusted in case of slow-growing respiratory mutants or different cell types. The goal is to use cultures 70–80% confluent for [^{35}S]-methionine labeling.
24. Optional step: cells can be incubated in media containing 40 μg/mL chloramphenicol for 24 h prior to metabolic labeling. Chloramphenicol is a specific reversible mitoribosome inhibitor. Inhibition of mitochondrial translation leads to the accumulation in mitochondria of nuclear-encoded translation factors, which in turn produce a spike of mitochondrial translation once the inhibition is removed [12].
25. HEK293T cells do not attach well to plasticware and can be easily washed off. Perform all washes carefully, avoid adding the buffer directly on the cells, but add it against the vessel wall.
26. Emetine is an irreversible inhibitor of cytoplasmic translation and can only be used for pulse-only experiments. Anysomicin

and cycloheximide are reversible inhibitors of cytosolic translation and can be used for pulse-chase experiments.

27. To properly determine the rate by which the different translation products are synthesized, it is essential to observe a linear increase in signal intensity over increasing pulse time. We recommend performing a time course including 10, 20, and 30 min pulses. Longer times may be needed for a partially deficient translation mutant.
28. Optional step: for pulse-only experiments, cells can be incubated in 1 mL complete DMEM medium for 5 min at 37 °C in a CO_2 incubator after the pulse with [^{35}S]-methionine to favor the complete synthesis of newly synthesized polypeptides.
29. To follow the kinetics of newly synthesized protein degradation, perform a time course with multiple chase times from 1 to 24 h, depending on the cells used and the purpose of the analysis.
30. For example, add 2.2 mL 1 M sucrose buffer if you have started with 7.5 mL of T-K-Mg buffer.
31. If using a small Teflon/glass homogenizer to resuspend the mitochondrial pellet, avoid generating any suction that could break the mitochondria.
32. The volume of the ST buffer can be adjusted (0.2–0.5 mL) depending on the size of the mitochondrial pellet.
33. Optional: as negative control it is possible to add a sample treated with the mitoribosome inhibitor chloramphenicol (CAP) (*see* Fig. 2). Pretreat 1 mg mitochondria in 1 mL translation buffer with 0.5 mg/mL for 1 h at 37 °C prior the addition of [^{35}S]-methionine.
34. Optional: background due to protein synthesis of contaminant cytoplasmic ribosomes is normally negligible. However, it is possible to add 100 μg/mL emetine to the translation buffer to inhibit cytoplasmic ribosomes.
35. We recommend storing samples at −80 °C for no longer than 1 month to still have a detectable S^{35} signal.
36. To obtain straight distinct bands, avoid loading the samples in the external wells at the two sides of the gel. Additionally, load 20 μL of 1× Laemmli buffer in any well remained empty.
37. Electrophoresis can be run at 120–140 V for 3–6 h, depending on gel length. Always avoid overheating the gel. Longer gels (~21 cm in length) can be run at 80–90 V overnight until the front dye reaches ~1 cm from the bottom of the gel.
38. We use a semidry electrotransfer apparatus and perform the transfer for 3 h at 300 mA. The transfer efficiency can be

verified by checking that no prestained molecular weight marker remains in the gel after transfer.

39. The dry nitrocellulose membrane can be also directly exposed to the X-ray film. However, it can easily attach to the film, making it difficult to remove. The cellophane wrap avoids this from occurring, but, on the other hand, it slightly reduces signal intensity. Because of the very low ^{35}S radiation penetration, any material thicker than the cellophane wrap will partially or completely screen the signal.
40. For autoradiography, exposition times vary from 1 h to overnight for yeast whole cell labeling and *in organello* experiments and from 1 to 7 days for mammalian cell labeling. For image acquisition and quantification, a valuable alternative to autoradiography is the use of a Phosphorimager system.
41. For total protein normalization, an alternative to Western blot is represented by the staining of the membrane with Ponceau S dye (Ponceau staining solution: 3% (w/v) TCA, 2 mg/mL Ponceau S, Acid Red 112).

Acknowledgments

Our work is supported by the US Army Research Office (grant W911NF-21-1-0359 to FF) and the Florida Department of Health (grant 22B12 to FF).

References

1. Herrmann JM, Woellhaf MW, Bonnefoy N (2013) Control of protein synthesis in yeast mitochondria: the concept of translational activators. Biochim Biophys Acta (BBA) Mol Cell Res 1833(2):286–294
2. Fontanesi F (2013) Mechanisms of mitochondrial translational regulation. IUBMB Life 65(5):397–408
3. Singh V, Itoh Y, Huynen MA et al (2022) Activation mechanism of mitochondrial translation by LRPPRC-SLIRP. bioRxiv. 2022: 2022.06.20.496763
4. Kummer E, Ban N (2021) Mechanisms and regulation of protein synthesis in mitochondria. Nat Rev Mol Cell Biol 22(5):307–325
5. Wallis JW, Chrebet G, Brodsky G et al (1989) A hyper-recombination mutation in S. cerevisiae identifies a novel eukaryotic topoisomerase. Cell 58(2):409–419
6. Herrmann JM, Stuart RA, Craig EA et al (1994) Mitochondrial heat shock protein 70, a molecular chaperone for proteins encoded by mitochondrial DNA. J Cell Biol 127(4):893–902
7. Enriquez JA, Attardi G (1996) Analysis of aminoacylation of human mitochondrial tRNAs. Methods Enzymol 264:183–196
8. Maiti P, Antonicka H, Gingras A-C et al (2020) Human GTPBP5 (MTG2) fuels mitoribosome large subunit maturation by facilitating 16S rRNA methylation. Nucleic Acids Res 48(14): 7924–7943
9. Fernandez-Vizarra E, Ferrin G, Perez-Martos A et al (2010) Isolation of mitochondria for biogenetical studies: an update. Mitochondrion 10(3):253–262
10. Côté C, Poirier J, Boulet D (1989) Expression of the mammalian mitochondrial genome. Stability of mitochondrial translation products as a function of membrane potential. J Biol Chem 264(15):8487–8490
11. Edgar D, Shabalina I, Camara Y et al (2009) Random point mutations with major effects on protein-coding genes are the driving force behind premature aging in mtDNA mutator mice. Cell Metab 10(2):131–138
12. Leary SC, Sasarman F (2009) Oxidative phosphorylation: synthesis of mitochondrially

encoded proteins and assembly of individual structural subunits into functional holoenzyme complexes. Methods Mol Biol 554:143–162

13. Hobbie SN, Akshay S, Kalapala SK et al (2008) Genetic analysis of interactions with eukaryotic rRNA identify the mitoribosome as target in aminoglycoside ototoxicity. Proc Natl Acad Sci U S A 105(52):20888–20893

14. Mays JN, Camacho-Villasana Y, Garcia-Villegas R et al (2019) The mitoribosome-specific protein mS38 is preferentially required for synthesis of cytochrome c oxidase subunits. Nucleic Acids Res 47(11):5746–5760

15. Seshadri SR, Banarjee C, Barros MH et al (2020) The translational activator Sov1 coordinates mitochondrial gene expression with mitoribosome biogenesis. Nucleic Acids Res 48(12):6759–6774

16. Barrientos A, Zambrano A, Tzagoloff A (2004) Mss51p and Cox14p jointly regulate mitochondrial Cox1p expression in Saccharomyces cerevisiae. EMBO J 23(17):3472–3482

17. Poutre CG, Fox TD (1987) PET111, a Saccharomyces cerevisiae nuclear gene required for translation of the mitochondrial mRNA encoding cytochrome c oxidase subunit II. Genetics 115(4):637–647

18. Manthey GM, McEwen JE (1995) The product of the nuclear gene PET309 is required for translation of mature mRNA and stability or production of intron-containing RNAs derived from the mitochondrial COX1 locus of Saccharomyces cerevisiae. EMBO J 14(16):4031–4043

19. Maiti P, Kim H-J, Tu Y-T et al (2018) Human GTPBP10 is required for mitoribosome maturation. Nucleic Acids Res 46(21):11423–11437

Chapter 13

Assembly of the Mitochondrial Translation Initiation Complex

Cristina Remes, Minh Duc Nguyen, Henrik Spahr, Martin Ng, Clark Fritsch, Arpan Bhattacharya, Hong Li, Barry Cooperman, and Joanna Rorbach

Abstract

Mitochondria maintain their own translational machinery that is responsible for the synthesis of essential components of the oxidative phosphorylation system. The mammalian mitochondrial translation system differs significantly from its cytosolic and bacterial counterparts. Here, we describe detailed protocols for efficient in vitro reconstitution of the mammalian mitochondrial translation initiation complex, which can be further used for mechanistic analyses of different aspects of mitochondrial translation.

Key words Mitoribosome, Translation, Translation initiation, tRNA purification, Translation factors

1 Introduction

Mitochondria are specialized eukaryotic organelles with important roles in energy production, cellular homeostasis, and metabolism. Mitochondrial protein synthesis is achieved by specialized mitoribosomes, which are structurally divergent from their bacterial ancestor and highly specialized for co-translational insertion of the synthesized proteins into the inner mitochondrial membrane. The initiation of protein synthesis in mitochondria differs significantly from the bacterial and eukaryotic counterparts. Mammalian mitochondrial transcripts are leaderless, with very short (1–3 nucleotides) or absent 5′-untranslated regions (UTRs). In addition, initiation factor 1, which is essential in the bacterial and cytosolic translational machineries in eukaryotic systems, is absent in mitochondria, while the mitochondrial initiation factors 2 (mtIF2) and 3 (mtIF3), both homologous to bacterial initiation factors, have acquired mitochondria-specific extensions and insertions [1, 2]. *In vitro* reconstituted translation systems have been broadly used to dissect the mechanism of the distinct steps in prokaryotic and eukaryotic translation, but such information

Antoni Barrientos and Flavia Fontanesi (eds.), *The Mitoribosome: Methods and Protocols*,
Methods in Molecular Biology, vol. 2661, https://doi.org/10.1007/978-1-0716-3171-3_13,

regarding the mitochondrial system is currently lacking. While recent cryo-EM studies have uncovered important features of the mitochondrial translation apparatus [3–7], detailed biophysical characterization, including kinetics of different stages of mitochondrial translation, is missing. As mitochondrial gene expression has been implicated in many pathophysiological processes [8], understanding mechanistic details of mitochondrial protein synthesis is necessary for the future development of therapeutics targeting this process. Here, we present protocols for purification of the translation initiation machinery, using human mitoribosomes, *E. coli* fMet-tRNA^{fMet}, mitochondrial translation initiation factors mtIF2 and mtIF3, and leaderless mRNAs. In addition, we describe a method for double labelling of the mitoribosome, which can be applied to characterize the dynamic interplay of the two mitoribosomal subunits during different stages of translation, using single-molecule techniques.

2 Materials

All solutions should be prepared with DEPC-treated water (*see* **Note 1**), water, and molecular grade reagents.

2.1 Purification of 55S Ribosomes

2.1.1 Purification of Mitochondria from HEK Cells

1. HEK293S TetR GnTI^{-} (*see* **Note 2**).
2. Dulbecco's Modified Eagle Medium (DMEM) supplemented with 10% FBS, 100 I.U./mL Penicillin, 100 μg/mL Streptomycin.
3. Freestyle 293 Expression Medium supplemented with 5% FBS, 100 I.U./mL Penicillin, 100 μg/mL Streptomycin.
4. Phosphate Buffered Saline (PBS).
5. Lysis buffer: 25 mM HEPES-KOH, pH 7.5, 150 mM KCl, 20 mM $Mg(OAc)_2$, 2% Triton X-100, 2 mM DTT, protease inhibitors, RNase inhibitors.
6. MIB buffer: 25 mM HEPES/KOH pH 7.5, 50 mM KCl, 20 mM $Mg(OAc)_2$, 2 mM DTT, protease inhibitors.
7. SM4 buffer: 25 mM HEPES/KOH pH 7.5, 50 mM KCl, 20 mM $Mg(OAc)_2$, 2 mM DTT, 281 mM sucrose, 844 mM mannitol, protease inhibitors.

2.1.2 Purification of Crude Mitoribosomes

1. MIBSM buffer: 3 volumes MIB: 1 volume SM4.
2. 20 mg/mL Digitonin solution in MIBSM buffer (*see* **Notes 3** and **4**).
3. Lysis buffer: 25 mM HEPES/KOH pH 7.5, 50 mM KCl, 20 mM $Mg(OAc)_2$, 2% Triton X-100, protease inhibitors, RNase inhibitors.

4. Ribosome buffer: 25 mM HEPES/KOH pH 7.5, 50 mM KCl, 20 mM $Mg(OAc)_2$, 2 mM DTT, protease inhibitors, RNase inhibitors.
5. 10% sucrose in ribosome buffer, supplemented with protease inhibitors, RNase inhibitors.
6. 30% sucrose in ribosome buffer, supplemented with protease inhibitors, RNase inhibitors.

2.2 Double Labelling of Mitoribosomes

2.2.1 Fluorescent Labelling of Mitoribosomes

1. Cy3-NHS ester (*see* **Note 5**).
2. Cy5-NHS ester (*see* **Note 5**).
3. Labelling buffer: 50 mM HEPES-KOH, pH 7.5, 100 mM KCl, 10 mM $Mg(OAc)_2$.
4. Sucrose cushion/labelling buffer: 50 mM HEPES-KOH, pH 7.5, 100 mM KCl, 10 mM $Mg(OAc)_2$, 1.1 M sucrose.

2.2.2 Dissociation of Labelled Mitoribosomes

1. Dissociation buffer: 50 mM HEPES/KOH pH 7.5, 300 mM KCl, 5 mM $MgCl_2$, 1 mM DTT, protease inhibitors, RNase inhibitors.
2. 10% sucrose in dissociation buffer, supplemented with protease inhibitors, RNase inhibitors.
3. 30% sucrose in dissociation buffer, supplemented with protease inhibitors, RNase inhibitors.

2.2.3 Reassociation of Ribosomal Subunits Labelled with Different Fluorophores

1. Reassociation buffer: 50 mM HEPES/KOH pH 7.5, 50 mM KCl, 20 mM $Mg(OAc)_2$, 1 mM DTT, supplemented with protease inhibitors, RNase inhibitors.
2. 10% sucrose in reassociation buffer, supplemented with protease inhibitors, RNase inhibitors.
3. 30% sucrose in reassociation buffer, supplemented with protease inhibitors, RNase inhibitors.

2.3 Preparation of E. coli fMet-tRNA^{fMet}

2.3.1 Preparation of Immobilized 3′-Biotinylated Oligonucleotide Matrix

1. 10 mM Tris–HCl pH 7.6.
2. 3′-biotinylated oligonucleotide: (5′) ATGAGCCCGACGAGC TACCAGGCTGCTC/bio/ (3′).
3. Pierce™ High Capacity Streptavidin Agarose.

2.3.2 Preparation of Immobilized 3′-Amine Oligonucleotide Matrix

All buffers should be prepared fresh before use and stored at 4 °C.

1. Coupling buffer: 0.2 M $NaHCO_3$, 0.5 M NaCl, pH 8.3.
2. Wash buffer 1: 1 mM HCl.
3. Quench buffer: 0.1 M Tris–HCl, pH 8.5.
4. Wash buffer 2: 0.1 M NaOAc, 0.5 M NaCl, pH 5.0.

5. N-Hydroxysuccinimidyl-Sepharose 4 Fast Flow.
6. 3′-amine oligonucleotide: (5′) ATGAGCCCGACGAGCTAC CAGGCTGCTC/am/ (3′).

2.3.3 Isolation of E. coli *fMet-tRNA^{fMet} from Bulk* E. coli *tRNA*

1. *E. coli* bulk tRNAs.
2. 10 mM Tris–HCl pH 7.6.
3. 1 M $MgCl_2$.
4. 2× annealing buffer: 20 mM Tris–HCl, pH 7.6, 1.8 M Tetramethylammonium (TMA) Chloride, 0.2 mM EDTA.
5. NaOAc 3 M, pH 5.2.
6. 1 M Tris–HCl, pH 9.0.
7. Ethanol, 200 proof.

2.3.4 Isolation of Bulk E. coli *Synthetase*

1. *E. coli* MRE600 cells.
2. S100 lysis buffer: 20 mM Tris–HCl pH 7.6, 100 mM NH_4Cl, 10 mM $Mg(OAc)_2$, 0.5 mM EDTA, 3 mM β-Mercaptoethanol (BME).
3. S100 buffer: 20 mM Tris–HCl, pH 7.6, 10 mM $MgCl_2$, 30 mM NaCl, 0.5 mM EDTA, 6 mM BME.
4. S100 sucrose cushion: 20 mM Tris–HCl pH 7.5, 500 mM NH_4Cl, 10 mM $Mg(OAc)_2$, 0.5 mM EDTA, 1.1 M sucrose, 3 mM BME.

2.3.5 Charging and Formylation of E. coli *tRNA^{fMet}*

1. 100 mM Methionine (cold amino acid).
2. 10.25 mCi/mL [S35] Met (hot amino acid).
3. *E. coli* bulk synthetase.
4. Aminoacylation and formylation buffer: 100 mM Tris–HCl, pH 7.6, 7 mM ATP, 20 mM $MgCl_2$, 10 mM KCl, 0.004 units/μL Pyrophosphatase, 1 mM EDTA, 20 mM Folinic Acid, 7 mM BME.
5. NaOAc 3 M pH 5.2.
6. Ethanol, 200 proof.
7. Phenol saturated with Tris–HCl, pH 4.3.
8. Chloroform.
9. FPLC Wash buffer: 50 mM NaOAc pH 5.0.
10. FPLC elution buffer: gradient 0–1 M NaCl in 50 mM NaOAc pH 5.0.

2.4 Overexpression and Purification of Mitochondrial Translation Initiation Factors

1. SOC medium: 0.5% yeast extract, 2% tryptone, 10 mM NaCl, 2.5 mM KCl, 10 mM $MgCl_2$, 10 mM $Mg(SO_4)_2$, 20 mM glucose.
2. Kanamycin agar plates: 50 ug/mL Kanamycin in Miller LB Agar.

3. Lysis buffer: 20 mM Tris–HCl pH 7.5, 500 mM NaCl, 10% glycerol, 4 mM BME, 1 mM PMSF, 1× PIC.
4. Lysozyme, lyophilized powder.
5. Nickel-nitrilotriacetic acid (NTA) agarose.
6. Wash buffer: 20 mM Tris–HCl pH 7.5, 200 mM NaCl, 10% glycerol, 1 mM DTT, 20 mM Imidazole.
7. Elution buffer: 20 mM Tris–HCl pH 7.5, 200 mM NaCl, 10% glycerol, 1 mM DTT, 300 mM Imidazole.
8. Dialysis buffer: 20 mM Tris–HCl pH 7.5, 200 mM NaCl, 10% glycerol, 1 mM DTT, 0.5 mM EDTA.
9. Gel filtration buffer: 20 mM Tris–HCl pH 7.5, 200 mM NaCl, 1 mM DTT, 0.5 mM EDTA.
10. cDNA: Human *mtIF2* (amino acids 38–727).
11. cDNA: Human *mtIF3* (amino acids 32–278).
12. pET-24b vector (Novagen).
13. Rosetta 2 (DE3) competent cells.
14. Luria Broth (LB), MagicMedia (ThermoFisher).

2.5 Assembly of the Mitochondrial Translation Initiation Complex

1. Initiation buffer: 50 mM Tris–HCl pH 7.6, 30 mM KCl, 10 mM $MgCl_2$, 1 mM DTT, 0.1 mM spermine, 1 mM spermidine, 5 mM GTP.
2. 10% sucrose in initiation buffer lacking GTP.
3. 10% sucrose in initiation buffer lacking GTP.
4. Sucrose cushion/initiation buffer: 50 mM Tris–HCl pH 7.6, 30 mM KCl, 10 mM $MgCl_2$, 1 mM DTT, 0.1 mM spermine, 1 mM spermidine, 1.1 M sucrose.

3 Methods

3.1 Purification of Mitoribosomes from HEK Cells

3.1.1 Purification of Mitochondria from HEK Cells

1. Culture HEK293S TetR GnTI$^-$ cells in DMEM, at 37 °C and 5% CO_2. Scale up to nine T175 flasks.
2. When reaching 80% confluency, harvest the cells by centrifugation at 500 rcf for 10 min at 4 °C (*see* **Note 6**).
3. Resuspend the pelleted cells in 300 mL Freestyle 293 Expression Medium supplemented with 5% FBS, and transfer the cell culture into a 3 L Corning disposable spinner flask. Incubate at 37 °C and 5% CO_2 in an incubator containing a magnetic plate, stirring continuously at 100 rpm.
4. Monitor the density of the cell culture. When the density reaches 3×10^6 cells/mL (usually takes 2 days), split the cells by centrifugation at 500 rcf for 10 min and dissolve the pellet in

Freestyle 293 Expression Medium to a density of 1×10^6 cells/mL. Repeat the procedure until a total culture volume of 3 L.

5. When the density of the cell culture reaches 6×10^6 cells/mL, harvest the cells by centrifugation at 1000 rcf for 10 min at 4 °C.
6. Resuspend the cell pellet in 200 mL PBS and centrifuge at 1000 rcf for 5 min at 4 °C.
7. Resuspend the cell pellet in 180 mL cold MIB buffer and gently mix the cells with a magnetic stirrer for 20 min at 4 °C.
8. Measure the total volume of the resuspended cells and add a volume of SM4 buffer corresponding to 1/3 of the total volume.
9. Transfer the cells to a 250 mL Teflon/glass Dounce homogenizer on ice and gently homogenize the cells using a tight-fitting pestle.
10. Centrifuge the lysed cells at 1000 rcf for 10 min at 4 °C. Collect the supernatant and discard the pellet.
11. Centrifuge the supernatant at 10,000 rcf for 10 min at 4 °C and collect the pellet consisting of crude mitochondria. Gently dissolve the mitochondria in 5 mL cold MIBSM buffer.
12. Add 300 U of RNase-free DNase and incubate at 4 °C for 30 min.
13. Centrifuge at 10,000 rcf for 10 min at 4 °C. Discard the supernatant and dissolve the pellet in 5 mL MIBSM Buffer.
14. Snap-freeze the crude mitochondria and store at −80 °C or continue with purification of crude 55S ribosomes.

3.1.2 Sucrose Gradient Purification of 55S Monosomes

1. Add 50 μL of 10 mg/mL Digitonin to the 5 mL mitochondria and incubate on ice for 5 min.
2. Add 30 mL of MIBSM buffer to dilute the concentration of digitonin and centrifuge at 10,000 rcf for 10 min at 4 °C. Discard the supernatant.
3. Dissolve the pellet in 2 mL lysis buffer and incubate for 20 min on ice.
4. Centrifuge at 27,000 rcf for 45 min at 4 °C and discard the pellet.
5. Prepare a linear 10–30% sucrose density gradient using a Gradient Maker (Biocomp Instruments), according to the manufacturer's instruction.
6. Carefully overlay 0.5 mL supernatant on top of one sucrose gradient prepared in SW41 tubes (*see* **Notes** 7 and **8**) and centrifuge at 21,000 rpm (75,416 rcf) for 21 h at 4 °C.
7. Fractionate the gradients into 300 μL fractions by monitoring the absorbance at 254 nm, and collect the peak corresponding to the 55S monosome (fractions 20–24, Fig. 1a).

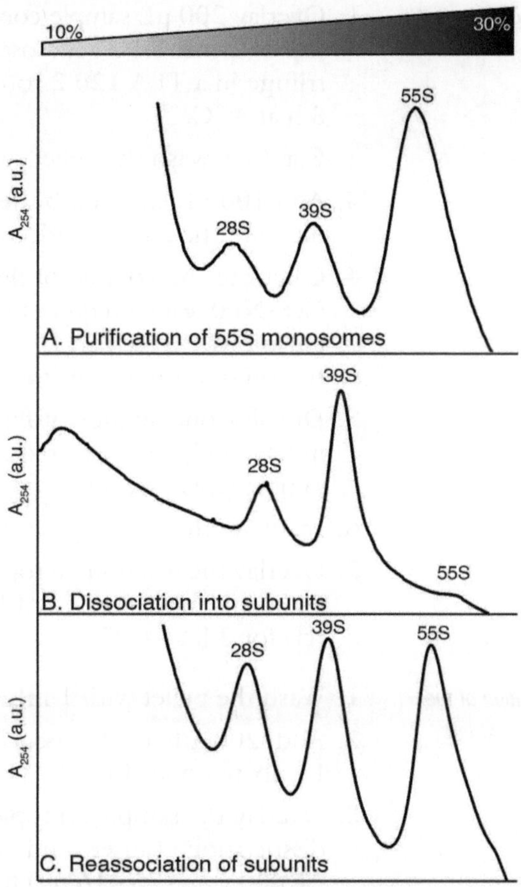

Fig. 1 Isolation of reassociated human 55S monosomes. (**a**) Purification of 55S monosomes through a 10–30% sucrose gradient. (**b**) Dissociation of the monosome into mitoribosomal subunits. (**c**) Reassociation of the mitoribosomal subunits and purification of the reassociated monosome through a sucrose gradient

8. Centrifuge in a TLA 120.2 rotor at 352,000 rcf for 16 h at 4 °C and discard the supernatant.
9. Carefully wash the pellet twice with 1 mL cold ribosome buffer.
10. Dissolve the pellet in 200 μL ribosome buffer. Aliquot the 55S monosomes to desired volumes, snap-freeze and store at −80 °C.

3.2 Double Labelling of Mitoribosomes

Here, we describe preparation of mitoribosomes labelled with Cy3 and Cy5 fluorophores at the large and small subunit, respectively, and vice versa. This is achieved by labelling of the full mitoribosome with a fluorescent dye (four to five fluorophores/monosome), dissociation into subunits and purification of labelled subunits, followed by reassociation of small and large subunits labelled with different fluorophores.

3.2.1 Labelling of the Full Monosome

1. Overlay 200 μL sample containing 500 μM mitoribosomes on top of 1 mL 1.1 M sucrose cushion/labelling buffer and centrifuge in a TLA 120.2 rotor at 90,000 rpm (351,955 rcf) for 3 h at 4 °C.
2. Carefully wash the pellet twice with 1 mL labelling buffer.
3. Add 100 μL labelling buffer and dissolve the pellet on ice and measure the concentration of mitoribosomes (*see* **Note 9**).
4. Calculate the volume of fluorescent dye (Cy3-NHS ester and Cy5-NHS ester) required to be added to the monosome sample in order to obtain 10 times excess in respect to the concentration of monosome (*see* **Note 10**).
5. Dissolve one aliquot of fluorescent dye in 10 μL DMSO and immediately add the required volume to the mitoribosome sample (*see* **Note 11**).
6. Incubate the labelling reaction at 37 °C for 45 min.
7. Overlay the reaction on top of sucrose cushion/labelling buffer and centrifuge in a TLA 120.2 rotor at 90,000 rpm (351,955 rcf) for 3 h at 4 °C.

3.2.2 Dissociation of the Labelled Monosome into Subunits

1. Wash the pellet with 1 mL cold dissociation buffer.
2. Add 200 μL cold Dissociation Buffer and dissolve on ice. Incubate on ice for 1 h.
3. Overlay the sample on top of cold 10–30% sucrose gradient in dissociation buffer and centrifuge in a SW41 rotor at 21,000 rpm (75,416 rcf) for 21 h at 4 °C.
4. Fractionate the sucrose gradient and collect the peaks corresponding to the small and large subunit into TLA120.2 tubes (*see* Fig. 1b).
5. Centrifuge in a TLA 120.2 rotor for 16 h at 90,000 rpm (351,955 rcf) at 4 °C.

3.2.3 Reassociation of Small and Large Subunits Labelled with Different Fluorophores

1. Wash the pellets containing labelled subunits twice with 500 μL cold reassociation buffer.
2. Add 50 μL reassociation buffer and dissolve the pellets on ice.
3. Measure the concentration of ribosomal subunits (*see* **Note 9**).
4. Mix equimolar amounts of large and small ribosomal subunits labelled with different fluorophores and incubate at 37 °C for 1 h.
5. Overlay the sample on top of 10–30% sucrose gradient in reassociation buffer and centrifuge in a SW41 rotor at 21,000 rpm (75,416 rcf) for 21 h at 4 °C.
6. Fractionate the sucrose gradient and collect the peak corresponding to the 55S monosome (*see* Fig. 1c).
7. Centrifuge in a TLA 120.2 rotor for 16 h at 90,000 rpm (351,955 rcf) at 4 °C.

3.3 Preparation of *E. coli* fMet-tRNA^{fMet}

Depending on the downstream application for the fMet-tRNA^{fMet}, we use two different methods for tRNA isolation from bulk *E. coli* tRNA. For ensemble co-sedimentation and filter-binding experiments, we purify tRNAs using a 3'-biotinylated oligonucleotide which is complementary to the target tRNA immobilized to streptavidin agarose matrix [9, 10] (*see* Subheading 3.3.1). For single-molecule FRET experiments, in which ribosomal complexes programmed with biotinylated mRNAs are specifically immobilized on a glass slide functionalized with streptavidin, tRNAs purified with biotinylated oligonucleotides are unsuitable due to biotin contamination. In this case, we bind the corresponding 3′-amine oligonucleotide to the matrix using N-hydroxysuccinimide-amine cross-linking [11] (*see* Subheading 3.3.2).

3.3.1 Preparation of Immobilized 3′-Biotinylated Oligonucleotide Matrix

1. Transfer 5 mL of streptavidin agarose slurry beads and 10 mL Tris–HCl 10 mM pH 7.5 into a 50 mL conical tube and mix gently.
2. Attach the Steriflip filter, and using the vacuum source, remove the buffer. Wash the beads two more times with 10 mL Tris–HCl 10 mM pH 7.5.
3. Dissolve 600 nmol 3′-biotinylated DNA-oligonucleotide in 1 mL Tris–HCl 10 mM pH 7.5.
4. Add 5 mL Tris–HCl 10 mM pH 7.5 and the dissolved 3′-biotinylated DNA-oligonucleotide to the washed streptavidin beads and incubate on a roller in the cold room overnight.
5. Remove the unbound 3′biotinylated DNA-oligonucleotide and wash the streptavidin beads with 10 mL Tris–HCl 10 mM pH 7.5 three times.
6. Dissolve the beads in 10 mL Tris–HCl 10 mM pH 7.5 and store them at 4 °C until use.

3.3.2 Preparation of Immobilized 3′-Amine Oligonucleotide Matrix (See Note 12)

1. Transfer 15 mL N-Hydroxysuccinimidyl-Sepharose 4 Fast Flow and 30 mL cold wash buffer 1 into a 50 mL conical tube and mix gently.
2. Centrifuge at 4000 rpm (3500 rcf) for 1 min at 4 °C and carefully remove the supernatant.
3. Wash the beads five more times with 30 mL cold wash buffer 1 by repeating **steps 1** and **2** five times.
4. Add 15 mL cold coupling buffer to the beads and mix.
5. Centrifuge at 4000 rpm (3500 rcf) for 1 min at 4 °C and carefully remove the supernatant.
6. Repeat **steps 4** and **5** three times.
7. Add 15 mL cold coupling buffer and mix gently.

8. Dissolve 500 nmol 3′-amine modified oligonucleotide in 1 mL coupling buffer, add it to the activated beads and vortex gently.
9. Incubate at room temperature for 2 h and in the cold room overnight, mixing the tube continuously on a roller.
10. Remove the unbound 3′-amine oligonucleotide by vacuum filtration.
11. Add 30 mL cold quench buffer to the beads and mix. Remove the buffer by vacuum filtration. Repeat this step three more times.
12. Add 30 mL cold wash buffer 2, mix and remove the buffer by vacuum filtration. Repeat this step twice.
13. Add 30 mL cold quench buffer, mix and remove the buffer by vacuum filtration. Repeat this step twice.
14. Repeat **steps 12** and **13** twice.
15. Add 30 mL quench buffer and incubate in the cold room overnight, mixing the tube continuously on a roller.
16. Vacuum-filter the supernatant and dissolve the beads in 30 mL 10 mM Tris–HCl pH 7.5. Store at 4 °C until use.

3.3.3 Isolation of E. coli tRNA^{fMet} from Bulk E. coli tRNA

1. Dissolve 100 mg bulk *E. coli* tRNA in 10 mL DEPC-treated water.
2. Incubate the bulk *E. coli* tRNA solution at 80 °C for 10 min.
3. Incubate 10 mL 2× annealing buffer at 65 °C for 10 min and add it to the bulk *E. coli* tRNA solution. Incubate the mixture at 65 °C for 10 min.
4. Slowly lower the incubation temperature to 37 °C by turning off the heat block (this process usually takes 45 min). Then incubate for additional 30 min.
5. Wash away the unbound bulk *E. coli* tRNA with cold 10 mL Tris–HCl 10 mM pH 7.5. Repeat this step until the absorbance measured at 260 nm is lower than 0.1 a.u.
6. Add 10 mL Tris–HCl 10 mM pH 7.5 to the beads and incubate at 45 °C for 10 min. Collect the eluted tRNA by vacuum filtration.
7. Repeat the previous step by incubating at 55 and 65 °C.
8. Adjust the Mg^{2+} concentration in the eluted tRNA samples to 10 mM, using a 1 M $MgCl_2$ solution.
9. To the tRNA^{fMet} 65 °C elution sample, add 0.1 volume of 1 M Tris–HCl pH 9 and incubate at 37 °C for 3 h to hydrolyze the charged tRNA.
10. Add 0.1 volumes of NaOAc 3 M pH 5.2 and 2.5 volumes of cold ethanol and incubate at −80 °C for 1 h.

11. Centrifuge in a SS34 rotor at 14,000 rpm (23,425 rcf) for 30 min at 4 °C and remove the supernatant.
12. Dry the pellet in a desiccator for 15 min.
13. Dissolve the dry pellet in 200 μL DEPC-treated water and measure the concentration of the $tRNA^{fMet}$.

3.3.4 Isolation of Bulk E. coli Synthetase

1. Resuspend 20 g MRE600 *E. coli* cell pellet in 60 mL S100 lysis buffer.
2. Lyse the cells with two cycles of French press at 1000–1200 psi.
3. Centrifuge at 14,000 rpm (23,425 rcf) for 30 min in a SS34 rotor. Discard the pellet.
4. Centrifuge at 17,500 rpm (35,508 rcf) for 1 h in a Ti45 rotor. Discard the pellet.
5. In Ti45 tubes, overlay the supernatant on top of an equal volume of S100 sucrose cushion.
6. Centrifuge at 32,000 rpm (118,727 rcf) for 16 h at 4 °C in a Ti45 rotor. Discard the pellet.
7. Dialyze the supernatant against S100 buffer.
8. Load the sample on a 15 mL DEAE Sepharose Fast Flow column equilibrated with S100 buffer.
9. Elute with S100 buffer, monitoring the absorbance of the eluate at 260 nm (*see* Fig. 2a), and collect 5 mL fractions. Measure the A_{280}/A_{260} of each fraction and pull together the fractions for which A_{280}/A_{260} is greater than 1.5.
10. Aliquot the bulk *E. coli* synthetase into 0.5 mL fractions, snap-freeze and store at −80 °C.

3.3.5 Charging and Formylation of E. coli $tRNA^{fMet}$

1. Incubate 20 μM tRNA $tRNA^{fMet}$ with 200 μM Met nonradioactive (cold amino acid), [^{35}S]-Met (hot amino acid) (*see* **Note 13**), and 0.1 volumes of *E. coli* S100 bulk synthetase in aminoacylation and formylation buffer at 37 °C for 45 min.
2. Add one volume saturated phenol pH 4.3 and mix thoroughly by vortexing.
3. Centrifuge at 10,000 rcf for 5 min at 4 °C and collect the top aqueous part.
4. Add one volume of chloroform and vortex thoroughly.
5. Centrifuge at 10,000 rcf for 5 min at 4 °C and collect the top aqueous part.
6. Add 0.1 volumes of NaOAc 3 M pH 5.2 and 2.5 volumes of cold ethanol and incubate at −80 °C for 1 h.
7. Centrifuge at 11,000 rcf for 30 min at 4 °C and remove the supernatant.
8. Dry the pellet in a desiccator for 15 min.

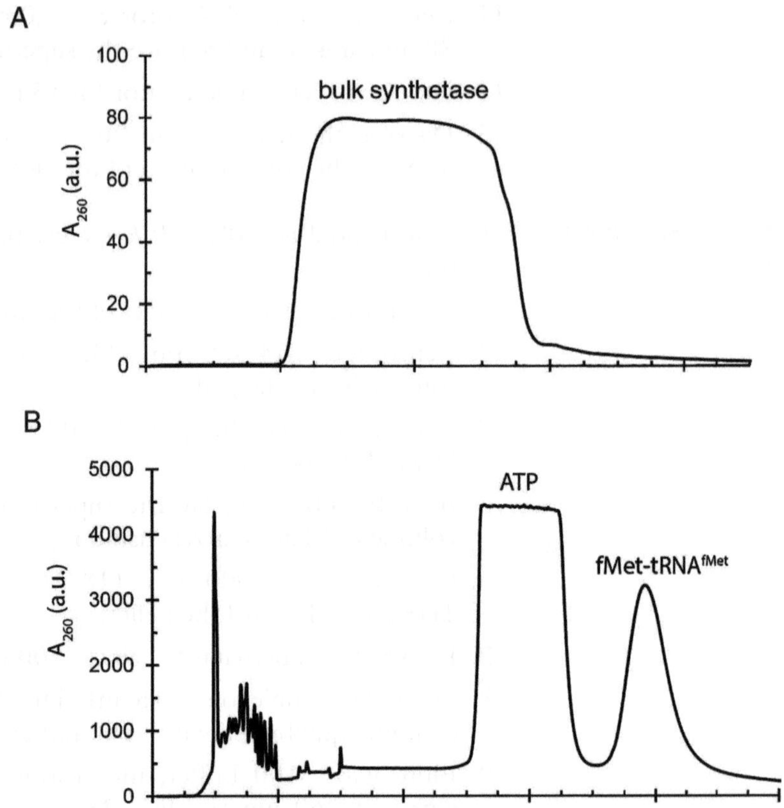

Fig. 2 (**a**) Purification of bulk *E. coli* synthetase through a DEAE Sepharose column. (**b**) Purification of charged fMet-tRNA^{Met} through a Q Sepharose column

9. Dissolve the dry pellet in 1 mL FPLC wash buffer.
10. Load the sample to a 1 mL Q-Sepharose resin, prewashed with 15 mL FPLC wash buffer.
11. Elute the fMet-tRNA^{fMet} with FPLC elution buffer.
12. Collect the peak corresponding to the fMet-tRNA^{fMet} (*see* Fig. 2b). Add 0.1 volumes of NaOAc 3 M pH 5.2 and 2.5 volumes of cold ethanol. Incubate at −80 °C for 1 h.
13. Centrifuge at 11,000 rcf for 30 min at 4 °C and remove the supernatant.
14. Dry the pellet in a desiccator for 15 min.
15. Dissolve the dry pellet in 200 μL DEPC-treated water.
16. Store the fMet-tRNA^{fMet} in aliquots at −80 °C.

3.4 Overexpression and Purification of Mitochondrial Translation Initiation Factors

3.4.1 Plasmid Transformation

1. Transfer ~50 ng of pET-24b plasmid containing mtIF2 or mtIF3 construct to 100 μL Rosetta 2 (DE3) competent cells.
2. Incubate the mixture on ice for 30 min, then heat shock at 42 °C for 60 s.
3. Place the mixture on ice for 5 min.
4. Add 500 μL of room temperature SOC media to the mixture.
5. Place the tube at 37 °C in a shaking incubator for 60 min.
6. Spread 100–200 μL of the mixture on the prewarmed Kanamycin agar plate.
7. Incubate overnight at 37 °C in the incubator.

3.4.2 Large-Scale Protein Expression

1. Pick a single colony from the plate and inoculate in 40 mL LB supplemented with 30 ng/mL of Kanamycin.
2. Incubate overnight in a shaking incubator at 220 rpm, at 37 °C.
3. Dilute 1:50 of LB culture media in MagicMedia (total 2 L of culture media) supplemented with 30 ng/mL of Kanamycin and incubate overnight (~16 h) at 25 °C in a shaking incubator (*see* **Note 14**).
4. Harvest the cells by centrifugation at 4000 rpm (3500 rcf) for 10 min at 4 °C and store the pellet at −80 °C until use.

3.4.3 Protein Purification

1. Resuspend the cell pellet in 200 mL of lysis buffer supplemented with 5 mg/mL of lysozyme and incubate at 4 °C for 1 h.
2. Add 500 Unit of Universal Nuclease into the mixture and incubate for additional 15 min at room temperature (*see* **Note 15**).
3. Centrifuge at 20,000 × *g* for 2 h at 4 °C to obtain the cell lysate and remove the cell debris.
4. Filter the supernatant with a 0.45 μm filter to remove cell debris and aggregated particles.
5. Add 5 mL of Nickel-NTA agarose pre-equilibrated with lysis buffer to the filtered cell lysate and incubate at 4 °C for 1.5 h.
6. Transfer the mixture into an open chromatography column for gravity flow through of unbound components.
7. Wash any remaining unbound or weakly bound proteins with 30 mL wash buffer (six times resin volume).
8. Elute the His-tagged proteins with 10 mL elution buffer.
9. Dialyze the eluate against dialysis buffer at 4 °C overnight to remove imidazole.
10. Collect the dialyzed sample and concentrate to 5 mL.

11. Load the concentrated protein sample into a size-exclusion chromatography system via a HiLoad 16/600 Superdex 200 pg gel filtration column (GE Healthcare) pre-equilibrated with the gel filtration buffer.
12. Pool the peak fractions containing the target proteins, concentrate to 1~2 mg/mL (measured by nanodrop or BCA assay), aliquot, flash-freeze and store at −80 °C.
13. Check the purity and protein concentration throughout purification via SDS/PAGE and compare with known amounts of BSA.

3.5 Assembly of the Mitochondrial Translation Initiation Complex

1. Prepare 200 μL initiation reaction containing 200 nM mitoribosomes, 240 nM mtIF2, 240 nM mtIF3, 240 nM fMet-$tRNA^{fMet}$, 1 μM mRNA (*see* **Note 16**) in initiation buffer. Incubate at 37 °C for 30 min and on ice for 15 min.
2. Overlay the reaction on top of one 10–30% sucrose gradient and centrifuge at 21,000 rpm (75,416 rcf) for 21 h at 4 °C in a SW41 rotor.
3. Fractionate the gradient into 300 μL fractions and collect the peak corresponding to the 55S monosome (fractions 20–24).
4. Concentrate the sample to 50 μL using a Vivaspin 500 centrifugal concentrator.
5. Measure the concentration of mitoribosomes (*see* **Note 9**) and determine the concentration of initiator tRNA bound to the initiation complex using the specific activity of the fMet-$tRNA^{fMet}$.

4 Notes

1. To prepare DEPC-treated water, add 1 mL of 0.1% Diethylpyrocarbonate (DEPC) to 1000 mL MQ water, mix for 3 h and autoclave. Let cool to room temperature before use.
2. HEK293S-derived cells grow as adherent monolayers on tissue culture-treated flasks and are also suitable for growing in cell suspension conditions.
3. Digitonin is toxic if inhaled and should be handled under a hood. The digitonin solution should be prepared immediately before use.
4. To dissolve digitonin, heat the required volume of MIB to a temperature just before boiling and add digitonin by continuously stirring on a magnetic plate.
5. For reproducible results, NHS ester dyes should be handled in a glovebox. To prepare aliquots, dyes should be dissolved in DMSO, aliquoted, lyophilized, and stored at −80 °C.

6. HEK293S TetR GnTI^{-} cells grow as adherent monolayers, but do not stick strongly to the culture plates. To split the cells, wash carefully the cell layer with 10 mL room temperature PBS, add 10 mL room temperature PBS, and gently tap the cell culture dish to detach the cells. Slowly pipette the cell suspension and transfer to a 15 mL conical tube.
7. The sucrose gradients should be prepared using room temperature sucrose solutions and cooled in the cold room before use.
8. For reproducible sucrose gradients, a maximum volume of 0.5 mL should be loaded on top of a gradient prepared in a SW41 tube.
9. The concentration of monosomes and mitoribosomal subunits is determined using the conversion:

   $1\ A_{260} = 32$ pmol 55S

   $1\ A_{260} = 77$ pmol 28S

   $1\ A_{260} = 55$ pmol 39S
10. Depending on the downstream application, ribosomes can be labelled with a low or high number of fluorophores. The labelling ratio can be optimized by varying the excess of fluorescent dye added in the labelling reaction and the reaction time. We obtained a labelling ratio of 0.8, 2.3, 5 dyes/monosome by incubation with an excess of fluorophore relative to the monosome of 1, 5, 10, respectively.
11. The half-time of NHS esters is approximately 10 min at neutral pH. The dye should be dissolved in labeling buffer immediately before use.
12. For best oligo binding efficiency, it is recommended to perform the washing steps quickly, maintaining a temperature of 4 °C throughout the procedure.
13. For our purposes, the fMet-tRNA^{fMet} was charged with a specific activity of 1500 cpm/pmol. The specific activity can be optimized by varying the ratio of hot/cold amino acid added in the charging reaction.
14. The MagicMedia (Thermo Fisher) is an *E.coli* expression medium, which provides higher protein yields with much less hands-on time (i.e., no OD monitoring and induction steps). The traditional protocol using IPTG-induced system can still be used as an alternative approach.
15. Alternative methods, such as sonication can be used to lyse the cells.
16. We obtained reproducible initiation activity of mitoribosomes on leaderless mRNAs with sizes ranging from 30 to 60 nucleotides, containing no 3′-UTR and the start codon AUG.

References

1. Yassin AS, Haque ME, Datta PP et al (2011) Insertion domain within mammalian mitochondrial translation initiation factor 2 serves the role of eubacterial initiation factor 1. Proc Natl Acad Sci U S A 108:3918–3923. https://doi.org/10.1073/pnas.1017425108
2. Bhargava K, Spremulli LL (2005) Role of the N- and C-terminal extensions on the activity of mammalian mitochondrial translational initiation factor 3. Nucleic Acids Res 33:7011–7018. https://doi.org/10.1093/nar/gki1007
3. Aibara S, Singh V, Modelska A, Amunts A (2020) Structural basis of mitochondrial translation. elife. https://doi.org/10.7554/ELIFE.58362
4. Kummer E, Leibundgut M, Rackham O et al (2018) Unique features of mammalian mitochondrial translation initiation revealed by cryo-EM. Nature 560:263. https://doi.org/10.1038/s41586-018-0373-y
5. Khawaja A, Itoh Y, Remes C et al (2020) Distinct pre-initiation steps in human mitochondrial translation. Nat Commun 11:1–10. https://doi.org/10.1038/s41467-020-16503-2
6. Koripella RK, Sharma MR, Bhargava K et al (2020) Structures of the human mitochondrial ribosome bound to EF-G1 reveal distinct features of mitochondrial translation elongation. Nat Commun 11:3830. https://doi.org/10.1038/s41467-020-17715-2
7. Kummer E, Schubert KN, Schoenhut T et al (2021) Structural basis of translation termination, rescue, and recycling in mammalian mitochondria. Mol Cell 81:2566. https://doi.org/10.1016/j.molcel.2021.03.042
8. Wang F, Zhang D, Zhang D et al (2021) Mitochondrial protein translation: emerging roles and clinical significance in disease. Front Cell Dev Biol 9:675465
9. Barhoom S, Farrell I, Shai B et al (2013) Dicodon monitoring of protein synthesis (DiCoMPS) reveals levels of synthesis of a viral protein in single cells. Nucleic Acids Res 41:e177. https://doi.org/10.1093/NAR/GKT686
10. Liu J, Pampillo M, Guo F et al (2014) Monitoring collagen synthesis in fibroblasts using fluorescently labeled tRNA pairs. J Cell Physiol 229:1121–1129. https://doi.org/10.1002/JCP.24630
11. Cui Z, Stein V, Tnimov Z et al (2015) Semisynthetic tRNA complement mediates in vitro protein synthesis. J Am Chem Soc 137:4404–4413. https://doi.org/10.1021/JA5131963/SUPPL_FILE/JA5131963_SI_001.PDF

Chapter 14

Reconstitution of Mammalian Mitochondrial Translation System Capable of Long Polypeptide Synthesis

Muhoon Lee and Nono Takeuchi-Tomita

Abstract

Mammalian mitochondria have their own dedicated protein synthesis system, which produces 13 essential subunits of the oxidative phosphorylation complexes. Here, we describe the in vitro reconstitution of the mammalian mitochondrial translation system, utilizing purified recombinant mitochondrial translation factors, 55S ribosomes from pig liver mitochondria, and a heterologous yeast tRNA mixture. The system is capable of translating leaderless mRNAs encoding model proteins, such as nanoluciferase with a molecular weight of 19 kDa, and is readily applicable for in vitro evaluations of mRNAs and nascent peptide chain sequences, as well as factors and small molecules that affect mitochondrial translation.

Key words In vitro translation, Reconstituted translation system, Mammalian mitochondria, 55S ribosome, Leaderless mRNA

1 Introduction

The protein synthesis system in mammalian mitochondria utilizes specialized mitochondrial ribosomes, tRNAs, and translation factors, which have peculiarly diverged from their bacterial counterparts to accomplish mitochondrial-specific tasks, such as exclusively synthesizing the hydrophobic membrane subunits of the respiratory chain complexes. Mitochondrial 55S ribosomes are characterized by reduced rRNA contents and the acquisition of numerous ribosomal proteins, with unique mitochondrial ribosomal proteins facilitating the synthesis of membrane proteins and their co-translational membrane insertion [1–3]. The 22 species of mt-tRNAs are generally shorter than the canonical tRNAs, and some even lack invariant nucleotides [4]. Somewhat different translation factors are utilized in mitochondria as compared to bacteria [2, 5, 6], such as two initiation factors (IF-2mt, IF-3mt), three

Antoni Barrientos and Flavia Fontanesi (eds.), *The Mitoribosome: Methods and Protocols*, Methods in Molecular Biology, vol. 2661, https://doi.org/10.1007/978-1-0716-3171-3_14,

elongation factors (EF-Tumt, EF-Tsmt, EF-G1mt), and three termination and ribosome recycling factors (RF-1Lmt/mtRF1a, EF-G2mt, RRFmt). These factors retain the conserved core structures of their bacterial counterparts, but often possess extra insertion or terminal extension sequences with unknown functions. The homolog of bacterial IF-1 is missing, but the insertion sequence of IF-2mt replaces its role [7]. The dual functions of bacterial EF-G, translocation and ribosome recycling, are performed separately by two homologs, EF-G1mt and EF-G2mt, respectively [8]. There is no mitochondrial homolog of bacterial RF3. The 13 proteins encoded by mtDNA are translated from nine monocistronic mRNAs and two dicistronic mRNAs with overlapping reading frames. Except for the two downstream reading frames in the dicistronic mRNAs, the remaining 11 reading frames are all encoded on leaderless mRNAs, which lack the 5′ leader sequence to promote mRNA recruitment to the ribosome [9]. A nonuniversal genetic code is used for mitochondrial mRNAs [4].

In addition to the recently developing innovative technologies of mtDNA editing [10] and mitochondrial gene silencing [11], the *in vitro* mitochondrial translation system is useful for studying mitochondrial protein synthesis dissecting the unique molecular interactions between mitochondrial ribosomes, mRNAs, tRNAs, and translation factors. We have recently developed a reconstituted mammalian mitochondrial translation system, using purified recombinant mitochondrial translation factors, 55S ribosomes from pig liver mitochondria, and a yeast tRNA mixture [5]. We used a heterologous yeast tRNA mixture because it is currently difficult to prepare a sufficient amount of mitochondrial tRNAs. This system is capable of correctly initiating translation from the start codon of the leaderless mRNA, and synthesizing long polypeptides with lengths that extend through the peptide tunnel, and is useful for analyzing the mechanism of translation initiation, and the interactions between the nascent peptide chain and the mitochondrial ribosome. Moreover, in combination with translational activators, assembly factors, and co-translational membrane insertion machineries, such as proteoliposomes with integrated mitoribosome membrane receptors, this mitochondrial translation system will lead the way toward the elucidation of the translation mechanism in mammalian mitochondria. Here, we provide the protocols for factor preparation, system manipulation, and translation product detection. The system is readily applicable for monitoring translation regulation by specific mRNA or nascent peptide chain sequences, and for *in vitro* evaluations of small molecules that target the mitochondrial translation apparatus (Fig. 1).

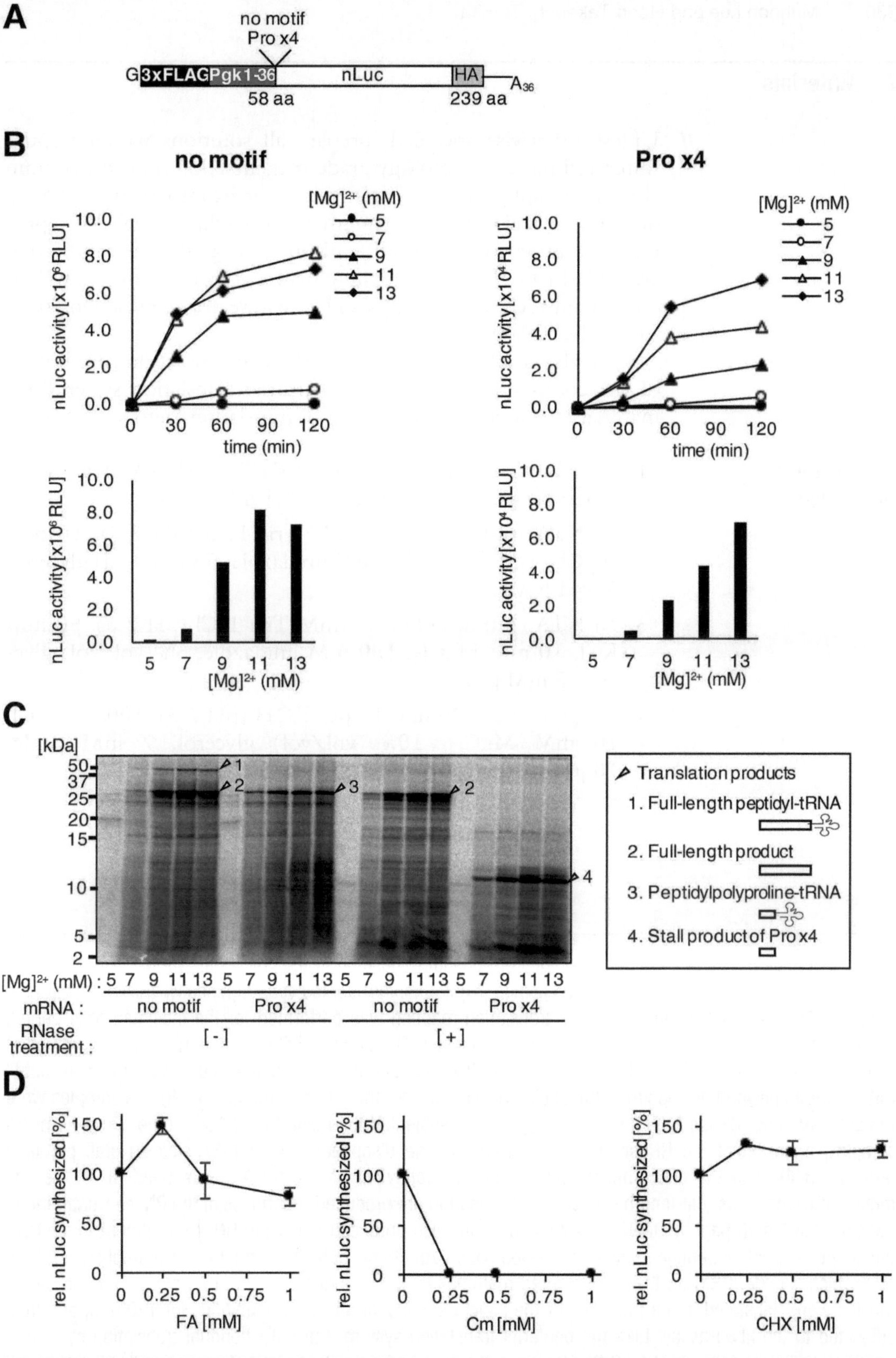

Fig. 1 Effects of the polyproline-ribosome stalling sequence and antibiotics on nanoluciferase synthesis in the reconstituted mammalian mitochondrial translation system. (**a**) Schematic of the mRNAs. In the "Prox4" mRNA, the polyproline sequence is inserted in front of the nanoluciferase (nLuc) sequence, while there is no

2 Materials

Unless otherwise specified, prepare all solutions with ultrapure water and molecular biology grade reagents. Solutions for column chromatography are filter-sterilized. β-mercaptoethanol (β-Me), dithiothreitol (DTT), phenylmethylsulfonyl fluoride (PMSF), protease inhibitor cocktail, GDP and other reagents when indicated, are added to buffers immediately before use. Reagents, as well as equipment such as pipette tips and glassware, for *in vitro* translation should be RNase-free.

See also Table 1 "Summary of the expression and purification of the components for the reconstituted translation system" for materials required for preparation of translation factors.

2.1 IF-2mt Preparation

1. Lysis buffer: 50 mM Tris–HCl (pH 7.5), 10 mM $MgCl_2$, 5% (vol/vol) glycerol, 5 mM β-Me, 0.1 mM PMSF.
2. Ni-NTA wash buffer: 50 mM Tris–HCl (pH 7.5), 10 mM $MgCl_2$, 1 M NH_4Cl, 10 mM imidazole, 5% (vol/vol) glycerol, 5 mM β-Me.
3. Ni-NTA elution buffer: 50 mM Tris–HCl (pH 7.5), 50 mM KCl, 10 mM $MgCl_2$, 150 mM imidazole, 5% (vol/vol) glycerol, 5 mM β-Me.
4. Dialysis buffer: 20 mM Hepes-KOH (pH 7.5), 100 mM KCl, 10 mM $MgCl_2$, 10% (vol/vol) glycerol, 7 mM β-Me, 10 μM GDP.
5. HiTrap Q A buffer: 20 mM Hepes-KOH (pH 7.5), 10 mM $MgCl_2$, 10% (vol/vol) glycerol, 7 mM β-Me, 10 μM GDP.

Fig. 1 (continued) insertion sequence at the corresponding position in the "no motif" mRNA. (**b**) "No motif" mRNA and "Prox4" mRNA were translated using [^{35}S]Methionine-labeled yeast aminoacyl-tRNA mix, with the indicated Mg^{2+} concentrations. After the translation reaction was performed for the indicated time period, aliquots were subjected to the nLuc assay (upper panels). The nLuc activities at the 120 min reaction point were plotted (lower panels). Note that "Prox4" mRNA provides only 1% of the nLuc activity of "no motif" mRNA, due to polyproline-mediated translation arrest. (**c**) After the 120 min reaction in (**b**), the samples were treated with either RNase A ([+]) or nothing ([−]), and subjected to Tricine SDS-PAGE. In the "Pro x4" mRNA samples, omission of the RNase treatment results in the disappearance of tRNA-cleaved stall products (arrowhead 4), with the concomitant appearance of peptidylpolyproline-tRNAs (arrowhead 3). In the "no motif" mRNA samples, full-length products (arrowhead 2) are produced regardless of the RNase treatment. In addition, full-length peptidyl-tRNAs (arrowhead 1) are also produced when the RNase treatment is omitted, due to the incomplete peptide release at the stop codon. (**d**) Effects of fusidic acid (FA), chloramphenicol (Cm), and cycloheximide (CHX) on the reconstituted mammalian mitochondrial translation system. The "no motif" mRNAs were translated in the presence of the indicated concentrations of antibiotics for 120 min, and then subjected to the nLuc assay. Like the bacterial translation system, the mitochondrial translation system is sensitive to Cm and resistant to CHX. However, the system is resistant to FA, because both EF-G1mt and EF-G2mt are resistant to FA, unlike bacterial EF-G [5]

Table 1
Summary of the expression and purification of the components for the reconstituted translation system

Component (origin)	IF-2mt (bovine)	IF-3mt (human)	EF-G1mt (human)	EF-G2mt (human)	EF-Tumt (bovine)	EF-Tsmt (bovine)	RF-1Lmt (human)	RRFmt (human)
1. Expression vector	pQE-BMIF2	pSUMO15b-HMIF3	pSUMO15b-HMEFG1	pSUMO15b-HMEFG2	pET24c-BMtu	pET24c-BMts	pET15b-HMRFIL	pET15b-HMRRF
2. *E. coli* host strain	BL21pRARE	Rosetta(DE3) pLyS	Rosetta(DE3) pLyS	Rosetta(DE3) pLyS	BL21(DE3)	Rosetta(DE3) pLyS	Rosetta(DE3) pLyS	Rosetta (DE3) pLyS
3. Culture								
Scale	6 L	6 L	6 L	6 L	2 L	2 L	2 L	2 L
Medium	2xYT	2xYT	2xYT	2xYT	2xYT	2xYT	2xYT	2xYT
Antibiotics	Amp, Cm	Amp, Cm	Amp, Cm	Amp, Cm	Kan	Kan, Cm	Amp, Cm	Amp, Cm
4. Induction								
A_{600}	0.5–0.8	0.5–0.8	0.5–0.8	0.5–0.8	0.5–0.8	0.5–0.8	0.5	0.5–0.8
IPTG (mM)	0.05	0.05	0.1	0.1	0.1	0.05	0.1	0.1
Temperature	18 °C	18 °C	18 °C	18 °C	18 °C	37 °C	18 °C	18 °C
Time	Overnight (16–20 h)	Overnight (16–20 h)	Overnight (16–20 h)	Overnight (16–20 h)	Overnight (16–20 h)	6 h	Overnight (16–20 h)	Overnight (16–20 h)
5. 1st-step purification	Ni-NTA (3 mL resin)	Ni-NTA (3 mL resin)	Ni-NTA (3 mL resin)	Ni-NTA (3 mL resin)	Ni-NTA (4 mL resin)	Ni-NTA (2 mL resin)	Ni-NTA (4 mL resin)	Ni-NTA (4 mL resin)
6. His-tag cleavage	–	+	+	+	–	–	–	+
Protease	–	Ulp1	Ulp1	Ulp1	–	–	–	Thrombin
Terminus for His-tag attachment	C	(N, cleaved)	(N, cleaved)	(N, cleaved)	C	C	N	(N, cleaved)

(continued)

Table 1
(continued)

Component (origin)	IF-2mt (bovine)	IF-3mt (human)	EF-G1mt (human)	EF-G2mt (human)	EF-Tumt (bovine)	EF-Tsmt (bovine)	RF-1Lmt (human)	RRFmt (human)
7. 2nd-step purification								
Column volume	HiTrap Q 20 mL	HiTrap SP 10 mL	HiTrap Q 20 mL	HiTrap Q 20 mL	HiTrap Q 20 mL	HiTrap Q 20 mL	HiTrap Q 20 mL	Q-Sepharose FF 2 mL (open column)
Elution	Linear gradient (100–500 mM KCl)	Linear gradient (100–700 mM KCl)	Linear gradient (100–300 mM KCl)	Linear gradient (100–300 mM KCl)	Linear gradient (50–300 mM KCl)	Linear gradient (40–350 mM KCl)	Linear gradient (50–300 mM KCl)	One step (100 mM KCl)
KCl concentration of protein elution	~220 mM	~350 mM	~260 mM	~280 mM	~200 mM	~150 mM	~150 mM	100 mM
9. Yield (mg/L culture)	1.4	0.6	1.2	1.2	17	6.7	12.5	15
10. Reference	This work[a]	This work[a]	This work[a]	This work[a]	[5]	[5]	[5]	[5]
11. Note								
Expression vector used in [5]	pET32a-BMIF2	pET29b-HMIF3	pET15b-HMEFG1	pET15b-HMEFG2				

[a]The purified factor shows essentially the same activity as the one described in [5]

6. HiTrap Q B buffer: 20 mM Hepes-KOH (pH 7.5), 1 M KCl, 10 mM $MgCl_2$, 10% (vol/vol) glycerol, 7 mM β-Me, 10 μM GDP.
7. Stock buffer: 20 mM Hepes-KOH (pH 7.5), 100 mM KCl, 10% (vol/vol) glycerol, 7 mM β-Me, 10 μM GDP.

2.2 IF-3mt Preparation

1. Lysis buffer: 50 mM Tris–HCl (pH 7.5), 100 mM KCl, 10% (vol/vol) glycerol, 5 mM β-Me, 0.1 mM PMSF.
2. Ni-NTA wash buffer: 50 mM Tris–HCl (pH 7.5), 1 M NH_4Cl, 10 mM imidazole, 10% (vol/vol) glycerol, 5 mM β-Me.
3. Ni-NTA elution buffer: 50 mM Tris–HCl (pH 7.5), 100 mM KCl, 150 mM imidazole, 10% (vol/vol) glycerol, 5 mM β-Me.
4. His-tag cleavage buffer: 20 mM Hepes-KOH (pH 7.5), 20 mM KCl, 5 mM $MgCl_2$, 950 mM NH_4Cl, 25 mM imidazole, 10% (vol/vol) glycerol, 5 mM β-Me.
5. Dialysis buffer: 20 mM Hepes-KOH (pH 7.5), 100 mM KCl, 0.1 mM EDTA, 10% (vol/vol) glycerol, 7 mM β-Me.
6. HiTrap SP A buffer: 20 mM Hepes-KOH (pH 7.5), 0.1 mM EDTA, 10% (vol/vol) glycerol, 7 mM β-Me.
7. HiTrap SP B buffer: 20 mM Hepes-KOH (pH 7.5), 0.1 mM EDTA, 1 M KCl, 10% (vol/vol) glycerol, 7 mM β-Me.
8. Stock buffer: 20 mM Hepes-KOH (pH 7.5), 100 mM KCl, 10% (vol/vol) glycerol, 7 mM β-Me.
9. Ulp1 protease: the Ulp1 active domain (Δ1-402) was purified as previously reported [12].

2.3 EF-G1mt Preparation

1. Lysis buffer: 50 mM Tris–HCl (pH 7.5), 100 mM KCl, 10% (vol/vol) glycerol, 5 mM β-Me, 0.1 mM PMSF.
2. Ni-NTA wash buffer: 50 mM Tris–HCl (pH 7.5), 1 M NH_4Cl, 10 mM imidazole, 10% (vol/vol) glycerol, 5 mM β-Me.
3. Ni-NTA Elution buffer: 50 mM Tris–HCl (pH 7.5), 100 mM KCl, 200 mM imidazole, 10% (vol/vol) glycerol, 5 mM β-Me.
4. His-tag cleavage buffer: 50 mM Hepes-KOH (pH 7.5), 20 mM KCl, 10 mM $MgCl_2$, 950 mM NH_4Cl, 25 mM imidazole, 10% (vol/vol) glycerol, 5 mM β-Me.
5. Dialysis buffer: 20 mM Hepes-KOH (pH 7.5), 100 mM KCl, 0.1 mM EDTA, 10% (vol/vol) glycerol, 7 mM β-Me.
6. HiTrap Q A buffer: 20 mM Hepes-KOH (pH 7.5), 0.1 mM EDTA, 10% (vol/vol) glycerol, 7 mM β-Me.
7. HiTrap Q B buffer: 20 mM Hepes-KOH (pH 7.5), 0.1 mM EDTA, 1 M KCl, 10% (vol/vol) glycerol, 7 mM β-Me.
8. Stock buffer: 20 mM Hepes-KOH (pH 7.5), 100 mM KCl, 10% (vol/vol) glycerol, 7 mM β-Me.
9. Ulp1 protease: the Ulp1 active domain (Δ1-402) was purified as previously reported [12].

2.4 EF-G2mt Preparation

1. Lysis buffer: 50 mM Tris–HCl (pH 7.5), 100 mM KCl, 10% (vol/vol) glycerol, 5 mM β-Me, 0.1 mM PMSF.
2. Ni-NTA wash buffer: 50 mM Tris–HCl (pH 7.5), 1 M NH_4Cl, 10 mM imidazole, 10% (vol/vol) glycerol, 5 mM β-Me.
3. Ni-NTA elution buffer: 50 mM Tris–HCl (pH 7.5), 100 mM KCl, 200 mM imidazole, 10% (vol/vol) glycerol, 5 mM β-Me.
4. His-tag cleavage buffer: 50 mM Hepes-KOH (pH 7.5), 20 mM KCl, 10 mM $MgCl_2$, 950 mM NH_4Cl, 25 mM imidazole, 10% (vol/vol) glycerol, 5 mM β-Me.
5. Dialysis buffer: 20 mM Hepes-KOH (pH 7.5), 100 mM KCl, 0.1 mM EDTA, 10% (vol/vol) glycerol, 5 mM β-Me.
6. HiTrap Q A buffer: 20 mM Hepes-KOH (pH 7.5), 0.1 mM EDTA, 10% (vol/vol) glycerol, 7 mM β-Me.
7. HiTrap Q B buffer: 20 mM Hepes-KOH (pH 7.5), 0.1 mM EDTA, 1 M KCl, 10% (vol/vol) glycerol, 7 mM β-Me.
8. Stock buffer: 20 mM Hepes-KOH (pH 7.5), 100 mM KCl, 10% (vol/vol) glycerol, 7 mM β-Me.
9. Ulp1 protease: Ulp1 active domain (Δ1-402) was purified as previously reported [12].

2.5 EF-Tumt Preparation

1. Lysis buffer: 50 mM Tris–HCl (pH 7.5), 100 mM KCl, 5 mM $MgCl_2$, 10% (vol/vol) glycerol, 5 mM β-Me, 0.1 mM PMSF, 10 μM GDP.
2. Ni-NTA wash buffer 1: 50 mM Tris–HCl (pH 7.5), 100 mM KCl, 5 mM $MgCl_2$, 1 M NH_4Cl, 10 mM imidazole, 10% (vol/vol) glycerol, 5 mM β-Me, 0.2 mM PMSF, 20 μM GDP.
3. Ni-NTA wash buffer 2: 50 mM Tris–HCl (pH 7.5), 40 mM KCl, 2 mM $MgCl_2$, 40 mM imidazole, 10% (vol/vol) glycerol, 5 mM β-Me, 20 μM GDP.
4. Ni-NTA elution buffer: 50 mM Tris–HCl (pH 7.5), 40 mM KCl, 2 mM $MgCl_2$, 250 mM imidazole, 10% (vol/vol) glycerol, 5 mM β-Me, 20 μM GDP.
5. Dialysis buffer: 25 mM Hepes-KOH (pH 7.5), 50 mM KCl, 10 mM $MgCl_2$, 10% (vol/vol) glycerol, 7 mM β-Me.
6. HiTrap Q A buffer: 25 mM Hepes-KOH (pH 7.5), 10 mM $MgCl_2$, 10% (vol/vol) glycerol, 7 mM β-Me.
7. HiTrap Q B buffer: 25 mM Hepes-KOH (pH 7.5), 1 M KCl, 10 mM $MgCl_2$, 10% (vol/vol) glycerol, 7 mM β-Me.
8. Stock buffer: 20 mM Hepes-KOH (pH 7.5), 100 mM KCl, 10% (vol/vol) glycerol, 7 mM β-Me.

2.6 EF-Tsmt Preparation

1. Lysis buffer: 50 mM Tris–HCl (pH 7.5), 100 mM KCl, 5 mM $MgCl_2$, 10% (vol/vol) glycerol, 5 mM β-Me, 0.1 mM PMSF, 20 μM GDP.
2. Ni-NTA wash buffer 1: 50 mM Tris–HCl (pH 7.5), 100 mM KCl, 5 mM $MgCl_2$, 1 M NH_4Cl, 10 mM imidazole, 10% (vol/vol) glycerol, 5 mM β-Me, 0.2 mM PMSF, 20 μM GDP.
3. Ni-NTA wash buffer 2: 50 mM Tris–HCl (pH 7.5), 40 mM KCl, 2 mM $MgCl_2$, 40 mM imidazole, 10% (vol/vol) glycerol, 5 mM β-Me, 20 μM GDP.
4. Ni-NTA elution buffer: 50 mM Tris–HCl (pH 7.5), 40 mM KCl, 2 mM $MgCl_2$, 250 mM imidazole, 10% (vol/vol) glycerol, 5 mM β-Me, 20 μM GDP.
5. Dialysis buffer: 20 mM Hepes-KOH (pH 7.5), 40 mM KCl, 2 mM $MgCl_2$, 10% (vol/vol) glycerol, 7 mM β-Me.
6. HiTrap Q A buffer: 20 mM Hepes-KOH (pH 7.5), 2 mM $MgCl_2$, 10% (vol/vol) glycerol, 7 mM β-Me.
7. HiTrap Q B buffer: 20 mM Hepes-KOH (pH 7.5), 1 M KCl, 2 mM $MgCl_2$, 10% (vol/vol) glycerol, 7 mM β-Me.
8. Stock buffer: 20 mM Hepes-KOH (pH 7.5), 100 mM KCl, 10% (vol/vol) glycerol, 7 mM β-Me.

2.7 RF-1Lmt Preparation

1. Lysis buffer: 50 mM Tris–HCl (pH 7.5), 100 mM KCl, 10% (vol/vol) glycerol, 5 mM β-Me, 0.1 mM PMSF.
2. Ni-NTA Wash buffer: 50 mM Tris–HCl (pH 7.5), 1 M NH_4Cl, 10 mM imidazole, 10% (vol/vol) glycerol, 5 mM β-Me.
3. Ni-NTA Elution buffer: 50 mM Tris–HCl (pH 7.5), 100 mM KCl, 200 mM imidazole, 10% (vol/vol) glycerol, 5 mM β-Me.
4. Dialysis buffer: 20 mM Hepes-KOH (pH 7.5), 50 mM KCl, 10% (vol/vol) glycerol, 7 mM β-Me.
5. HiTrap Q A buffer: 20 mM Hepes-KOH (pH 7.5), 10% (vol/vol) glycerol, 7 mM β-Me.
6. HiTrap Q B buffer: 20 mM Hepes-KOH (pH 7.5), 1 M KCl, 10% (vol/vol) glycerol, 7 mM β-Me.
7. Stock buffer: 20 mM Hepes-KOH (pH 7.5), 100 mM KCl, 10% (vol/vol) glycerol, 7 mM β-Me.

2.8 RRFmt Preparation

1. Lysis buffer: 50 mM Tris–HCl (pH 7.5), 100 mM KCl, 10% (vol/vol) glycerol, 5 mM β-Me, 0.1 mM PMSF.
2. Ni-NTA wash buffer: 50 mM Tris–HCl (pH 7.5), 1 M NH_4Cl, 10 mM imidazole, 10% (vol/vol) glycerol, 5 mM β-Me.
3. Ni-NTA elution buffer: 50 mM Tris–HCl (pH 7.5), 100 mM KCl, 150 mM imidazole, 10% (vol/vol) glycerol, 5 mM β-Me.

4. His-tag cleavage buffer 1: 20 mM Tris–HCl (pH 7.5), 150 mM KCl, 5 mM $CaCl_2$, 5% (vol/vol) glycerol, 5 mM β-Me.
5. His-tag cleavage buffer 2: 20 mM Hepes-KOH (pH 7.5), 1 M NH_4Cl, 25 mM imidazole, 10% (vol/vol) glycerol, 5 mM β-Me.
6. Dialysis buffer: 20 mM Hepes-KOH (pH 7.5), 100 mM KCl, 10% (vol/vol) glycerol, 7 mM β-Me.
7. Stock buffer: same as dialysis buffer.
8. Thrombin protease (Cytiva).

2.9 55S Ribosome Preparation

1. Extraction buffer: 0.26 M sucrose, 15 mM Tris–HCl (pH 7.5), 40 mM KCl, 15 mM $MgCl_2$, 0.8 mM EDTA (pH 8.0), 1.6% (vol/vol) Triton X-100, 6 mM β-Me, 0.05 mM spermine, 0.05 mM spermidine, 0.1 mM PMSF. Add Triton X-100, β-Me, spermine, spermidine, and PMSF immediately before use.
2. Sucrose cushion buffer (with Triton X-100): 1 M (34% (wt/vol)) sucrose, 20 mM Tris–HCl (pH 7.5), 100 mM KCl, 20 mM $MgCl_2$, 1.0% (vol/vol) Triton X-100, 6 mM β-Me. Add Triton X-100 and β-Me immediately before use.
3. Sucrose cushion buffer (without Triton X-100): 1 M (34% (wt/vol)) sucrose, 20 mM Tris–HCl (pH 7.5), 100 mM KCl, 20 mM $MgCl_2$, 6 mM β-Me. Add β-Me immediately before use.
4. 5× Re-association buffer: 75 mM Tris–HCl (pH 7.5), 500 mM KCl, 100 mM $MgCl_2$.
5. 1× Re-association buffer: 15 mM Tris–HCl (pH 7.5), 100 mM KCl, 20 mM $MgCl_2$, 6 mM β-Me. Mix 10 mL of 5× re-association buffer and 23.14 μL of 14 M β-Me, and then add RNase-free water up to 50 mL. Add β-Me immediately before use.
6. 1× Re-association buffer containing 200 mM KCl: Mix 10 mL of 5× re-association buffer, 0.75 g KCl, and 23.14 μL of 14 M β-Me, and then add RNase-free water up to 50 mL. Add β-Me immediately before use.
7. 6% (wt/vol) Sucrose gradient buffer: Mix 40 mL of 5× re-association buffer, 12 g sucrose, and 92.57 μL of 14 M β-Me, and then add RNase-free water up to 200 mL. Add β-Me immediately before use.
8. 38% (wt/vol) Sucrose gradient buffer: Mix 40 mL of 5× re-association buffer, 76 g sucrose, and 92.57 μL of 14 M β-Me, and then add RNase-free water up to 200 mL. Add β-Me immediately before use.

2.10 In Vitro Translation

1. 1.5 mM 18 a.a. (without Met and Cys): Mix 1.34 mg of Ala, 2.61 mg of Arg, 2.25 mg of Asn, 2.00 mg of Asp, 2.19 mg of Gln, 2.21 mg of Glu, 1.13 mg of Gly, 2.33 mg of His, 1.97 mg of Ile, 1.97 mg of Leu, 2.19 mg of Lys, 2.48 mg of Phe, 1.73 mg of Pro, 1.58 mg of Ser, 1.79 mg of Thr, 3.06 mg of Trp, 2.72 mg of Tyr, and 1.76 mg of Val, and then add 0.1 M Hepes-KOH (pH 7.5) up to 10 mL. Sterilize the solution by filtration. Store in approximately single-use aliquots (20 μL) at −80 °C.
2. 10 mM Cysteine (Cys): To 12.1 mg of Cys, add RNase-free water up to 10 mL. Sterilize the solution by filtration. Store in approximately single-use aliquots (20 μL) at −80 °C.
3. 20 mM ATP, GTP: Mix 200 μL of 100 mM ATP, 200 μL of 100 mM GTP, and 600 μL of RNase-free water. Sterilize the solution by filtration. Store in approximately single-use aliquots (20 μL) at −80 °C.
4. 12% PURE buffer (without Mg and SPD): Mix 1.5 mL of 1 M Hepes-KOH (pH 7.5), 1.5 mL of 2 M L-Glutamic acid potassium salt monohydrate, 30 μL of 1 M DTT, and 570 μL of RNase-free water. Sterilize the solution by filtration. Store in approximately single-use aliquots (20 μL) at −80 °C.
5. 1 M Creatine phosphate: To 3.27 g of creatine phosphate, add RNase-free water up to 10 mL. Sterilize the solution by filtration. Store in approximately single-use aliquots (20 μL) at −80 °C.
6. 1 μg/μL 10-formyl-5,6,7,8-tetrahydrofolic acid: Dissolve 25 mg folinic acid (calcium salt) (Sigma, Cat. No.: F7878) in 2 mL of 50 mM 2-mercaptoethanol. Add 220 μL hydrochloric acid, and incubate the solution for 3 h at room temperature. Dilute the solution to 1 mg/mL with water, and clarify by filtration. Store in approximately single-use aliquots (20 μL) at −20 °C.
7. 250 mM $Mg(OAc)_2$: To 536 mg of magnesium acetate tetrahydrate, add RNase-free water up to 10 mL. Sterilize the solution by filtration. Store in approximately single-use aliquots (20 μL) at −80 °C.
8. 10 mM Spermine (SP): Mix 100 μL of 1 M spermine and 9.9 mL of RNase-free water. Sterilize the solution by filtration. Store in approximately single-use aliquots (20 μL) at −80 °C.
9. 1 mM Methionine (Met): To 1.49 mg of Met, add RNase-free water up to 10 mL. Sterilize the solution by filtration. Store in approximately single-use aliquots (20 μL) at −80 °C.
10. Creatine kinase, myokinase, nucleoside-diphosphate kinase, pyrophosphatase: These enzymes were purified as previously reported [13].

11. MTF: *E. coli* MTF is purified as previously reported [13].
12. Purified mitochondrial translation factors, ribosomes.
13. mRNA: see "Sequences of template DNAs used for the translation reaction in Fig. 1" (Table 2).
14. Precharged yeast aminoacyl-tRNA mixtures: we previously described the detailed methods for the preparation of the yeast tRNA mixture and the tRNA charging with a yeast S100 fraction [14].
15. [^{35}S]Methionine (PerkinElmer, NEG009A).

2.11 Analysis of Translation Products

1. 0.1 mg/mL RNaseA: Mix 200 μL of 5 mg/mL RNase A and 9.8 mL of buffer (20 mM Hepes-KOH (pH 7.5), 100 mM KOAc (pH 7.5), 2 mM $Mg(OAc)_2$). Store in aliquots (1 mL) at −30 °C.
2. Nano-Glo Luciferase Assay System (Nano-Glo Luciferase Assay Substrate, Nano-Glo Luciferase Assay Buffer) (Promega).
3. 5 mg/mL RNaseA: To 7.5 mg of RNaseA, add buffer (20 mM Hepes-KOH (pH 7.5), 100 mM KOAc (pH 7.5), 2 mM Mg $(OAc)_2$) up to 1.5 mL. Store in aliquots (50 μL) at −30 °C.
4. 5× LiDS: To a mixture of 5% (wt/vol) lithium dodecyl sulfate (LiDS), 25% (vol/vol) glycerol, 100 mM Tris–HCl (pH 6.8), 0.025% (wt/vol) bromophenol blue (BPB), and 875 mM β-Me, add RNase-free water up to 10 mL. Store in aliquots at −30 °C.

3 Methods

3.1 Preparation of Translation Factors (IF-2mt, IF-3mt, EF-G1mt, EF-G2mt, EF-Tumt, EF-Tsmt, RF-1Lmt and RRFmt)

See also Table 1 "Summary of the expression and purification of the components for the reconstituted translation system," which includes all of the required information for the following protocol.

3.1.1 Cell Culture

1. Transform the expression vector (Table 1, row 1) into an *Escherichia coli* host strain (Table 1, row 2).
2. Grow cells in 2xYT broth supplemented with antibiotics (Table 1, row 3) at 37 °C.
3. When the culture reaches the indicated A_{600}, induce protein expression by adding IPTG, and then grow the cells for the indicated time period at the indicated temperature (Table 1, row 4).
4. Harvest cells by centrifugation. Unless the harvested cells are immediately processed for purification, freeze the cell pellets with liquid nitrogen and store at −80 °C.

Table 2
Sequences of template DNAs used for translation reactions in Fig. 1

[shaded], T7 promoter; **ATG**, initiation Met codon; ____, 3xFLAG; ___, pgk1-36; [box], polyproline; - - - -, HA; **A**x36, polyA36.

No motif: T7_Met_3XFLAG_PGK1-36 _nLuc_HA_A36
GGGCCTAATACGACTCACTATAG**ATG**GATTATAAAGATCACGACGGTGATTATAAAGATCA TGATATTGATTATAAAGATGACGATGATAAAGAATTATCTTCAAAGTTGTCTGTCCAAGAT TTGGACTTGAAGGACAAGCGTGTCTTCATCAGAGTTGACTTCAACGTCCCATTGGACGG TAAGAAGATCACTTCTACGGTTTTCACCTTGGAAGATTTCGTTGGTGATTGGAGACAAAC TGCTGGTTACAATTTGGATCAAGTCTTGGAACAAGGTGGTGTCTCTTCTTTGTTTCAAAA CTTGGGTGTTTCCGTTACCCCAATCCAAAGAATAGTTTTGTCTGGTGAAAACGGTTTGAA GATCGATATCCATGTTATCATCCCATACGAAGGTTTGTCAGGTGATCAAATGGGTCAAATC GAAAAGATCTTCAAGGTTGTTTACCCAGTTGATGATCACCACTTTAAGGTTATCTTGCACT ACGGTACTTTGGTCATTGATGGTGTTACTCCAAACATGATCGATTACTTTGGTAGACCTTA CGAAGGTATTGCTGTTTTCGATGGTAAGAAGATTACTGTCACTGGTACTTTGTGGAACGG TAACAAAATTATCGACGAAAGATTGATCAACCCAGACGGTTCTTTGTTGTTCAGAGTTAC TATTAACGGTGTTACCGGTTGGAGATTGTGCGAAAGAATTTTGGCTTACCCATACGACGT CCCAGACTACGCGTAA**AAAAAAAAAAAAAAAAAAAAAAAAAAAAAAAAAAAA**
Pro x4: T7_Met_3XFLAG_PGK1-36_P4 _nLuc_HA_A36
GGGCCTAATACGACTCACTATAG**ATG**GATTATAAAGATCACGACGGTGATTATAAAGATCA TGATATTGATTATAAAGATGACGATGATAAAGAATTATCTTCAAAGTTGTCTGTCCAAGAT TTGGACTTGAAGGACAAGCGTGTCTTCATCAGAGTTGACTTCAACGTCCCATTGGACGG TAAGAAGATCACTTCTCCTCCCCCACCGACGGTTTTCACCTTGGAAGATTTCGTTGGTGA TTGGAGACAAACTGCTGGTTACAATTTGGATCAAGTCTTGGAACAAGGTGGTGTCTCTTC TTTGTTTCAAAACTTGGGTGTTTCCGTTACCCCAATCCAAAGAATAGTTTTGTCTGGTGA AAACGGTTTGAAGATCGATATCCATGTTATCATCCCATACGAAGGTTTGTCAGGTGATCAA ATGGGTCAAATCGAAAAGATCTTCAAGGTTGTTTACCCAGTTGATGATCACCACTTTAAG GTTATCTTGCACTACGGTACTTTGGTCATTGATGGTGTTACTCCAAACATGATCGATTACTT TGGTAGACCTTACGAAGGTATTGCTGTTTTCGATGGTAAGAAGATTACTGTCACTGGTAC TTTGTGGAACGGTAACAAAATTATCGACGAAAGATTGATCAACCCAGACGGTTCTTTGTT GTTCAGAGTTACTATTAACGGTGTTACCGGTTGGAGATTGTGCGAAAGAATTTTGGCTTA CCCATACGACGTCCCAGACTACGCGTAA**AAAAAAAAAAAAAAAAAAAAAAAAAAAAA** **AAAAAAA**

3.1.2 First-Step Purification by Ni-NTA Agarose (Table 1, Row 5)

1. Resuspend the cells with 4 mL of Lysis buffer per 1 g of cells, and disrupt them by sonication. Remove cell debris by centrifugation at 10,000 × *g*.
2. Apply the supernatant to the indicated volume of Ni-NTA agarose resin in a column.
3. For EF-Tumt and EF-Tsmt: wash the column with 100 column volumes of Ni-NTA wash buffer 1 and then with 10 column volumes of Ni-NTA wash buffer 2.

   For the other proteins: wash the column with 100 column volumes of Ni-NTA wash buffer.
4. Elute the protein with 5 column volumes of Ni-NTA elution buffer, and analyze the eluted proteins by SDS-PAGE.

3.1.3 Histidine-Tag Cleavage (Table 1, Row 6)

1. For IF-2mt, EF-Tumt and EF-Tsmt: go directly to Subheading 3.1.4, **step 1**.

   For IF-3mt, EF-G1mt, and EF-G2mt: add Ulp1 protease to the sample (final approximately 3.1 μg/mL), and dialyze against 1 L of His-tag cleavage buffer at 4 °C overnight, with a buffer change after 1 h. Analyze the histidine-tag cleavage by SDS-PAGE.

   For RRFmt: add thrombin protease to the sample (final approximately 2 U/mL), and dialyze the sample against 1 L of His-tag cleavage buffer 1 at 4 °C overnight, with a buffer change after 1 h. After analyzing the histidine-tag cleavage by SDS-PAGE, add PMSF (final 0.2 mM) to stop the reaction.
2. For IF-3mt, EF-G1mt and EF-G2mt: incubate the sample and Ni-NTA agarose resin (same volume as Subheading 3.1.2, **step 2**) equilibrated with His-tag cleavage buffer on a rotator at 4 °C for approximately 30 min.

   For RRFmt: add solid NH_4Cl (final 1 M) and 2.5 M imidazole (final 25 mM) to the sample, and then incubate the sample and Ni-NTA agarose resin (same volume as Subheading 3.1.2, **step 2**), equilibrated with His-tag cleavage buffer 2, on a rotator at 4 °C for approximately 30 min.
3. Transfer the slurry into an empty column, and collect the flow-through fractions in a clean tube.
4. For IF-3mt, EF-G1mt, and EF-G2mt: wash with two column volumes of His-tag cleavage buffer, and combine the wash fraction with the flow-through fraction.

   For RRFmt: wash with two column volumes of His-tag cleavage buffer 2, and combine the wash fraction with the flow-through fraction.
5. Analyze the His-tag removal by SDS-PAGE.

3.1.4 Second-Step Purification by Ion-exchange Column (Table 1, Row 7)

1. Dialyze the sample against 1 L of dialysis buffer at 4 °C, either overnight with a buffer change after 1 h, or for 4 h with a buffer change after 2 h.
2. For the proteins except RRFmt: load the sample onto the indicated ion-exchange column. After extensive washing, elute with the indicated linear gradient of A-buffer to B-buffer (with the compositions indicated in Table 1, row 7) in 15 column volumes. Analyze the fractions by SDS-PAGE and pool the protein containing fractions.

   For RRFmt: incubate the sample with 2 mL of Q-Sepharose FF resin equilibrated with stock buffer on a rotator at 4 °C for approximately 30 min. Load the slurry into an empty column, and collect the flow-through fractions in a clean tube. Wash with two column volumes of stock buffer, and combine the wash fraction with the flow-through fraction.
3. For the proteins except RRFmt: dialyze the sample against 1 L of stock buffer at 4 °C overnight, with a buffer change after 1 h.

   For RRFmt: skip this step, and go directly to **step 4**.
4. Concentrate the purified protein to approximately 5 mg/mL using Amicon Ultra Centrifugal Filters (10 kDa cut-off) (MERCK).
5. Determine the concentration of the purified protein by a Bradford assay using BSA as the standard, and check the purity by SDS-PAGE analysis.
6. Aliquot the purified factors, fast-freeze them in liquid nitrogen, and store them at −80 °C.

3.2 55S Ribosome Preparation

55S ribosomes are purified from crude ribosomes prepared from porcine liver mitoplasts, essentially according to the protocol in [15] with minor modifications. See [15] for the mitoplast preparation.

3.2.1 Crude Ribosome Preparation (from 30 g of Mitoplasts)

1. Remove approximately 30 g of frozen mitoplast pellets from the −80 °C storage, and resuspend the pellets with extraction buffer (final 0.1 g-mitoplast/mL-extraction buffer) in a 500 mL or 1 L beaker.
2. Transfer the mitoplast suspension to a glass homogenizer tube, and homogenize the sample with about five strokes of a Teflon pestle at low speed (<300 rpm).
3. Adjust the volume of the sample to 330 mL with extraction buffer and stir the sample for 15 min at 4 °C to complete the lysis of the mitoplasts.
4. Centrifuge the lysed mitoplasts for 45 min at 51,000 × *g* (21,000 rpm, Type 45 Ti, BECKMAN COULTER) at 4 °C.

5. Transfer the supernatant (approximately 300 mL) to a clean 500 mL graduated cylinder, and discard the pellet. Leave a little supernatant rather than transferring the supernatant completely, to avoid contamination from the pellet. Measure the exact volume of the supernatant.
6. Transfer the sample to a 500 mL beaker, and slowly add solid KCl to a 300 mM final concentration (190 mg KCl per 10 mL sample), while mixing gently with a magnetic stirrer at 4 °C. Remember that the extraction buffer already contains 40 mM KCl.
7. Divide the sample (approximately 50 mL each) into six 45 Ti ultracentrifuge tubes (BECKMAN COULTER), and place 11 mL of sucrose cushion buffer (with Triton X-100) into the very bottom of the tube using a syringe.
8. Centrifuge the sample at 142,000 × *g* (35,000 rpm, Type 45 Ti, BECKMAN COULTER) at 4 °C for 22 h.
9. Collect two-thirds of the supernatant, approximately 40 mL in each tube, in 50 mL Falcon tubes as S100, fast-freeze the samples in liquid nitrogen, and store them at −80 °C.
10. Completely remove the remaining supernatant by aspiration. The pellets are generally small and yellow-brown.
11. Add approximately 200 μL of 1× re-association buffer per tube and dissolve the pellets by gentle shaking for approximately 1 h at 4 °C.
12. Collect the supernatants in a 15 mL tube.
13. Repeat **steps 11** and **12** twice.
14. If portions of the pellets still remain, rinse out the tubes with an additional 100 μL of 1× re-association buffer, and combine it with the original sample.
15. Dispense approximately 1 mL of the sample into 1.5 mL tubes, and centrifuge at 14,000 × *g* for 5 min at 4 °C.
16. Collect the supernatants in a 15 mL tube.
17. Add 200 μL of 1× re-association buffer per tube, and resuspend the pellet by pipetting. Centrifuge the samples at 14,000 × *g* for 5 min at 4 °C, and combine the supernatants with the sample from **step 16**. Repeat this step at least 3 times, until the pellets disappear.
18. Add the sample to 40 mL of 1× re-association buffer containing 200 mM KCl, and mix well by gentle pipetting.
19. Transfer the sample into a 45 Ti ultracentrifuge tube, and place 20 mL of sucrose cushion buffer (without Triton X-100) into the very bottom of the tube using a syringe.
20. Centrifuge the sample at 142,000 × *g* (35,000 rpm, Type 45 Ti, BECKMAN COULTER) for 20 h at 4 °C.

21. Remove the supernatant by aspiration.
22. Add approximately 200 μL of 1× re-association buffer to the ultracentrifuge tube and dissolve the pellet by gentle shaking for approximately 1 h at 4 °C.
23. Transfer the sample to a 1.5 mL tube. For the subsequent steps, the use of low protein binding tubes is recommended.
24. Repeat **step 22**.
25. Combine the sample with the sample from **step 23**.
26. Rinse out the ultracentrifuge tube with an additional 100 μL of 1× re-association buffer. Resuspend all of the residual pellet by pipetting, even if it does not dissolve completely, and combine the suspension with the original sample.
27. Centrifuge the sample at 12,000 rpm for 5 min at 4 °C.
28. Transfer the supernatant to a new 1.5 mL tube.
29. Add 100 μL of 1× re-association buffer to the pellet, and resuspend the pellet by pipetting. Centrifuge the samples at 12,000 rpm for 5 min at 4 °C, and combine the supernatant with the sample from **step 28**. Repeat this step until there is no more clear material from the pellets, which become opaque debris.
30. Centrifuge the sample at 12,000 rpm for 10 min at 4 °C, and transfer the supernatant to a new 1.5 mL tube.
31. Determine the concentration of the sample by measuring its UV absorption spectrum and reading the A_{260}. Generally, the yield of crude ribosomes is about 50 A_{260} from 30 g mitoplasts. To subsequently obtain good quality 55S ribosomes, the A_{260}/A_{280} ratio here should be greater than 1.4.
32. Fast-freeze the purified crude ribosomes in liquid nitrogen, and store them at −80 °C.

3.2.2 55S Ribosome Preparation (from Crude Ribosomes)

1. Remove approximately 70 A_{260} of crude ribosomes from the −80 °C freezer, and thaw slowly. Centrifuge the sample at 12,000 rpm for 10 min at 4 °C.
2. Transfer the supernatant to a new 1.5 mL tube, and adjust its volume to 570 μL with 1× re-association buffer.
3. Add 30 μL of 10 mM puromycin to give a final concentration of 0.5 mM, and incubate the sample at 27 °C for 15 min.
4. Centrifuge the sample at 12,000 rpm for 10 min at 4 °C, and transfer the supernatant to a new 1.5 mL tube.
5. Load the entire sample onto a single linear 36 mL gradient of 6–38% sucrose in 1× re-association buffer.
6. Centrifuge the gradient in a swinging bucket rotor at 72,000 × *g* (20,000 rpm, SW 28, BECKMAN COULTER) for 16 h at 4 °C.

7. Recover the gradient from top to bottom, using a density gradient fractionator (Towa Labo) at a flow rate of 0.05 min/fraction (300 μL/fraction), while monitoring the absorbance at 260 nm. Fractions containing 55S monosomes are generally located about two-thirds into the gradient, and recovered at about 3.6 min.
8. Pool the fractions containing 55S monosomes, and add 1× re-association buffer up to 60 mL.
9. Transfer the sample into a 45 Ti ultracentrifuge tube (BECKMAN COULTER), and centrifuge the sample at 61,000 × *g* (23,000 rpm, Type 45 Ti, BECKMAN COULTER) for 27 h at 4 °C.
10. Immediately after centrifugation is complete, discard the supernatant. Remove any excess liquid from the pellet using an aspirator. Note that the pellet at this point is loose and will quickly dissolve.
11. Add 100 μL of 1× re-association buffer to the ultracentrifuge tube, and dissolve the pellet by gentle shaking for approximately 1 h at 4 °C.
12. Transfer the sample to a new 1.5 mL tube. For the subsequent steps, the use of low protein binding tubes is recommended.
13. Repeat **step 11**.
14. Combine this sample with the sample from **step 12**. If pellets still remain, rinse out the ultracentrifuge tube with an additional 50 μL of 1× re-association buffer, and combine it with the original sample.
15. Centrifuge the sample at 12,000 rpm for 5 min at 4 °C.
16. Transfer the supernatant to a new 1.5 mL tube.
17. Determine the concentration of the sample by measuring the UV absorption spectrum of the sample and reading the A_{260}. The yield of 55S ribosomes is variable, but generally about 500–1000 pmol are obtained from 70 A_{260} units of crude ribosomes, estimating that the 55S ribosomes at 1 A_{260} correspond to 32 pmol. Good quality 55S ribosomes usually have an A_{260}/A_{280} ratio greater than 1.7.
18. Prepare aliquots of the purified 55S ribosomes, freeze them in liquid nitrogen, and store them at −80 °C.

3.3 In Vitro Translation

This section describes a standard protocol for the translation of mRNA encoding nanoluciferase (nLuc). We recommend the separation of the translation reaction from transcription and aminoacylation, so that the Mg^{2+} and polyamine concentrations in the reaction can be controlled (*see* **Notes 1–4**). Therefore, *in vitro* transcribed mRNAs and a pre-charged yeast aminoacyl-tRNA mixture are used for the reaction. Prepare the desired mRNAs by *in vitro* transcription using T7 RNA polymerase and the template DNAs, according to the standard protocols (*see* **Note 5**, and "Sequences of template DNAs used for the translation reaction in Fig. 1" (Table 2)). We described the detailed procedures for the pre-charged yeast aminoacyl-tRNA mixture elsewhere [14].

1. Prepare the "buffer mixture," "enzyme mixture," "aminoacyl-tRNA mixture," and "mRNA solution" (total 10 μL) on ice, as follows (*see* **Note 6**):

	Components	Volume (μL)
Buffer mixture	1.5 mM 18 a.a. (without Met and Cys)	1
	10 mM Cys	0.1
	20 mM ATP, GTP	0.5
	12% PURE buffer (without Mg and SPD)	1.2
	1 M Creatine phosphate	0.2
	1 μg/μL 10-formyl-5,6,7,8-tetrahydrofolic acid	0.1
	250 mM $Mg(OAc)_2$[a]	(Final 11 mM)
	10 mM SP	0.1
	20 mM SPD[b]	0
	1 mM Met	0.5
Enzyme mixture	Creatine kinase	(Final 100 nM)
	Myokinase	(Final 20 nM)
	Nucleoside-diphosphate kinase	(Final 15 nM)
	Pyrophosphatase	(Final 15 nM)
	IF-2mt	(Final 0.5 μM)
	IF-3mt	(Final 1 μM)
	EF-G1mt	(Final 0.5 μM)
	EF-G2mt	(Final 0.5 μM)
	EF-Tumt	(Final 5 μM)
	EF-Tsmt	(Final 1 μM)
	RF-1Lmt	(Final 0.5 μM)
	RRFmt	(Final 0.5 μM)
	MTF(*E. coli*)	(Final 5 μM)
	55S ribosome	(Final 0.2 μM)
mRNA solution	mRNA[c]	(Final 0.025–0.1 μM)
Aminoacyl-tRNA mixture	Aminoacyl-tRNA[d]	(0.045 A_{260} units)
RNase-free water	RNase-free water	(Up to 10 μL)

[a] *See* **Notes 1**, **2**, and **3**
[b] *See* **Notes 1**, **3**, and **4**
[c] *See* **Note 5**
[d] *See* **Note 7**

2. Mix "buffer mixture" and "RNase-free water".
3. Add "enzyme mixture" to the mixture tube on ice, gently tap the mixture tube and centrifuge the mixture.

4. Add "aminoacyl-tRNA mixture" on ice, gently tap the mixture tube and centrifuge the mixture.
5. Add "mRNA solution" on ice, gently tap the mixture tube and centrifuge the mixture.
6. Incubate the reaction mixture at 37 °C for an appropriate duration (3 h).
7. At an appropriate time point, place the sample on ice to stop the reaction.

3.4 Analysis of Translation Products

3.4.1 Measurement of Nanoluciferase Activity

1. After the translation reaction, mix 2 μL of the translation reaction sample and 18 μL of 0.1 mg/mL RNase A, to stop the reaction, on ice (*see* **Note 8**).
2. Prepare a Nano-Glo Substrate-Buffer mixture, which contains Nano-Glo Luciferase Assay Substrate (Promega) and Nano-Glo Luciferase Assay Buffer (Promega) in a ratio of 1:50. Mix the sample from **step 1** (20 μL) and 20 μL of the Nano-Glo Substrate-Buffer mixture on ice.
3. Dispense the sample into a white 96-well half-area plate and incubate for 17 min at room temperature (27 °C).
4. Measure the nanoluciferase activity using a GloMax plate reader (Promega) (PROTOCOLS: Nano-Glo/Integration-Time: 0.3 s).

3.4.2 Detection of ^{35}S-Met-Labeled Proteins by SDS-PAGE or Tricine-PAGE

1. After the translation reaction, mix the translation reaction sample (10 μL) with 0.4 μL of 5 mg/mL RNase A, and then incubate the mixture for 10 min at 30 °C. Thereafter, add 2.5 μL of 5× LiDS dye to the sample and mix well (*see* **Note 9**).
2. Apply the resulting sample onto a 15% SDS-PAGE gel or Tricine-PAGE gel (*see* **Note 10**) and run it at 200 V (constant voltage) for approximately 2 h.
3. Immerse the gel in 10% (vol/vol) acetic acid-10% (vol/vol) ethanol and shake at room temperature for 30 min.
4. Repeat **step 3**.
5. Immerse the gel in 10% (vol/vol) acetic acid-10% (vol/vol) ethanol-3% (vol/vol) glycerol, and then shake at room temperature for 1 h. Overnight shaking is also possible.
6. Transfer the gel onto a piece of filter paper, place plastic wrap over the gel, and dry with gel dryer.
7. Expose the gel on an imaging plate for an appropriate time (approximately 4–6 h) and analyze using a Bio-Analyzing System (BAS-5000 Fujifilm). A filter paper spotted with a dilution series of [^{35}S] Met (4.0×10^{-4}–7.0×10^{-2} pmol) should also be analyzed, in order to obtain a standard curve for [^{35}S] Met.

4 Notes

1. The optimal Mg^{2+} and polyamine concentrations for the translation reaction depend on the mRNA sequence. For example, Fig. 1b shows the nLuc synthesis under different Mg^{2+} concentration conditions, and accordingly, the Mg^{2+} and polyamine concentrations for nLuc synthesis, [Mg^{2+}/Spermidine (SPD)/Spermine (SP)] (mM), is usually set as [11/0/0.1].
2. The following components are dissolved in stock buffer containing the indicated Mg^{2+} concentrations in parentheses: creatine kinase (10 mM), myokinase (10 mM), nucleoside-diphosphate kinase (10 mM), pyrophosphatase (10 mM), *E. coli* MTF (10 mM), 55S ribosome (20 mM) [5, 13]. The volume of 250 mM $Mg(OAc)_2$ to be added to the reaction is determined by taking into account the concentration of Mg^{2+} derived from these components, such that the final concentration of Mg^{2+} in the reaction is 11 mM.
3. An aminoacylation and/or transcription-translation coupled reaction is also possible by modifying the reaction mixture, with the Mg^{2+} and polyamine concentrations in the reaction, [Mg^{2+}/Spermidine (SPD)/Spermine (SP)] (mM), set as [13/2/0.1], and adding 2 mM ATP, 2 mM GTP, 1 mM CTP, and 1 mM UTP to the reaction. When aminoacylation is coupled, 30–300 units of each *E. coli* ARS [13] and 0.56 A_{260} units *E. coli* tRNA mixture are included, instead of the yeast aminoacyl-tRNA mixture. For the transcription-translation coupled reaction, 30 nM T7 RNA polymerase and 10 nM template DNA are included, instead of the mRNA. However, also *see* **Note 4** for the effect of spermidine on the translation reaction.
4. Figure 1c shows that the insertion of a ribosome arrest sequence, in this case a polyproline-encoding sequence, at a specific position in the mRNA gives rise to the formation of a stalled ribosome nascent peptide chain complex (RNC). We note that the omission of spermidine in the translation reaction is critical for the RNC formation, as the nascent peptide chain is substantially hydrolyzed and released from the stalled ribosome in the presence of spermidine, although the mechanism for this spontaneous peptide release is presently unclear [5].
5. The template DNA of the objective leaderless mRNA should contain a T7 promotor, a single G, the ORF of the protein, and $polyA_{36}$. As examples, the sequences of the template DNAs used for the translation reaction in Fig. 1 are shown in Table 2. A single G is inserted between the T7 promotor and the ORF of the protein to promote the transcription of the

leaderless mRNA by T7RNA polymerase. Since a yeast tRNA mixture is utilized in the translation reaction, the universal genetic code, rather than the mitochondrial genetic code, is utilized for the ORF of the protein. To mimic the mitochondrial mRNAs [9], polyA_{36} is usually included. We note, however, that polyA_{36} has a minimal effect on the translation efficiency, at least for the nLuc mRNA.

6. To reduce manipulation errors, the components are combined as a mixture and added in one step. For example, a mixture for multiple samples in **steps 2–4** in Subheading 3.3 is prepared in one tube and aliquoted in **step 5** in Subheading 3.3 into the mRNA solution previously prepared in different tubes.
7. For the detection of [^{35}S]Met-labeled proteins by SDS-PAGE or Tricine-PAGE, prepare the aminoacyl-tRNA mixture using [^{35}S]Met, according to [14]. When performing translation reactions using [^{35}S]Met-labeled aminoacyl-tRNAs and further analysis of the translation products, dispose of the used chips and microtubes with respect to your lab procedure for radioactive waste.
8. When analyzing the time course of nanoluciferase synthesis, withdraw 2 μL from the reaction solution at the desired time point. At this time, gently open the lid of the reaction tube, and do not remove the tube from the heat block, in order to minimize the temperature change of the sample. Gently remove 2 μL from the very top of the reaction solution. Do not disturb the reaction solution by pipetting.
9. When analyzing peptidyl-tRNAs (Fig. 1c, for example), the samples are directly mixed with 5 × LiDS dye, without the incubation with RNase A, and subjected to SDS-PAGE analysis.
10. When analyzing short peptides (<10 kDa), Tricine SDS-PAGE analysis is recommended.

Acknowledgements

We would like to thank all the laboratory members who were involved in this project. This work was supported in part by grants to N.T. from MEXT/JSPS Grant-in-Aid for Scientific Research (18K06054), The Naito Foundation, The Hamaguchi Foundation for the Advancement of Biochemistry, The Koyanagi Foundation, and The TERUMO LIFE SCIENCE FOUNDATION.

References

1. Aibara S, Singh V, Modelska A et al (2020) Structural basis of mitochondrial translation. elife 9:e58362
2. Kummer E, Ban NE (2021) Mechanisms and regulation of protein synthesis in mitochondria. Nat Rev Mol Cell Biol 22:307–325
3. Itoh Y, Andrell J, Choi A et al (2021) Mechanism of membrane-tethered mitochondrial protein synthesis. Science 371:846–849
4. Suzuki T, Nagao A, Suzuki T (2011) Human mitochondrial tRNAs: biogenesis, function, structural aspects, and diseases. Annu Rev Genet 45:299–329
5. Lee M, Matsunaga N, Akabane S et al (2021) Reconstitution of mammalian mitochondrial translation system capable of correct initiation and long polypeptide synthesis from leaderless mRNA. Nucleic Acids Res 49:371–382
6. Christian BE, Spremulli LL (2012) Mechanism of protein biosynthesis in mammalian mitochondria. Biochim Biophys Acta 1819:1035–1054
7. Ayyub SA, Varshney U (2020) Translation initiation in mammalian mitochondria- a prokaryotic perspective. RNA Biol 17:165–175
8. Tsuboi M, Morita H, Nozaki Y et al (2009) EF-G2mt is an exclusive recycling factor in mammalian mitochondrial protein synthesis. Mol Cell 35:502–510
9. D'Souza AR, Minczuk M (2018) Mitochondrial transcription and translation: overview. Essays Biochem 62:309–320
10. Silva-Pinheiro P, Minczuk M (2022) The potential of mitochondrial genome engineering. Nat Rev Genet 23:199–214
11. Cruz-Zaragoza LD, Dennerlein S, Linden A et al (2021) An in vitro system to silence mitochondrial gene expression. Cell 184:5824–5837
12. Tamaru D, Amikura K, Shimizu Y et al (2018) Reconstitution of 30S ribosomal subunits in vitro using ribosome biogenesis factors. RNA 24:1512–1519
13. Shimizu Y, Ueda T (2010) PURE technology. Methods Mol Biol 607:11–21
14. Nagai R, Xu YC, Liu C et al (2021) In vitro reconstitution of yeast translation system capable of synthesizing long polypeptide and recapitulating programmed ribosome stalling. Method Protoc 4:45
15. Spremulli LL (2007) Large-scale isolation of mitochondrial ribosomes from mammalian tissues. Methods Mol Biol 372:265–275

Chapter 15

Human Mitoribosome Profiling: A Re-engineered Approach Tailored to Study Mitochondrial Translation

Iliana Soto, Mary Couvillion, and L. Stirling Churchman

Abstract

To understand the human mitochondrial translation process, tools are required to dissect this system at a global scale. The mechanisms and regulation of translation in mitochondria are different from those in the cytosol, and mitochondrial ribosomes have distinct biochemical properties. In this chapter, we describe in detail the modifications we have made to the ribosome profiling approach to adapt it to the unique characteristics of the human mitochondrial ribosome. This approach maximizes the fraction of mitochondrial ribosomes recovered, providing a snapshot of the mitochondrial translation landscape with minimal bias. We also describe the use of mouse lysate as an internal spike-in control for normalization, allowing quantification of global changes in translation across samples. Finally, we outline the bioinformatic pipelines to process the raw reads and identify mitoribosome A sites in the absence of untranslated regions flanking open reading frames. This method offers a subcodon-resolution time-sensitive global approach to explore the mitochondrial translation process in human cells.

Key words Human mitochondrial translation, Mitochondrial ribosome profiling

1 Introduction

The mechanisms and regulation of mitochondrial translation have been challenging to study, and gaps exist in our knowledge despite the plethora of pathologies caused by mutations in the mitochondrial translation machinery [1, 2]. The mitochondrial translation response to environmental and developmental changes is currently unexplored because tools have not been developed for the quantitative, time-resolved studies of mitochondrial translation in vivo. In recent years, mitochondrial translation has been studied using ribosome profiling conditions that had previously been tailored to study translation in the cytosol [3]. Mitochondrial ribosomes (mitoribosomes) are entirely distinct from cytoplasmic ribosomes (cytoribosomes) and have different biochemical properties. They have a higher protein to RNA ratio, as a result of a contraction of ribosomal RNA and the acquisition of several additional proteins, which

Antoni Barrientos and Flavia Fontanesi (eds.), *The Mitoribosome: Methods and Protocols*, Methods in Molecular Biology, vol. 2661, https://doi.org/10.1007/978-1-0716-3171-3_15,

facilitate the particular requirements of protein synthesis in mitochondria [4], requiring the re-engineering of existing ribosome profiling methods for the study of mitochondrial translation. The improved protocol we present in this chapter offers a powerful tool to better study mitochondrial gene expression and to gain insight into the translation process itself.

Ribosome profiling has previously been applied to study mitochondrial translation in human cells [5–7]. Despite high-sequencing depth (50–150 million reads per library), these approaches display incomplete coverage over mitochondrial open reading frames, potentially due to capture inefficiencies [8]. Here we present a method optimized at each step (*see* Fig. 1) to enrich mitochondrial ribosome-protected fragments (mitoRPFs) over the more abundant cytosolic ribosome-protected fragments (cytoRPFs), achieving 97–100% coverage over all mitochondrial codons in multiple cell types from a sequence depth of 50–150 million reads. First, because the subunits of the mitoribosome tend to dissociate more readily than those of the cytoribosome and must be released from the mitochondrial membrane, we use relatively low salt and high magnesium conditions along with a specialized detergent (lauryl maltoside) to preserve the integrity of mitochondrial polysomes while fully extracting them from the mitochondrial

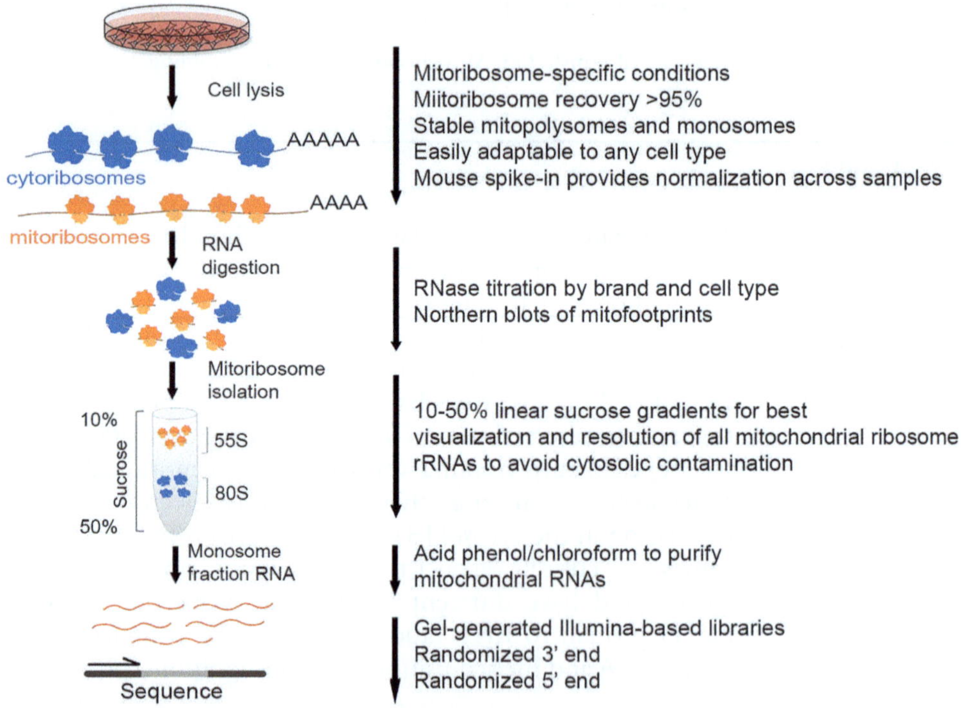

Fig. 1 Schematic of the human mitoribosome profiling approach. We describe the modifications to each step of the approach, from sample collection to building and sequencing libraries

inner membrane [9]. Second, we titrated RNase I and found that higher concentrations should be used compared with cytosolic ribosome profiling, likely because of the enhanced membrane solubilization and absence of lysate clarification step. Additionally, the mitoribosome contains less structural RNA than the cytoribosome making it more resistant to degradation by RNase treatment. Third, the gradient used for fractionation has the resolution to separate mitochondrial ribosomes (mitomonosomes, 55S) from cytosolic ribosomes (cytomonosomes, 80S). Immunoblotting is required to track these separate fractions as mitoribosomes are not abundant enough to produce a peak of absorbance at 260 nm. Fourth, we include a unique molecular identifier (UMI) in order to detect and remove duplicate reads that result from PCR amplification, producing an accurate quantitative measure of mitoribosome occupancy across the mitochondrial transcriptome.

Developmental, environmental, and genetic perturbations can cause drastic changes in mitochondrial gene expression, and with only 13 protein-coding genes, these changes are often global. We describe here the use of mouse lysate to serve as a spike-in control so that these global changes can be observed.

Complementary to the improvement of sample generation by ribosome profiling, we also designed novel bioinformatic tools. Although many computational packages are available for the analysis of ribosome profiling data, the peculiarities of mammalian mitochondrial transcripts require special attention. In particular, the absence of untranslated regions in most mRNAs makes A site determination challenging. Here, we provide the main steps for data analysis including the use of mouse mitoRPFs as spike-in controls, with special focus on determining the offset for A site transformation. The optimization of these steps from sample collection to data analysis provides a robust method to monitor translation in human mitochondria and limit bias.

2 Materials

All solutions should be prepared with nuclease-free water and only nuclease-free tubes should be used.

2.1 Cell Growth

1. Complete Dulbecco's Modified Eagle Medium (DMEM), high glucose formulation, containing sodium pyruvate.
2. Fetal bovine serum (FBS).
3. Cosmic calf serum.
4. HeLa S3 and NIH3T3 mouse fibroblasts cell lines.

2.2 Cell Lysis and Sample Collection

1. 1× PBS pH 7.4.
2. 5% Lauryl maltoside stock solution in nuclease-free water.
3. 5 M NH_4Cl stock solution in nuclease-free water.
4. 1 M $MgCl_2$ stock solution in nuclease-free water.
5. 0.1 M DTT stock solution in nuclease-free water.
6. 1 M Tris–HCl pH 7.5 stock solution in nuclease-free water.
7. 50× EDTA-free protease inhibitor cocktail stock solution in nuclease-free water.
8. *Lysis buffer*: 0.25% Lauryl maltoside, 50 mM NH_4Cl, 20 mM $MgCl_2$, 0.5 mM DTT, 10 mM Tris, pH 7.5, 1× EDTA-free protease inhibitor cocktail (*see* **Note 1**).
9. Loose-fit glass/glass 5 mL homogenizer.

2.3 RNase Digestion

1. RNase I.
2. SuperaseIn.
3. NanoDrop instrument.

2.4 Mitochondrial Ribosome Isolation

2.4.1 Sucrose Gradient Sedimentation

1. Ultracentrifuge tubes.
2. Ultracentrifuge tubes, rubber stoppers, and 10 cm cannula (Biocomp).
3. 60% filter-vacuumed sucrose stock in nuclease-free water.
4. 5% Lauryl maltoside stock solution in nuclease-free water.
5. 5 M NH_4Cl stock solution in nuclease-free water.
6. 1 M $MgCl_2$ stock solution in nuclease-free water.
7. 0.5 M EDTA, pH:8 stock solution in nuclease-free water.
8. 0.1 M DTT stock solution in nuclease-free water.
9. 1 M Tris pH 7.5 stock solution in nuclease-free water.
10. 50× EDTA-free protease inhibitor cocktail stock solution in nuclease-free water.
11. *50% sucrose mix*: 50% sucrose, 0.1% lauryl maltoside, 50 mM NH_4Cl, 10 mM $MgCl_2$, 0.5 mM DTT, 10 mM Tris, pH 7.5, 1× EDTA-free protease inhibitor cocktail (*see* **Note 1**).
12. *10% sucrose mix*: 10% sucrose, 0.1% lauryl maltoside, 50 mM NH_4Cl, 10 mM $MgCl_2$, 0.5 mM DTT, 10 mM Tris, pH 7.5, 1× EDTA-free protease inhibitor cocktail (*see* **Note 1**).
13. Beckman Ultracentrifuge and SW 41 Ti Swinging Bucket Rotor.
14. Biocomp sucrose gradient master and fractionator instrument.

2.4.2 Mitochondrial Ribosome Visualization in Sucrose Gradients

1. *5× sample loading buffer for Western blots*: 4% SDS, 250 mM Tris pH 6.8, 250 mM DTT, 30% glycerol, and 0.6 mg/mL bromophenol blue (Aliquot and store at −20 °C).
2. 15% Bis-Tris polyacrylamide precast gels.
3. 20% MOPS western running buffer (1 M MOPS, 1 M Tris, 2% SDS, 20 mM EDTA).
4. Pre-stained protein ladder.
5. MRPL12 (Proteintech, 14795-1-AP) and MRPS18-B (Proteintech, 16139-1-AP) antibodies.
6. Skim milk powder.
7. *Western washing buffer:* 10 mM Tris pH:7.5, 0.25 mM EDTA, 150 mM NaCl, 0.1% Tween-20.
8. *Western transfer buffer:* 14 g/L Glycine, 3 g/L Tris-base, 20% Methanol.
9. Nitrocellulose membrane.
10. LiCor secondary rabbit antibody.
11. LiCor instrument.

2.5 Mitochondrial Footprint Isolation

1. Acid Phenol:Chloroform:IAA (125:24:1), pH 4.5.
2. 100% Isopropanol.
3. 3 M NaOAc stock in nuclease-free water.
4. GenElute LPA Linear Polyacrylamide 25 mg/mL stock solution.
5. 70% EtOH stock in nuclease-free water.

2.6 Mitochondrial Footprint Size Purification

1. *2× Urea loading buffer*: 0.03% Urea, 6.5 mM EDTA, 1 mM Tris pH 7.5, and 0.6 mg/mL of bromophenol blue in nuclease-free water.
2. Low molecular weight DNA ladder.
3. 28 nt RNA oligo [5′AGU CAC UUA GCG AUG UAC ACU GAC UGU G/3Phos/].
4. 10× TBE buffer (0.89 M tris, 0.89 M borate, and 0.02 M EDTA).
5. 15% polyacrylamide TBE-Urea precast gels.
6. SYBR Gold stain.
7. Blue light cutting surface.
8. 18G needle.
9. CoStar 1× spin columns.
10. 100% Isopropanol.
11. 3 M NaOAc stock in nuclease-free water.

12. GenElute LPA Linear Polyacrylamide 25 mg/mL stock solution.
13. 70% EtOH stock in nuclease-free water.

2.7 Library Preparation

In order to monitor all steps of the library construction, use a purified RNA oligo of similar size, for example, 28 nucleotides (Table 1).

2.7.1 RNA Dephosphorylation

1. 10× T4 PNK buffer.
2. T4 PNK.

2.7.2 3′ End Linker Ligation

1. Linker N10: 5′ pre-adenylated, 3′ blocked at 1 μg/μL [/5rApp/**NNNNNNNNNN**CTGTAGGCACCATCAAT/3ddC/].
2. 50% PEG 8000.
3. 10× T4 RNL2 buffer.
4. T4 RNL2.

2.7.3 Purification of 3′ End Ligation Product

1. 100% Isopropanol.
2. GenElute LPA Linear Polyacrylamide 25 mg/mL stock solution.
3. 3 M NaOAc stock in nuclease-free water.
4. 70% EtOH stock in nuclease-free water.
5. 10% TBE-Urea precast gel.
6. SYBR Gold stain.
7. *2× Urea loading buffer*: 0.03% Urea, 6.5 mM EDTA, 1 mM Tris pH 7.5 and 0.6 mg/mL of bromophenol blue in nuclease-free water.
8. Low molecular weight DNA ladder.
9. Blue light cutting surface.
10. 10 mM Tris–HCl pH 8.0.

2.7.4 Reverse Transcription

1. 1.25 μM RT Random Primer [/5Phos/**NNNN**GATCGTCGGACTGTAGAACTCTGAACCTGTC/iSp18/CACTCA/iSp18/CAAGCAGAAGACGGCATACGAGATATTGATGGTGCCTACAG].
2. 5× First Strand (FS) buffer.
3. 10 mM dNTPs.
4. 0.1 M DTT.
5. SuperaseIn.
6. SuperScriptIII.
7. 1 N NaOH.

Table 1
Oligos sequences and application in protocol

Oligo	Sequence	Notes
oGAB11	5′ AGU CAC UUA GCG AUG UAC ACU GAC UGU G/3Phos/	RNA oligo used as control for library preparation.
N10 3′ Linker	/5rApp/NNNNNNNNNNCTGTAGGCACCATCAAT/3ddC/	DNA oligo 5′ pre-adenylated, 3′ blocked. Introduces 10 random bases (N) to the 3′ end.
RT 4N Primer	5Phos/NNNNGATCGTCGGACTGTAGAACTCTGAACCTGTC/iSp18/CA CTCA/iSp18/CAAGCAGAAGACGGCATACGAGATATT GATGGTGCCTA CAG	DNA oligo for reverse transcription reaction. Introduces 4 random bases (N) to the 5′ end.
PCR Forward Primer	CAAGCAGAAGACGGCATACGAGATATTGATGG	DNA oligo for PCR amplification of libraries. Amplifying 5′ → 3′.
PCR Reverse Primer (Indexed)	AATGATACGGCGACCACCGAGATCTACACGATCGGAAGAGCACAC GTCTGAACTCCAGTCACNNNNNNCGACAGGTTCAGAGTTC	DNA oligo for PCR amplification of libraries. Amplifying 3′ → 5′ introducing Illumina index (N).
Custom Sequencing Primer	CGACAGGTTCAGAGTTCTACAGTCCGACGATC	DNA oligo for sequencing libraries.

2.7.5 Purification of Reverse Transcription Product

1. 100% Isopropanol.
2. GenElute LPA Linear Polyacrylamide 25 mg/mL stock solution.
3. 3 M NaOAc stock in nuclease-free water.
4. 70% EtOH stock in nuclease-free water.
5. 10% TBE-Urea precast gel.
6. SYBR Gold stain.
7. *2× Urea loading buffer*: 0.03% Urea, 6.5 mM EDTA, 1 mM Tris pH 7.5 and 0.6 mg/mL of bromophenol blue in nuclease-free water
8. Low molecular weight DNA ladder.
9. Blue light cutting surface.

2.7.6 Circularization

1. 10× CircLigase buffer.
2. 1 mM ATP.
3. 50 mM $MnCl_2$.
4. CircLigase enzyme.

2.7.7 PCR Amplification

1. 5× Phusion buffer.
2. 10 mM dNTPs.
3. 20 μM Forward Primer [CAAGCAGAAGACGGCATAC GAGATATTGATGG].
4. 20 μM Reverse Primer (*Illumina Index*) [AATGATACGGC-GACCACCGAGATCTACACGATCGGAAGAGCACAC GTCTGAACTCCAGTCAC**NNNNNN**CGACAGGTTC AGAGTTC].
5. 2 U/μL Phusion Polymerase.

2.7.8 PCR Product Purification

1. 8% TBE precast gel.
2. *10× DNA loading buffer:* 400 μL/mL of 100% Glycerol, 0.5% SDS, 10 mM EDTA pH 8 and 5 mg/mL of bromophenol blue.
3. 100% Isopropanol.
4. GenElute LPA Linear Polyacrylamide 25 mg/mL stock solution.
5. 3 M NaOAc stock in nuclease-free water.
6. 70% EtOH stock in nuclease-free water.
7. SYBR Gold stain.
8. 100 bp DNA ladder.
9. Blue light cutting surface.

2.8 Library Quality Controls and Quantitation

1. Qubit quantification reagent for high-sensitivity DNA.
2. BioA regents for high-sensitivity DNA.
3. Qubit instrument.
4. BioA instrument.

2.9 Library Sequencing

1. 100 μM Custom sequencing primer in nuclease-free water [CGACAGGTTCAGAGTTCTACAGTCCGACGATC].
2. Illumina platform sequencing [we used a NextSeq 500 instrument].

2.10 Data Analysis

1. High-performance computing cluster or equivalent computing power.
2. Programs installed: Python, R, cutadapt, STAR, samtools, jvarkit.

3 Methods

3.1 Cell Growth

Cell cultures should be grown in dishes to facilitate quick access during lysis. Plan to collect at least 5×10^6 cells for each sample to have enough material to perform mitoribosome profiling. Grow human cell lines in DMEM supplemented with 10% FBS. To prepare spike-in lysates, culture mice fibroblasts in the same way as human cells, but use 10% Cosmic serum in cultures instead of FBS. Plan to generate enough mouse lysate to spike in all samples of the experiment (95%:5%, human:mouse by A260 values). Grow all cultures in the absence of streptomycin or penicillin to avoid any antibiotic effects on the mitoribosome.

1. Grow cell cultures in 15 cm dishes containing 35 mL of FBS or Cosmic serum-supplemented medium.
2. Allow cells to reach 75–80% confluency in a 37 °C, 5% CO_2 incubator.

3.2 Cell Lysis and Sample Collection

When collecting cell samples for ribosome profiling it is critical to minimize the manipulation time and to keep the samples cold to avoid disrupting ribosome-RNA interactions. Collect and freeze human and mouse samples independently. They will be mixed before RNase digestion described in Subheading 3.3.

1. Lysis and sample collection can be done either in the main lab space or in a cell culture room hood, if a vacuum line is available. Place plate on ice, remove media using vacuum line, and quickly wash plate once with 10 mL ice-cold 1× PBS, pH 7.4. Remove as much liquid as possible.

2. Add 600 μL (*see* **Note 2**) of ice-cold lysis buffer, scrape cells with cell lifter, and transfer to a loose-fit glass/glass 5 mL dounce homogenizer. Dounce 5–6 times while on ice. Move homogenized lysates to 1.5 mL RNase-free tubes in either 450 μL (human), or 200 μL (mouse) aliquots, flash-freeze tubes in a styrofoam bucket containing liquid nitrogen. Move aliquots to −80°C freezer for storage (*see* **Note 3**).

3.3 RNase Digestion

Careful titration of RNase I treatment is necessary to balance between the enrichment of tight footprints and not disrupting the RNA-containing mitoribosomes. Mouse lysate is added to the human lysate prior to this step so they are processed in parallel. RNase I efficiency varies from brand to brand and between lots of the same brand, requiring its titration for the specific available enzyme and cell line used. The conditions presented here are specific for HeLa cell lysates digested with New England Biolabs RNase If enzyme.

1. Thaw human and mice whole cell lysates on ice.
2. Measure RNA concentration of lysates using NanoDrop. For each sample, use A260 estimates to mix human and mouse lysates in 95:5% (human: mouse) ratios (*see* **Note 4**).
3. To 450 μL of mouse-spiked human lysate, add RNase I so that the enzyme Units/A260 is 0.04 Units/ng/μL. Digest at room temperature for 30 min without rotation. To stop digestion, add 0.180 U/μL of SuperaseIn and place tube on ice (*see* **Note 5**).
4. Clarify lysates by centrifugation at 12,500 RCF for 5 min at 4 °C. Transfer clarified supernatant to a clean tube. Keep on ice. Use in Subheading 3.4. Save 15 μL of clarified lysate as input control for immunoblotting in Subheading 3.5.

3.4 Mitochondrial Ribosome Isolation

In mammalian cells, the sedimentation coefficient of the mitoribosome is significantly different from that of the cytoribosome allowing the use of sucrose gradients for purification. Using a linear gradient of wide range (10–50%), instead of a cushion, offers the resolution of all species of the mitoribosome: the small and large subunits, monosomes, and polysomes. Such separation is critical in avoiding unwanted RNA contaminants in the libraries.

3.4.1 Sucrose Gradient Sedimentation

1. Prepare enough 10% and 50% sucrose mixes to use 6 mL of each per gradient (*see* **Note 1**). Secure 12 mL ultracentrifuge tubes on a rack and layer 10% sucrose mix first at the bottom of the tubes. Using a long needle cannula, go to the bottom of the tube and place 6 mL of the 50% sucrose mix under the 10% layer, being careful not to introduce bubbles. Cap tubes using

Biocomp rubber stoppers, being careful to seal the tubes without any air or bubbles at the top.

2. Use Biocomp Sucrose Gradient Master instrument to mix gradients following a program to form linear sucrose concentration gradients from 10% to 50%. Fourteen cycles of time and angles:

   S01. 0:05/85.0/35

   S02. 0:01/77.0/00

   S03. 0:04/86.0/35

   S04. 0:01/77.0/00

   S05. 0:03/86.5/35

   S06. 0:01/77.0/00

   S07. 0:05/85.0/35

   S08. 0:01/77.0/00

   S09. 0:04/86.0/35

   S10. 0:01/77.0/00

   S11. 0:03/86.5/35

   S12. 0:01/77.0/00

   S13. 0:20/81.0/14

   S14. 0:07/86.0/20

3. Once ready, gradients can be used within 1 h of mixing, kept at 4 °C until samples are ready to load (*see* **Note 6**).
4. To load samples, uncap tubes and remove volume from top of gradient equal to volume of clarified lysate to be loaded (*see* **Note 7**). Carefully load up to 450 μL of sample with a 1 mL tip placed against the tube wall to avoid disturbing the gradient.
5. Centrifuge gradients using SW 41 Ti rotor in an ultracentrifuge at 40,000 RPM for 3 h at 4 °C. This time and speed will capture all mitoribosome species from the small subunit to polysomes.
6. Fractionate gradients using the automated Biocomp fractionator system into 12 equal-volume fractions. Cytosolic ribosomes are detected using a 254 nm UV filter, and traces are saved and recorded as csv files (*see* Fig. 2).
7. Save 40 μL of each fraction to be used in immunoblot experiments (*see* Subheading 3.4.2) (*see* **Note 8**). Freeze remainder of fractions at −80 °C until mitomonosome fraction has been confirmed by immunoblot analysis.

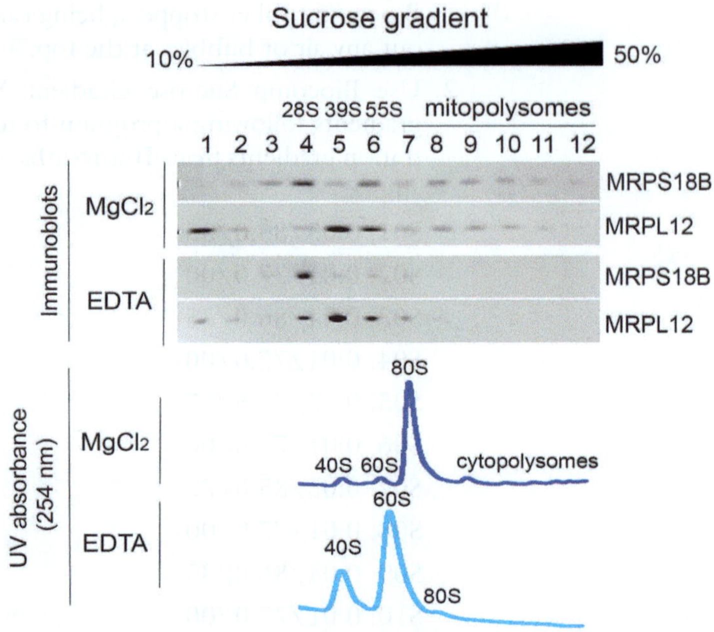

Fig. 2 Sedimentation of mito- and cytoribosomes in 10–50% linear sucrose gradients. HeLa S3 cells were lysed in the presence of either $MgCl_2$ or EDTA. Clarified lysates were loaded onto linear 10–50% sucrose gradients. Detection of mitoribosome (top panel) in each fraction was achieved by Western blots against proteins of the large (MRPL12) and small subunit (MRPS18B). We followed the sedimentation of cytoribosomes by their UV absorbance at 254 nM (bottom panel)

3.4.2 Mitochondrial Ribosome Visualization

Cytosolic ribosomes significantly outnumber mitoribosomes, making the detection of mitoribosomes in whole cell lysates challenging using UV absorbance. We used immunoblot analysis targeting both subunits of the mitoribosome to visualize mitoribosomes sedimentation in sucrose gradients and to monitor their integrity.

1. Prepare immunoblot samples by adding 10 μL of 5× loading buffer to the 40 μL sample saved in Subheading 3.4.1, **step 6**. Mix well and resolve on a 10% 15-well Bis-Tris polyacrylamide gel (*see* **Note 9**). Load 20 μL of sample/well and 5 μL of pre-stained protein ladder standard.
2. Run the gel until blue dye reaches the bottom, then transfer at 400 mA for 1 h and 15 min to a nitrocellulose membrane using western transfer buffer at 4 °C.
3. Block the nitrocellulose membrane in 5% skim milk powder in western washing buffer for 30 min.
4. Remove milk and add MRPL12 and MRPS18B antibodies each at 1:1000 v/v, diluted in 5% milk dissolved in western washing buffer. Incubate overnight at 4 °C on a shaking platform (*see* **Note 10**).

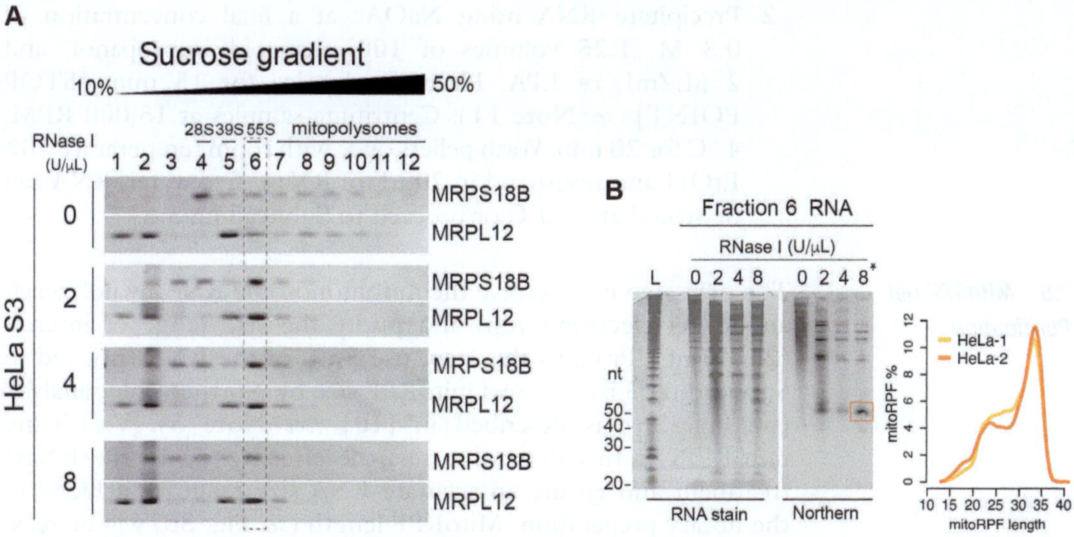

Fig. 3 Titration of RNase I to generate mitoribosome footprints. Increasing concentrations (0, 2, 4, and 8 Units/μL) of the nuclease were used to digest 450 μL of HeLa S3 lysate at room temperature for 30 min. Digested lysates were further clarified and loaded onto 10–50% linear sucrose gradients. (**a**) Detection of mitoribosomes in each fraction was achieved by Western blots against proteins of the large (MRPL12) and small subunit (MRPS18B). Dotted lined box indicates mitomonosome fraction. (**b**) RNA from the mitomonosome fraction (fraction 6) was further analyzed by Northern blot (bottom panel), showing a robust and compact footprint for the highest RNase I amount used (8 Units/μL), denoted with an (*) as the concentration used in subsequent sample generation. The mitoRPF length distribution from this sample (marked inside the box) and a replicate are shown to the right

5. Wash the membrane three times for 15 min at room temperature using western washing buffer. Add secondary LiCor antibody (rabbit-680), diluted to 1:10,000 volumes in 5% milk in western washing buffer. Incubate for 1 h at room temperature.
6. Remove the secondary antibody and wash the membrane twice for 15 min at room temperature with western washing buffer. Scan membrane using a LiCor instrument set to 700 nm wavelength (rabbit-680) (*see* Figs. 2 and 3a).

3.5 Mitochondrial RPF Isolation

MitoRPF isolation was performed exclusively from the identified mitomonosome fraction in Subheading 3.4. After sucrose gradient fractionation, the selected fraction can either be flash frozen in liquid nitrogen and stored at −80 °C for up to a week, or directly extracted for RNA isolation as explained below.

1. Mitoribosome footprints were recovered from the monosome-enriched fraction. To each sample, add equal volume of acid phenol/chloroform, pH 4.5. Vortex vigorously for 20 s and centrifuge at 4 °C for 10 min at 15,000 RPM. Transfer aqueous top layer to a clean RNase-free microcentrifuge tube.

2. Precipitate RNA using NaOAc at a final concentration of 0.3 M, 1.25 volumes of 100% ice-cold isopropanol, and 2 μL/mL of LPA. Place on dry ice for 15 min. [STOP POINT] (*see* **Note 11**). Centrifuge samples at 15,000 RPM, 4 °C for 20 min. Wash pellets once with room temperature 70% EtOH and resuspend in 20 μL of RNase-free water. RNA can be stored at −80°C or proceed to Subheading 3.6.

3.6 MitoRPF Gel Purification

The next step is to resolve the mitomonosome RNA by polyacrylamide gel electrophoresis and purify the size range of interest (25–40 nt). Prior to this step, use 5 μL of the RNA (isolated in Subheading 3.5) to assess mitoRPF size by Northern blot analysis (*see* Fig. 3b) as described in [10] (*see* **Note 5**). Visualizing mitoRFPs at this step will assess the effectiveness of the RNase treatment and ensure an accurate RNA size range is included in the library preparation. MitoRPF length (*see* Fig. 3b) will be reassessed bioinformatically after sequencing the libraries (*see* Subheading 3.10).

1. Combine 15 μL of RNA from Subheading 3.5 with 20 μL 2× Urea RNA loading buffer.
2. Prepare 0.5 μg of low molecular weight DNA ladder in 20 μL of 1× Urea RNA loading buffer.
3. Denature samples and ladder 90 s at 80 °C, immediately followed by 90 s on ice.
4. Pre-run a 15% TBE-Urea polyacrylamide gel in 1× RNase-free TBE buffer 15 min at 200 V (*see* **Note 12**). Load samples and 10 μL of ladder onto gel and run for 1.5 h at 200 V or until the front of the gel reaches bottom.
5. Stain gel with SYBR Gold (1:10,000 in RNase-free 1× TBE) for 2 min with shaking at room temperature. Rinse gel once in RNase-free water, place in a sheet protector, and image on Typhoon scanner. Use these scanning settings: 520 BP emission filter, blue (488) laser, and PMT voltage 400–600. If a Typhoon instrument is not available, image gel using UV transilluminator.
6. Prepare microcentrifuge tubes for crushing the gel slices. Pierce the bottom of a 0.5 mL tube using a 18G needle and nest it inside a 1.5 mL tube.
7. Place stained gel on a blue light cutting platform and excise sizes from 25 to 40 nt (*see* Fig. 4). Move gel pieces to the nested tube setup. Close smaller tube and leave bigger tube open. Centrifuge tubes at 14,000 RPM for 2 min. Gel slurry will be collected at the bottom of the 1.5 mL tube.
8. Remove the smaller tube and add 400 μL of RNase-free water to the gel slurry. Incubate at 70 °C for 15 min.

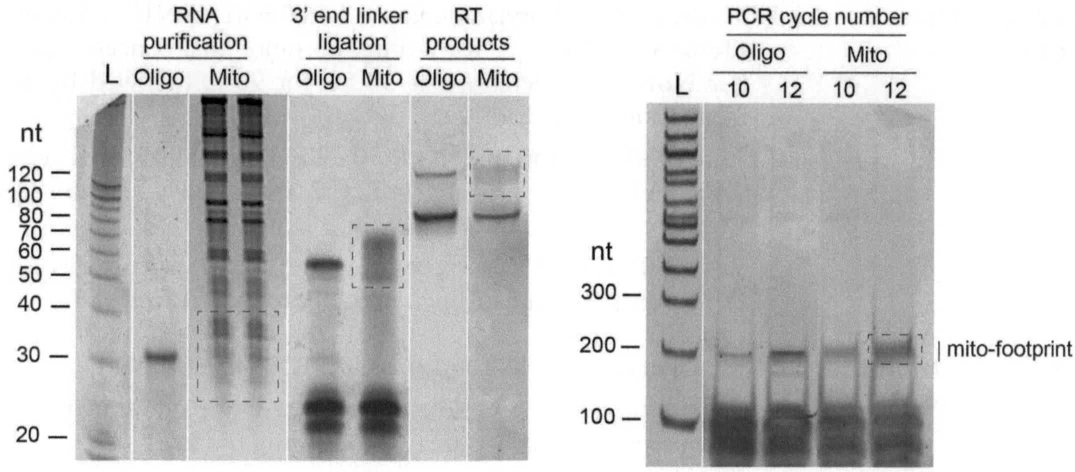

Fig. 4 Effect of using translation inhibitors in ribosome profiling. Comparison of relative synthesis rate values (tpm) for mtDNA-encoded OXPHOS subunits in cells exposed or not to ribosome inhibitors. Pearson correlation, r. HeLa cells were treated with cycloheximide (CHX, 100 μg/mL) and chloramphenicol (CAP, 100 μg/mL) for 15 and 5 min, respectively, to stall cytosolic and mitochondrial translation. After treatment, cells were washed with 1× PBS, pH 7.4 and subjected to lysis using our mitoribosome profiling conditions

9. Move gel slurry solution to a CoStar Spin column using a wide-bore tip and centrifuge at 14,000 RPM for 2 min at room temperature.
10. Move flow through to a clean tube and precipitate recovered RNA using 0.3 M NaOAc, 1.25 volumes of 100% cold isopropanol and 2 μL/mL of LPA. Incubate on dry ice for 15 min and centrifuge at 15,000 RPM for 20 min at 4 °C. [STOP POINT] (*see* **Note 11**).
11. Wash pellets with room temperature 70% EtOH and resuspend in 5 μL of RNase-free water.

3.7 Library Preparation

Libraries are prepared as described in [3], with a few modifications. Using only the RNA from the mitomonosome fraction provides a more purified sample, avoiding the need for rRNA depletion. We also introduce ten randomized nucleotides to the linker ligation at the 3′ end and 4 to the 5′ end. All intermediate products in the library preparation are visualized and purified using gels in order to monitor and quality control every step of the library generation process.

3.7.1 RNA Dephosphorylation

1. Denature the 5 μL of RNA from Subheading 3.5 for 90 s at 80 °C and immediately place sample on ice for 90 s.
2. Add 4.5 μl PNK mix: 0.95 μL 10× T4 PNK buffer, 2.55 μL RNase-free water, 1 μL T4 PNK. Incubate at 37 °C, 1 h.
3. Heat-inactivate T4 PNK by incubating at 70 °C for 10 min. Spin tube to collect possible condensation.

3.7.2 3′ End Linker Ligation

1. To the dephosphorylated sample, add 0.5 μL of N10 3′ Linker (Table 1) at 1 μg/μL for 0.5 μg, 85 pmol final concentration (*see* **Note 13**). Denature at 80 °C for 90 s, followed by an incubation on ice for 90 s.
2. Add 10 μL ligation mix: 8 μL 50% PEG 80008000, 1 μL 10× T4 RNA ligase buffer, 1 μL T4 RNL2, truncated (*see* **Note 14**). Incubate at room temperature for 3 h.
3. Add 400 μL of RNase-free water, and follow instructions for RNA precipitation explained in Subheading 3.6, **step 10**.

3.7.3 Purification of 3′ End Ligation Product

1. Resuspend pellets in 10 μL 1× Urea RNA load buffer.
2. Prepare 0.3 μg of low molecular weight DNA ladder in a final volume of 10 μL in 1× urea RNA loading buffer.
3. Denature samples and ladder at 80 °C for 90 s, followed by ice for 90 s.
4. Load samples and ladder on a 10% TBE-Urea polyacrylamide gel and run for 1 h at 200 V or until blue dye reaches the bottom of the gel.
5. Stain and image gel as described above in Subheading 3.6, **step 5**.
6. Excise ligated product (*see* Fig. 4) from 50 to 70 nt and perform gel extraction and isopropanol precipitation as explained in Subheading 3.6, **steps 6–11**. [STOP POINT] (*see* **Note 11**).
7. Resuspend ligated RNA in 10 L of 10 mM Tris, pH 8.0

3.7.4 Reverse Transcription

1. Transfer resuspended ligated RNA to a PCR tube. All subsequent incubation steps should be performed in a thermocycler.
2. Add 2 μL of 1.25 μM RT 4 N primer (Table 1).
3. Denature for 90 s at 80 °C and immediately place on ice for 90 s.
4. Add 8 μL reverse transcription mix: 4 μL 5× FS buffer, 1 μL 10 mM dNTPs, 1 μL 0.1 M DTT, 1 μL SuperaseIn, 1 μL Superscript III. Mix well. Incubate 30 min at 48 °C.
5. Add 2.2 μL of 1 N freshly diluted sodium hydroxide and incubate 20 min at 98 °C to degrade RNA.
6. Transfer to a 1.5 mL microcentrifuge tube and add 400 μL water. Precipitate sample as described above in Subheading 3.6, **step 10**.

3.7.5 Purification of Reverse Transcription Product

1. Resuspend pellet in 10 μL 1× Urea RNA load buffer.
2. Prepare 0.3 μg low molecular weight DNA ladder in a final volume of 10 μL 1× Urea RNA load buffer. Denature sample

and ladder and load on a 10% TBE-Urea polyacrylamide gel. Run gel for 1 h at 200 V or until blue dye reaches the bottom of the gel.

3. Stain and image gel as described above in Subheading 3.6, **step 5**.
4. Excise ligated product (*see* Fig. 4) from 90 to 110 nt and perform gel extraction and isopropanol precipitation as explained in Subheading 3.6, **steps 6–11**. [STOP POINT]. (*see* **Note 15**).
5. Resuspend purified RT product in 5 uL of nuclease-free water.

3.7.6 Circularization

1. To resuspended RT product, add 15 μL circularization mix: 10 μL water, 2 μL 10× circularization buffer, 1 μL 1 mM ATP, 1 μL 50 mM manganese chloride, 1 μL CircLigase.
2. Incubate 1 h at 60 °C.
3. Heat-inactivate CircLigase by incubating 10 min at 80 °C. Store circles at −20 °C.

3.7.7 PCR Amplification of Libraries

PCR Cycle Number Determination

1. Prepare a master mix reaction to test different PCR cycle numbers: Combine 5 μL of circles with 78.5 μL PCR mix: 56.3 μL water, 16.7 μL 5× Phusion HF buffer, 1.7 μL 10 mM dNTPs, 1.5 μL 20 μM oSMD2 forward primer (Table 1), 1.5 μL 20 μM indexed primer (Table 1), 0.8 μL 2 U/μL Phusion polymerase (*see* **Note 16**).
2. Aliquot 16.7 μL reaction mixture into each of five PCR tubes.
3. Program thermocycler 30 s at 98 °C and 14 cycles of: 15 s at 98 °C, 15 s at 65 °C, 7 s at 72 °C.
4. Remove samples from thermocycler during elongation step (72 °C) after 10 and 12 cycles (*see* Fig. 4).
5. To each sample, add 2.0 μL 10× DNA loading buffer.
6. Prepare 0.3 μg of 100 bp-DNA ladder in a final volume of 20 μL 1× DNA load buffer. Use 10 μL/gel.
7. Do not heat-denature samples, but load directly on a 8% TBE non-denaturing polyacrylamide gel. Run gel in 1× TBE for 45 min at 180 V. Pre-running the non-denaturing gel is optional.
8. Stain and image gel as previously described (Subheading 3.6, **step 5**).

PCR Amplification with Determined Cycle Number

1. Prepare PCR reaction mix as described in Subheading 3.7.7.1. Make enough master mix/sample to amplify eight reactions of 16.7 μL each.
2. Aliquot 16.7 μL reaction mixture into PCR tubes.

3. Set thermocycler program for 30 s at 98 °C for the number of cycles determined in Subheading 3.7.7.1 of: 15 s at 98 °C, 15 s at 65 °C, 7 s at 72 °C.

3.7.8 PCR Product Purification

1. Pool all reactions for each sample and resolve them in non-denaturing conditions as explained in Subheading 3.7.7.1
2. Excise amplified PCR product at around 180–190 nt, avoiding any lower product band resulting from unextended reverse transcription primer (*see* Fig. 4).
3. Perform overnight gel extraction: Place gel pieces in 1.5 mL microcentrifuge tube and add 650 μL of DNA gel extraction buffer. Incubate overnight at room temperature, rotating end over end.
4. Move liquid to a CoStar Spin column and centrifuge for 2 min at 14,000 RPM to remove any gel residue.
5. Precipitate libraries by adding 2 μL/mL of LPA and 800 μL of ice-cold 100% isopropanol. Incubate on dry ice for 15 min and spin at 15,000 RPM at 4 °C for 20 min. Wash pellet once with room temperature 70% EtOH. Resuspend pellets in 15 μL of 10 mM Tris, pH:8.5.

3.8 PCR Product Quality Controls and Quantitation

Assess library concentration and quality using Qubit Fluorometric Quantitation dsDNA kit (high sensitivity), and Bioanalyzer DNA high-sensitivity chip according to manufacturer's instructions.

3.9 Sequencing

When submitting samples for sequencing, supply custom "Read 1" sequencing primer: oNTI202 (Table 1).

3.10 Data Analysis

Human mitoRPFs are typically 30–34 nucleotides. The UMI is 14 nucleotides, so 60-cycle reads are sufficient for covering these plus enough of the 3′ adapter for identification and removal. For robust downstream analyses, plan for at least 30–50 million raw reads, although results vary depending on the cell type. Cytosolic rRNA typically composes 50–70% of all reads and mitochondrial rRNA around 10%. In mitochondria-rich cell types however the mitochondrial:cytosolic rRNA ratio may be increased and vice versa for mitochondria-poor cell types. Mitochondrial mRNA-mapping reads (the mitoRPFs) should ultimately make up 3–10% of the raw reads. PCR duplicates should be removed using the UMI and not a prebuilt tool that uses only alignment positions to identify them.

For more details, see https://github.com/churchmanlab/human-mitoribosome-profiling

3.10.1 Raw FASTQ Files Initial Processing

1. Remove adapter sequence (e.g., with Cutadapt https://github.com/marcelm/cutadapt). Discard reads shorter than 25 nucleotides after adapter removal. The 14-nt UMI will be

removed in the next step leaving a minimum of 11-nt RPFs. This short minimum length is used to allow retention of potential halfmers resulting from translation initiation on leaderless mRNAs. Discard reads in which the adapter was not identified because the UMI cannot be identified in these cases.

2. Remove four nucleotides from the 5′ end and ten nucleotides from the 3′ end. These together are the UMI, which is to be associated with the sequence identifier.
3. Remove the reads aligning to cytosolic and mitochondrial rRNAs.
4. Map the remaining reads with a splice aware alignment tool, for example, STAR (https://github.com/alexdobin/STAR) to the reference genome that includes nuclear chromosomes and the mitochondrial genome (*see* **Note 17**). If mouse lysate was added for spike-in use, the mouse genome should be concatenated with the human genome for alignment in order to get the most accurate results. However, spike-in read counts will be determined from a separate alignment step (*see* Subheading 3.10.1, **step 8**).
5. Remove true PCR duplicates by collapsing reads that have identical mapping coordinates, sequence, and UMI to a single occurrence.
6. Filter the deduplicated BAM file using samtools (https://github.com/samtools) to keep only mitochondrial mRNA mapping reads. The resulting BAM file will be the input for the following section.
7. In parallel to **steps 4–6**, determine the spike-in normalization factor: Filter the output of **step 3** for a minimum read length of 20 nucleotides.
8. Align to mouse mitochondrial mRNAs (not the entire mouse mitochondrial genome) allowing 0 mismatches. Either allow soft clipping or trim the 5′ terminal nucleotide which is often added, untemplated, by reverse transcriptase during library preparation.
9. Remove true PCR duplicates as in **step 5**.
10. The resulting read count is the spike-in normalization factor.

3.10.2 Transform the Read Signal to Ribosome A Site Positions

This section describes how to transform the read signal to ribosome A site positions (*see* **Note 18**). Only three mitochondrial genes have 5′ regions upstream of the start codon that are long enough to observe full-length mitoRPFs of initiating mitoribosomes. These genes are MT-ATP6 and MT-ND4, which both overlap with their upstream genes MT-ATP8 and MT-ND4L, respectively, in bicistronic transcripts. The third is MT-CO3, which is immediately downstream of MT-ATP6 and is inefficiently processed, leaving a

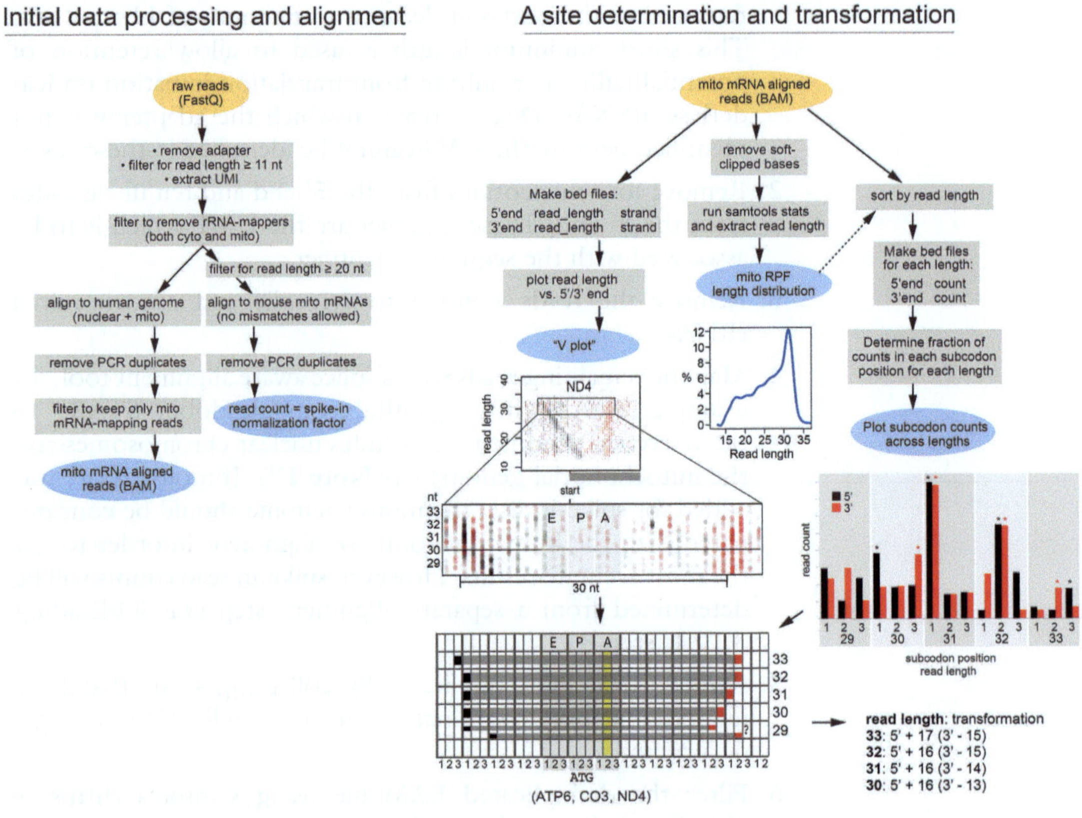

Fig. 5 Library preparation for mitoribosome profiling. All steps described in Subheading 3.7, including: mitoRFPs gel purification, 3′ end linker ligation, Reverse Transcription (RT) reaction and PCR amplification. Dotted lined boxes indicate areas in the gel that were included for product isolation. RNA and DNA products were run in either denaturing TBE-Urea acrylamide gels (mitoRFPs isolation, linker ligation, RT product), or non-denaturing TBE acrylamide gels (PCR amplification product). RNA and DNA products were visualized using SYBR Gold staining and their size estimated using a 10 bp DNA ladder (L). We included RNA oGAB11 (Oligo) (Table 1) as a technical control through all steps of the library preparation process to monitor the quality of the mitoRPF products (Mito)

tricistronic transcript (MT-ATP8/ATP6/CO3). These start codons are used to set the approximate register for the offset from the 5′ or 3′ end of reads to the A site. To determine the exact offset, the most frequent subcodon position of 5′ and 3′ ends for each read length is calculated from across all open reading frames. Notably, because trimming can occur from both ends, reads that are not the full length of 30–34 nucleotides will have an ambiguous register and their A site offset cannot be determined (e.g., 29 nt reads in Fig. 5). The input file for the following steps is the BAM file from **step 6** above: deduplicated human mitochondrial mRNA mappers.

1. First, plot the mitoRPF read length distribution. For an accurate output, soft-clipped bases must first be removed (*see* note in Subheading 3.10.1, **step 8**), which can be done using jvarkit (https://github.com/lindenb/jvarkit/blob/master/src/main/java/com/github/lindenb/jvarkit/tools/biostar/Biostar84452.java). Samtools stats will then output the read length distribution directly from the BAM file. See README.md and scripts at https://github.com/churchmanlab/human-mitoribosome-profiling/tree/main/1_AlignData.
2. Next make "V plots," which plot read length versus genomic position, to visualize mitoRPF 5′ and 3′ ends in relation to start codons. To do this, make two 3-column files: (1) 5′ position, read length, strand and (2) 3′ position, read length, strand. Again soft-clipped bases are ignored, unless they result from poly(A) tails. See README.md and scripts at https://github.com/churchmanlab/human-mitoribosome-profiling/tree/main/2_MakeLengthBeds.
3. Finally find the most frequent subcodon position (1st, 2nd, or 3rd) for both 5′ and 3′ ends of full-length reads. This is done individually for each read length. Make individual SAM files for each read length in the range of full-length as determined in **step 1** above (e.g., 29–33 nucleotides). From each of these make two 2-column files: (1) 5′ position, read count and (2) 3′ position, read count. Using these, determine the number of reads in each subcodon position across all open reading frames. See README.md and scripts at https://github.com/churchmanlab/human-mitoribosome-profiling/tree/main/3_CountFramePerLength.
4. Use the V plot at start codons and subcodon counts across lengths to determine the offset to the A site for each read length, as shown in Fig. 5. For example, take 30 nt reads (black horizontal line on the V plot in Fig. 5; also see the scale line below). The darkest red region (3′ ends) on this line is 16 nt downstream of the middle of the start codon, which is in the P site of an initiating ribosome. Now, looking at the subcodon position bar plot, we see that accordingly, 3′ ends of 30 nt reads are most frequently at the 3rd subcodon position (red asterisk) throughout the genome. The same process can be done using the 5′ ends of 30 nt reads and then for other read lengths. Occasionally, the precise positioning will be ambiguous on the V plot, which only represents a single locus. Therefore, it is important to always corroborate offsets using the subcodon bar plot, which uses data from across the entire transcriptome. Offsets can be taken from either the 5′ end or the 3′ end. Read signals are assigned to the center of the A site (subcodon position 2) so that even with one nucleotide of error

either way reads will still be assigned to the correct codon. To carry out A site transformation after determining offsets, see README.md and scripts at https://github.com/churchmanlab/human-mitoribosome-profiling/tree/main/4_AsiteTransformation.

4 Notes

1. $MgCl_2$ can be switched with 5 mM EDTA to assess polysome dissociation and ribosome subunit separation in sucrose gradients (*see* Fig. 2). Addition of antibiotics to buffers is optional. We did not detect differences in mitomonosome stability when omitting chloramphenicol and/or cycloheximide (*see* Fig. 6).
2. Make more lysis buffer than needed for volume adjustments. Adjust lysis buffer volume to RNA cell content. Use 600 μL of buffer for human fibroblasts and myoblasts, but 1 mL for HEK293T and HeLa S3 cells.
3. Cell lysates can be stored at −80 °C for up to 3–4 months. Do not freeze and thaw multiple times.
4. Mix enough sample volume to dedicate 450 μL of lysate for mitoribosome profiling and 400 μL for RNA-seq for RNA abundance control, so that both experiments share the same input source.
5. Enzyme concentration was estimated for the specific RNase brand and titrated for each cell line used. Footprint size after RNase digestion was assessed using Northern blot analysis as previously described [10]. Example: HeLa cells required 8 Units/μL of enzyme to digest 450 μL of lysate at 200 ng/μ L concentration estimated by absorbance at 260 nm, while human fibroblasts required 4.4 Units/μL.

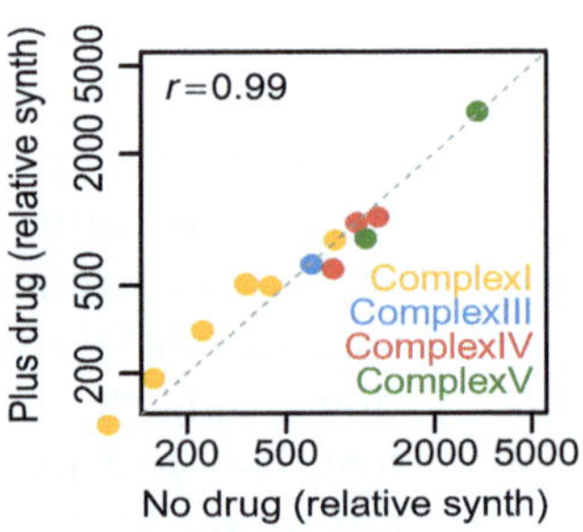

Fig. 6 Flow chart of the computational steps for the analysis of human mitoribosome profiling. Inputs and outputs are highlighted, and sample plots are shown to illustrate the process of determining the A site transformation, which can be done using either the 5′ or the 3′ read end

6. Keep mixes at room temperature during gradient mixing, but cool gradients afterwards prior to use.
7. Removing a small volume from the top of the gradient will prevent overflow during sample fractionation and will not significantly affect the resolution of the gradient.
8. With the time and speed parameters used above, the mitomonosome peak is expected in fraction 6. Depending on sample volume loaded, the mitomonosome may also be found in fraction 5, which makes immunoblotting a necessary step for accurate monosome identification.
9. Do not boil western samples before running.
10. MRPL12 running size is ~12 kDa, while MRPS18B is ~21 kDa, making their separation possible using the above conditions. Both antibodies show minimal background signal, which allows for their simultaneous use on the same membrane.
11. Store RNA at −80 °C in alcohol prior to resuspending.
12. Clear urea from wells with a syringe and needle before the pre-run and immediately before loading the samples. Load 20 μL/well and leave one empty well in between samples to avoid cross-contamination. To fit this volume, use 1.5 mm thick gels, or split the volume to fit 1 mm gels.
13. Linker should be present in at least fivefold molar excess compared to substrate RNA.
14. PEG 8000 is very viscous and care must be taken in preparing this reaction. It is advisable to prepare enough mix for at least one extra reaction.
15. RT sample can be stored in precipitation mix for weeks at −80 °C before moving to the next step. If the product band runs too close to the primer band, the sample will need to be repurified in a second gel run.
16. Perform PCR test with the index primer that will be used for each library.
17. It is important to allow alignment of reads with multiple mapping positions and retain them for downstream processing because of nuclear mitochondrial DNA segments present in the nuclear genome.
18. Compute the A-site offsets for each experiment. While each cell line generally has a consistent offset, there are small differences between cell lines, and differences in RNase treatment can have an effect.

References

1. Elson JL, Smith PM, Greaves LC et al (2015) The presence of highly disruptive 16S rRNA mutations in clinical samples indicates a wider role for mutations of the mitochondrial ribosome in human disease. Mitochondrion 25:17–27
2. Yarham JW, Elson JL, Blakely EL et al (2010) Mitochondrial tRNA mutations and disease. Wiley Interdiscip Rev RNA 1:304–324
3. Ingolia NT, Brar GA, Rouskin S et al (2012) The ribosome profiling strategy for monitoring translation in vivo by deep sequencing of ribosome-protected mRNA fragments. Nat Protoc 7:1534–1550
4. Amunts A, Brown A, Toots J et al (2015) Ribosome. The structure of the human mitochondrial ribosome. Science 348:95–98
5. Rooijers K, Loayza-Puch F, Nijtmans LG, Agami R (2013) Ribosome profiling reveals features of normal and disease-associated mitochondrial translation. Nat Commun 4:1–8
6. Pearce SF, Rorbach J, Van Haute L et al (2017) Maturation of selected human mitochondrial tRNAs requires deadenylation. Elife:6. https://doi.org/10.7554/eLife.27596
7. Pearce SF, Cipullo M, Chung B et al (2021) Mitoribosome profiling from human cell culture: a high resolution view of mitochondrial translation. Methods Mol Biol 2192:183–196
8. Soto I, Couvillion M, McShane E et al (2021) Balanced mitochondrial and cytosolic translatomes underlie the biogenesis of human respiratory complexes. bioRxiv 2021.05.31.446345
9. Englmeier R, Pfeffer S, Förster F (2017) Structure of the human mitochondrial ribosome studied in situ by cryoelectron tomography. Structure 25:1574–1581.e2
10. Couvillion MT, Churchman LS (2017) Mitochondrial ribosome (mitoribosome) profiling for monitoring mitochondrial translation in vivo. Curr Protoc Mol Biol 119: 4.28.1–4.28.25

Chapter 16

The $ARG8^m$ Reporter for the Study of Yeast Mitochondrial Translation

Daniel Flores-Mireles, Yolanda Camacho-Villasana, and Xochitl Pérez-Martínez

Abstract

Mitochondrial translation is an intricate process involving both general and mRNA-specific factors. In addition, in the yeast *Saccharomyces cerevisiae*, translation of mitochondrial mRNAs is coupled to assembly of nascent polypeptides into the membrane. $ARG8^m$ is a reporter gene widely used to study the mechanisms of yeast mitochondrial translation. This reporter is a recodified gene that uses the mitochondrial genetic code and is inserted at the desired *locus* in the mitochondrial genome. After deletion of the endogenous nuclear gene, this reporter produces Arg8, an enzyme necessary for arginine biosynthesis. Since Arg8 is a soluble protein with no relation to oxidative phosphorylation, it is a reliable reporter to study mitochondrial mRNAs translation and dissect translation form assembly processes. In this chapter, we explain how to insert the $ARG8^m$ reporter in the desired spot in the mitochondrial DNA, how to analyze Arg8 synthesis inside mitochondria, and how to follow steady-state levels of the protein. We also explain how to use it to find spontaneous suppressors of translation defects.

Key words *ARG8m*, ATP synthase, bc_1 complex, Cytochrome *c* oxidase, Mitochondria, Mitochondrial DNA, Suppressor, Respiratory complexes, Translation, Yeast

1 Introduction

S. cerevisiae mitochondrial translation can be studied in vivo by location of different reporter genes in the mitochondrial DNA (mtDNA). This can happen due to the ease to transform and manipulate yeast mitochondrial DNA [1]. This process must consider addition of mitochondrial 5′ and 3′ flanking regions to the reporter gene and recodification since the yeast mitochondrial genetic code differs from the universal genetic code (Reviewed in [2]). The reporters that have been inserted into the mtDNA are Arg8 [3], GFP [4, 5], and Barstar [6]. Arg8 is by far the most

Antoni Barrientos and Flavia Fontanesi (eds.), *The Mitoribosome: Methods and Protocols*, Methods in Molecular Biology, vol. 2661, https://doi.org/10.1007/978-1-0716-3171-3_16,

successfully used mitochondrial reporter, and today the studies of translation of all mtDNA-encoded mRNAs have used this reporter [3, 7–12].

Arg8 is the acetylornithine aminotransferase (EC 2.6.1.11), an enzyme that participates in ornithine and arginine production [13]. In yeast, this enzyme is encoded by the nuclear gene *ARG8*, and the protein is directed to mitochondria with the aid of a mitochondrial targeting sequence (MTS) that is cleaved once the protein is localized in mitochondria. To use this protein as a reporter gene for mitochondrial mRNAs translation and as selectable marker, Steele and collaborators [3] deleted the nuclear gene and inserted a recodified version in the mtDNA, named $ARG8^{m}$, that uses the yeast mitochondrial genetic code. The produced Arg8 protein has the exact same sequence as the nuclear one, and the MTS can be included or not in this sequence, depending on the particular question we want to address. Since Arg8 is a soluble protein that has no relation to the oxidative phosphorylation (OXPHOS) complexes, it has been demonstrated to be a reliable reporter to study translation of the mtDNA-encoded OXPHOS subunits and dissect it from posttranslational processes, like assembly.

This reporter can be fused in frame to a mitochondrial *orf* or a fragment of an *orf* to create chimeric proteins (*see* Fig. 1). In this context, one can analyze how the presence of the mitochondrial coding sequence of your favorite mitochondrial gene (*YFmG*) influences translation. $ARG8^{m}$ can replace the coding region of *YFmG* and preserve the original 5′ and 3′ untranslated regions. This mutation is useful to analyze the role of untranslated regions of the mRNA of interest on translation of the reporter. In addition, the *orf* of *YFmG* can be inserted at an ectopic site in the mitochondrial genome. This site is commonly used and locates at 285 nt upstream of the *COX2* locus. Ectopic *YFmG's orf* is flanked by the *COX2* UTRs since they are the shortest ones in the mtDNA and prevent mtDNA recombination with the original *COX2 locus*, leading to a stable mitochondrial genome [6]. In this context, it is possible to dissect the role of *YFmG's* untranslated regions on one side, and of *YFmG*'s *orf* on the other side. Both *YFmG*'s untranslated regions and coding regions are present in the same mtDNA molecule, but physically apart (*see* Fig. 1). Moreover, $ARG8^{m}$ have been inserted at this ectopic site to study stability of the mtDNA [14].

A detailed protocol for yeast mitochondrial transformation have been published [3]. Here we focused on how to transform and integrate $ARG8^{m}$ into the mitochondrial genome (*see* Subheading 3.1), and how to growth yeast cells on media lacking arginine to assess $ARG8^{m}$ mRNA translation (*see* Subheading 3.2). We also briefly explained how $ARG8^{m}$ translation can be assessed by (^{35}S)-Methionine labeling (*see* Subheading 3.3), how

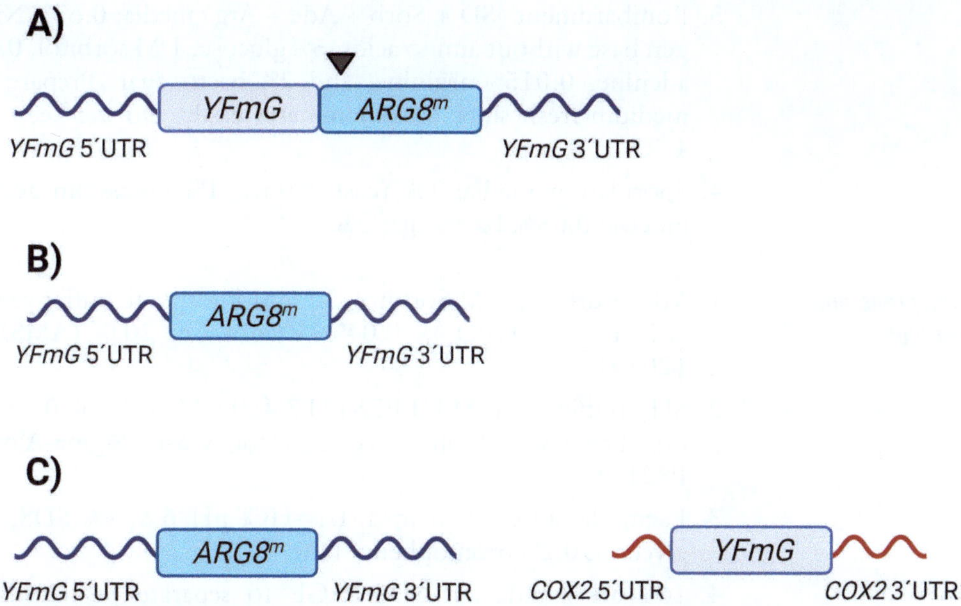

Fig. 1 The mitochondrial reporter gene *ARG8^m* can be inserted in the mtDNA in different forms. (**a**) The reporter can be fused in frame with the 3′ end of your favorite mitochondrial gene (*YFmG*). The stop codon of *YFmG* is thus replaced by the *ARG8^m* coding sequence. This reporter could also be fused to fragments of *YFmG* if the effect of a particular section of *YFmG* on *ARG8^m* translation is to be analyzed. The *ARG8^m* sequence can code for the Arg8 mitochondrial targeting sequence (▾), so once the $Yf_m g$-Arg8 precursor is produced, the Arg8 moiety can release from $Yf_m g$ region. (**b**) The *ARG8^m* reporter can completely replace the *YFmG* coding sequence. (**c**) The *ARG8^m* reporter completely replaces the *YFmG* coding sequence. In addition, *YFmG* coding sequence, flanked with ~73 and ~119 nucleotides from the *COX2* 5′ and 3′ UTRs, respectively, are inserted at an ectopic site of mtDNA, 285 nt upstream of the *COX2* coding sequence. (Created with BioRender.com)

to make quick and small mitochondrial preparations to study steady-state levels of the produced Arg8 protein (*see* Subheading 3.4), and how to use this reporter to find yeast suppressors that can help identifying new factors and pathways related to mitochondrial translation (*see* Subheading 3.5). Lastly, we included an example of how this reporter can be useful to differentiate between translation and posttranslational effects.

2 Materials

2.1 Media for Yeast Cell Growth

1. Complete media: 1% yeast extract, 2% bacto peptone, 2% glucose (for YPD), 2% Raffinose (for YPRaf), 2% galactose (for YPGal), or 3% glycerol/3% ethanol (for YPEG). For solid media, add 2% bacto agar.
2. Minimal media: 0.67% Nitrogen base without amino acids (BD Biosciences, Difco), 2% glucose, drop-out media lacking the desired amino acids, add according to the manufacturer's instructions (Formedium). For solid media, add 2% bacto agar.

3. Bombardment (SD + Sorb + Ade + Arg) media: 0.67% Nitrogen base without amino acids, 5% glucose, 1 M sorbitol, 0.01% adenine, 0.015% arginine, and 2% bacto agar. Prepare this medium fresh since it contaminates easily. Do *not* keep it at 4 °C.
4. Sporulation media: 1% Yeast extract, 1% Potassium acetate, glucose 0.05%, bacto agar 2%.

2.2 Mitopreps and Western blot

1. MPZ buffer: 1.2 M Sorbitol, 20 mM Phosphate buffer pH 7.4 (KH_2PO_4/K_2HPO_4), 0.04% Zymolyase 20T (AMSBIO, 120491-1).
2. SHP buffer: 20 mM HEPES pH 7.4, 0.6 M Sorbitol, 0.33 μL/mL Protease inhibitors cocktail for yeast (Sigma-Aldrich, P8215).
3. Laemmli buffer: 100 mM Tris–HCl pH 6.8, 4% SDS, 20% glycerol, 0.2% bromophenol blue.
4. 16% acrylamide for SDS-PAGE to separate proteins from *in vivo* or *in organello* translation assays (use a stock solution of 30% acrylamide, 0.3% bis-acrylamide): 388 mM Tris–HCl pH 8.8, 0.01% SDS, 0.05% Ammonium persulfate, 0.05% N, N, N′, N′-tetrametiletilendiamina.
5. 12% acrylamide (from a stock solution of 30% acrylamide, 0.8% bis-acrylamide) 388 mM Tris–HCl pH 8.8, 0.01% SDS, 0.05% Ammonium persulfate, 0.05% N, N, N′, N′-tetrametiletilendiamina.
6. Western blot blocking solution: 1% Fat-free milk, 50 mM Tris–HCl pH 7.6, 150 mM NaCl.
7. Western blot antibody solution: 0.1% Fat-free milk, 10 mM Tris–HCl pH 7.6, 100 mM NaCl, 1 mM EDTA-Na pH 8.0, 0.1% Tween-20.
8. We use a polyclonal antibody against Arg8 [3] diluted 1:10,000 and Peroxidase AffiniPure Goat Anti-Rabbit IgG diluted 1:2000 (Jackson Immuno Research, 111-035-144).
9. Western blot washing solution: 10 mM Tris–HCl pH 7.6, 100 mM NaCl, 1 mM EDTA-Na pH 8.0, 0.005% Tween-20.
10. Stripping solution: 0.7% 2-Mercaptoethanol, 62.5 mM Tris–HCl pH 6.8, 2% SDS.

2.3 Tetrad Dissection

1. Digestion buffer: 1.5 mg/mL Zymolyase 20T, 1 M sorbitol. Store aliquots at −20 °C.
2. Light microscope with micromanipulator (we use a Zeiss Scope A1 AX10 with a 10×/0.25 objective).

2.4 Mitochondrial Transformation

1. PDS-1000/ He™ Biolistic Particle Delivery System (BIO-RAD).
2. Helium tank grade 5 (99.99%).
3. Rupture discs 1350 Psi (BIO-RAD, 1652330).
4. Vacuum pump of 100 L/min capacity.
5. The most used strains have the following genotype: *Mata* or *Matα, ura3-52, arg8Δ::hisG, ade2-101, Kar1-1 (rho*^{0}*)*.

2.5 Serial Dilutions

1. Replica plater for 96-well plate, 8 × 6 array (Sigma-Aldrich, R2383).
2. Assay plate 96-well flat bottom (Corning, 3370).

3 Methods

3.1 Mitochondrial Transformation to Insert the ARG8^m Reporter

Generation of yeast strains containing the *ARG8*^{m} reporter inserted at a desired site in the mitochondrial genome requires four general steps: (1) Transformation of a *rho*^{0} strain (strain devoid of mtDNA) with a bacterial plasmid containing the *ARG8*^{m} reporter to generate a synthetic *rho*$^-$ strain (*see* Fig. 2a) (*see* **Note 1**). (2) Selection of transformants carrying the bacterial plasmid in mitochondria (correct synthetic *rho*$^-$ cells) (*see* Fig. 2b). (3) Transfer of the *ARG8*^{m} reporter from the synthetic *rho*$^-$ strain (donor) to the yeast strain with mtDNA where the reporter will be finally integrated (acceptor) through cytoduction. (4) Selection of the correct acceptor colonies (*see* Fig. 3).

All procedures are carried out under sterile conditions. Flame sterilization with a Bunsen burner is enough to prepare materials for transformation (bombardment step does not require a Bunsen burner). Al materials and reagents must be sterile.

3.1.1 Yeast Strain Preparation for Six Transformation Reactions by Microprojectile Bombardment

1. Grow cells on 2 mL culture of YPRaf or YPGal for two nights at 30 °C under constant agitation.
2. Inoculate 30 mL of fresh YPRaf or YPGal with 300 μL of culture. Grow overnight at 30 °C under constant agitation.
3. Centrifuge cells at 654 × *g* for 5 min at room temperature. Resuspend pellet in 600 μL of YPD.
4. Spread 100 μL on a petri dish containing bombardment medium using a glass cell spreader. Allow it to absorb and dry for at least 1–3 h before bombardment. Repeat procedure for a total of six plates.

3.1.2 Tungsten Particle Preparation

1. Sterilize 50 mg of tungsten particles (Bio-Rad Tungsten M-10 Microcarriers 1652266) in 1.5 mL microtube tube with 1.4 mL of 70% ethanol. Vortex and incubate at room temperature for 10 min.

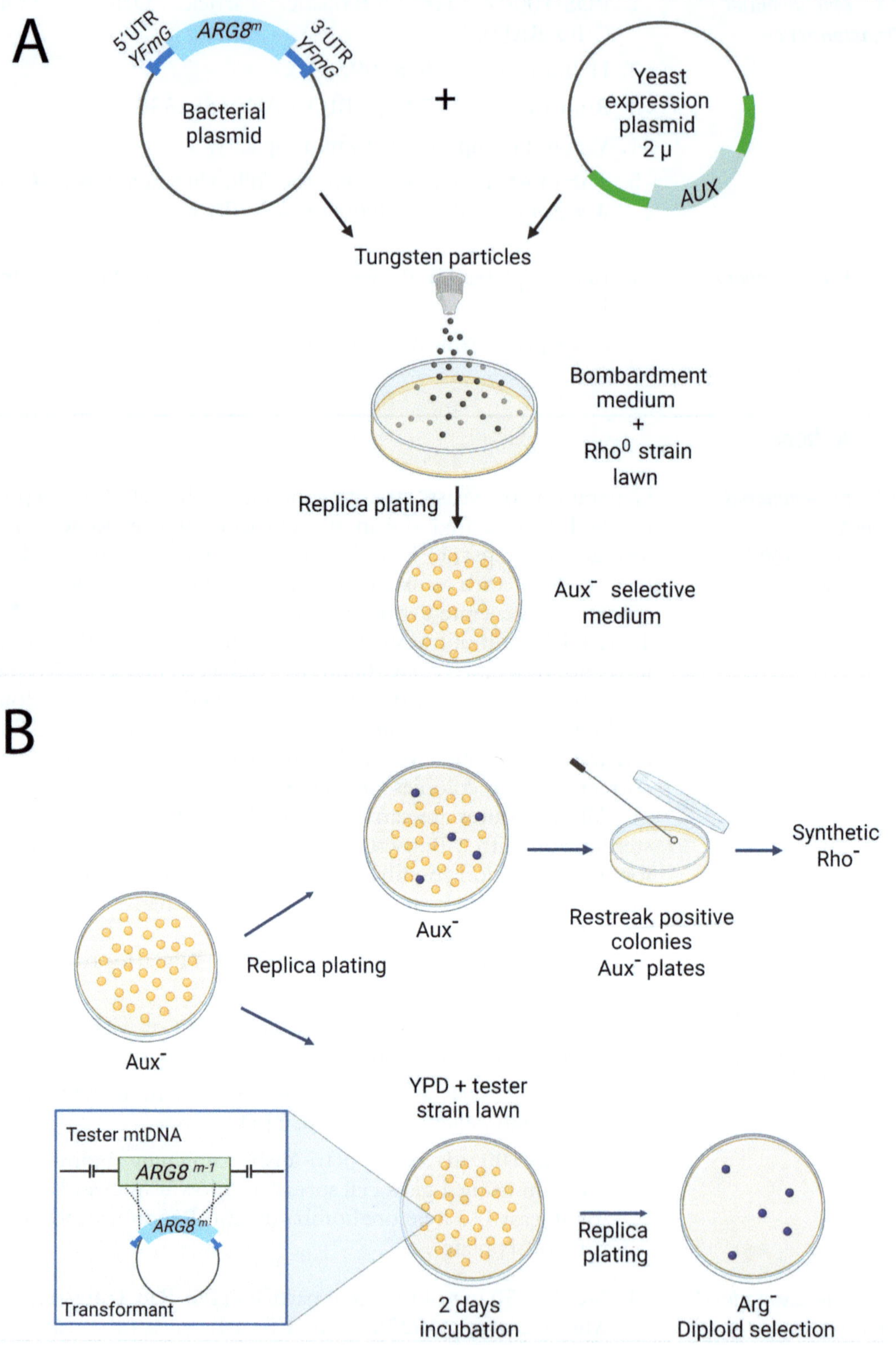

Fig. 2 Diagram of the procedure to (**a**) transform mitochondria with a bacterial plasmid containing *ARG8*^{m} and to (**b**) select the synthetic rho^{-} colonies after mitochondrial transformation. (Created with BioRender.com)

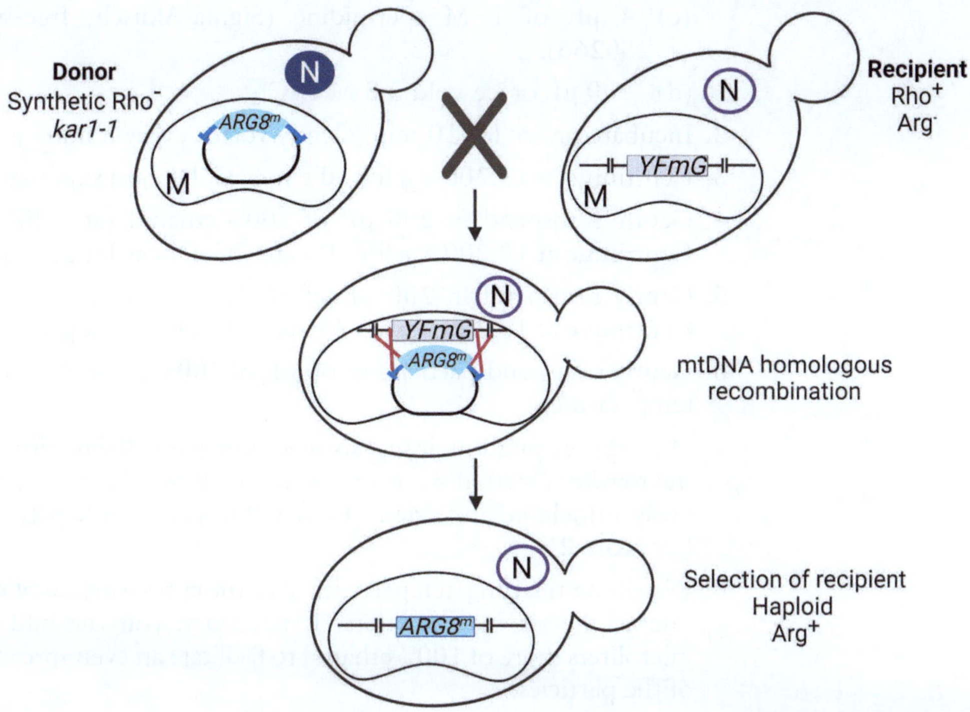

Fig. 3 During cytoduction, donor (synthetic rho$^-$) and recipient (strain of interest) cells mate, and cytosolic contents, including mitochondria (M) combine. Due to a *kar1* mutation, the nuclear fusion (N) is prevented. (Created with BioRender.com)

2. Centrifuge at 13,200 × *g* for 15 min.
3. Wash particles with 1.5 mL of sterile water and vortex. Centrifuge at 13,200 × *g* for 15 min. Discard supernatant.
4. Add 16.7 μL of glycerol/mg of particles. Store at −20 °C.

3.1.3 Precipitation of Plasmid DNA on Tungsten Particles

Two different bacterial plasmids are transformed into yeast cells: One plasmid is a 2 micron yeast expression plasmid (e.g., YEp351 or YEp352 with *LEU2* or *URA3* genes, respectively, for auxotrophy selection) and is used to select nuclear transformed cells. The second plasmid is a bacterial one (e.g., pBluescript, pGem, pUC) and contains the *ARG8ᵐ* sequence in the desired context and must be flanked by 100 or more bp of mitochondrial genome sequence on the 5′ and 3′ sides to insert into the desired locus by homologous recombination.

1. To a 1.5 mL microtube on ice, add the following reagents:
   (a) 5 μg of 2 micron plasmid (e.g., YEp351 or YEp352) and 15 μg of plasmid containing the *ARG8ᵐ* reporter, both plasmids in a total volume of maximum 51 μL in water.
   (b) 100 μL of tungsten particles glycerol suspension.

   (c) 4 μL of 1 M spermidine (Sigma-Aldrich, free-base, S0266).
   (d) 100 μL of ice-cold 2.5 M $CaCl_2$.
2. Incubate on ice for 10 min. Gently vortex every 3 min.
3. Centrifuge at 13,200 × *g* for 30 s at 4 °C. Discard supernatant.
4. Gently resuspend in 200 μL of 100% ethanol (at −20 °C). Centrifuge at 13,200 × *g* for 30 s at 4 °C. Discard supernatant.
5. Gently resuspend in 200 μL of 100% ethanol (at −20 °C). Centrifuge at 13,200 × *g* for 30 s at 4 °C. Discard supernatant.
6. Gently resuspend particles in 60 μL of 100% ethanol at room temperature.
7. During precipitation, insert six macrocarriers (*"flying discs"– do not sterilize them*) in six macrocarrier holders (which was previously autoclaved and dried). Place holders in a sterile petri dish (*see* **Note 2**).
8. Distribute the tungsten particles over the entire surface of each one of the six macrocarriers. If necessary, you can add few microliters more of 100% ethanol to facilitate an even spreading of the particles.
9. Let dry for 20 min until no liquid is visible.
10. In the meantime, sterilize the cavity of the microprojecter with 70% ethanol let stand for 10 min and then wipe it out and dry the cavity.

3.1.4 Bombardment

For microprojectile bombardment, carefully follow the manual. Conditions are:

1. Pressure for rupture discs: 1350 psi. Adjust pressure on the Helium tank at 2000 Psi (do not treat the rupture discs with isopropanol as suggested in the manual, it decreases the transformation rate dramatically).
2. Place the petri plate with the yeast strain lawn at the second level from bottom of the micro projector's cavity.
3. Apply vacuum: fire at 26–27 inches of Hg. Release vacuum immediately after bombardment.
4. After bombardment, remove parts of flying discs from the agar plate with sterile forceps.
5. Incubate plates for 6–7 days at 30 °C (*see* **Note 3**).
6. At the end, wash equipment with 70% ethanol. Wash microcarrier holders with 0.1% SDS, then with distilled water and finally with 100% ethanol, autoclave sterilize.
7. To select mitochondrial transformants, replica plate with sterile velveteen on two different media (*see* Fig. 2b): (1) Medium

lacking uracil (if transformed with YEp352) or leucine (if transformed with YEp351). This will be the master plate. (2) YPD medium with a 150 μL lawn of a YPD yeast culture carrying the allele *ARG8*$^{m-1}$. This is a tester strain that is unable to grow in medium lacking arginine. Only the diploid will grow after mating with a transformant colony carrying the plasmid in mitochondria.

8. Incubate for 2 days at 30 °C.
9. Replica plate the diploid plate from **step** 7 (YPD plate with diploids resulting from the cross between *ARG8*$^{m-1}$ tester and transformant colonies) on medium lacking arginine. Those colonies that contain the transformed plasmid in mitochondria (named synthetic *rho*$^{-}$) will be able to grow on media lacking arginine. Identify the positive colony on the master plate (*see* **Note 4**). Take single colonies form the master plate with a sterile toothpick and streak each one on a fresh plate lacking uracil or leucine accordingly.
10. Repeat replica plating on media lacking uracil or arginine (master plate) and on YPD with a lawn of the *ARG8*$^{m-1}$ allele (*see* **Note 5**).
11. Select one to six different synthetic *rho*$^{-}$ cells, grow overnight at 30 °C each one on 2 mL of YPD and prepare 25% glycerol stocks to preserve at −70 °C.

Table 1 presents the most frequently used tester strains to select synthetic *rho*$^{-}$ when the *ARG8*^{m} reporter is inserted on different mitochondrial *loci*.

3.1.5 Genetic Recombination to Integrate the ARG8^{m} Reporter in the Mitochondrial Genome

The next step for integrating the *ARG8*^{m} reporter on the desired *locus* in the mtDNA is to transfer the bacterial plasmid from the synthetic *rho*$^{-}$ mitochondria (donor) into the final yeast strain of interest (acceptor). This process is achieved by cytoduction, a mating-based method to introduce mitochondria from the *rho*$^{-}$ donor strain to the acceptor yeast strain. This method takes advantage of a *kar1-1* allele with limited nuclear fusion [15–17]. Mating the donor with the receptor allows cytoplasmic content to mix, and mitochondria from both strains are fused, allowing for mtDNA recombination. The resulting cytoductant contains the original nucleus of the acceptor. However, the *ARG8*^{m} reporter from the synthetic *rho*$^{-}$ integrates on the desired *locus* of the acceptor's mtDNA by homologous recombination. The only absolute requirement is that at least the donor or recipient must carry the *kar1-1* mutation (*see* Fig. 3).

1. Grow 2 mL of YPD cultures of donor and acceptor strains overnight at 30 °C with constant agitation.

Table 1
Tester strains most used to select synthetic rho⁻ cells

Mitochondrial gene where *ARG8^m* was inserted	Tester strain	Tester's mitochondrial relevant genotype	ARG8^m construct present in synthetic *rho*⁻	Medium to test diploid phenotype after mating synthetic *rho*⁻ and tester	Reference
COX2	NB40-3C	*rho*⁺, *cox2-62*	Ectopic[a] *cox2Δ::ARG8^m*	YPEG	[14]
COX2	NB71 or TF236	*rho*⁺, *cox3Δ::arg8^m-1*	*cox2Δ::ARG8^m* or *COX2::ARG8^m* [b]	-Arg	[8]
COX1	NB71	*rho*⁺, *cox1D369N*	*COX1(1-512)::pARG8^m* [c]	YPEG	[9]
COX1	L45	*rho*⁺, *cox3Δ::arg8^m-1*	*cox1Δ::ARG8^m*	-Arg	[9]
COX1	*L45*	*rho*⁺, *cox1D369N*	*cox2Δ::COX1*[a], *cox1Δ::ARG8^m* [d]	YPEG	[9]
COB	TF236 or NB71	*rho*⁺, *cox3Δ::arg8^m-1*	*cobΔ::ARG8^m*	-Arg	[28]
COB	NB160	*rho*⁺, *cox2-30*	*cox2Δ::COB*[a], *cobΔ::ARG8^m* [d]	YPEG	[25, 29]
COB	NB71	*rho*⁺, *cox3Δ::arg8^m-1*	*COB(1-352)::pARG8^m* [c]	-Arg	[28]
COX3	GW22	*rho*⁺, *cox3-421*	*cox3Δ::ARG8^m*	YPEG	[3]
ATP6	NB40-3	*rho*⁺, *cox2-62*[e]	*atp6Δ::ARG8^m*	YPEG	[30]
ATP8	NB40-3C	*rho*⁺, *cox2-62*[e]	*atp8Δ::ARG8^m*	YPEG	[11]
VAR1	DFS188[f]	*rho*⁺	*var1(1-20)::ARG8^m*	-Arg	[10]
ATP9	NB40-3C	*rho*⁺, *cox2-62*[e]	*atp9Δ::ARG8^m*	YPEG	[31]

[a]Ectopic insertion of the mitochondrial gene at 285 nt upstream of the *COX2 orf*. The insertion includes the 5′ and 3′ flanking regions of *COX2*. Importantly, the resultant mtDNA contains in addition the wild type, functional *COX2* gene in its original *locus*

[b]*ARG8^m* was fused in frame to different fragments of the *COX2* coding region

[c]The *ARG8^m* sequence has the MTS

[d]To create the strains with genes at the ectopic site in mtDNA, first it is necessary to make the complete replacement of the gene from the original locus with the *ARG8^m* reporter

[e]The plasmid contains the *COX2* sequence in addition to the *ARG8^m* sequence. *COX2* is used as a rescue gene to select transformants but is not going to be integrated on the final mtDNA. That is why a *cox2* mutant is used as tester strain

[f]After transformation, the synthetic rho⁻ was mated to DFS188 to generate the diploids that were used on the study. In other words, no tester strain was used, instead the construction was directly integrated on the diploid strain

2. Mix 750 μL of donor and 250 μL of acceptor on a sterile 1.5 mL microtube.
3. Centrifuge 1 min at 13,200 × *g* and discard supernatant.
4. With sterile scissors cut the tip of the micropipette tip. Gently resuspend the combined cells by pipetting up and down 2–3 times and place combined cells as a drop on a solid YPD plate.
5. Incubate 3–5 h at 30 °C to allow mating.
6. Look under the light microscope for the presence of shmoo formation in cells. Shmoo cells have a shape that is closer to a triangle with rounded corners.
7. If shmoo-shaped cells are present, take a small scoop of the mixed cells and resuspend on 2 mL YPD. Incubate for 2–3 h at 30 °C with constant agitation.
8. Dilute 10 μL of the culture in 990 μL of sterile water. Vortex briefly.
9. Spread 100 μL of the mix on a petri dish containing selection medium using a glass cell spreader. Make duplicates for each cytoduction.
10. Incubate at 30 °C for 2–3 days until appearance of single colonies.
11. Make replica plates with velvet on selection media for acceptors containing the desired mtDNA integration of the ARG8^{m} reporter gene and to discard the few diploids that are formed. Incubate overnight at 30 °C. Table 2 indicates an example of how to choose selective media.

Table 2
Example of selective media to use to select positive colonies after cytoduction

	Relevant genotype
Donor strain (synthetic rho^{-})	*Mata, kar1-1, ade2-101 (rho*$^{-}$*)*
Acceptor strain	*Matα, leu2-3,112 (rho*$^{+}$ *YFmG::ARG8*^{m}*)*
Medium to plaque cytoduction	-ade This medium selects against the donor strain. It will also be the master plate.
Medium to detect diploid	-leu Diploids are discarded from further purification.
Medium to detect correct acceptor	-arg If your cytoductant is unable to express the *ARG8*^{m} reporter, a tester strain with the *arg8*$^{m-1}$ allele can be used. Spread 100 μL of a YPD culture of the tester strain on YPD plates. Replica plate the master plate, allow to mate for 2 days at 30 °C. replica plate on medium lacking arginine. Incubate overnight at 30 °C.

12. Pick at least six different positive cytoductants and restreak on master medium (on the above example it would be medium lacking adenine) to obtain single colonies (*see* **Note 6**).
13. Repeat replica plating on all the above selective media.
14. Restreak and test positive colonies on the master medium as in points 12 and 13 for a final round of purification of positive cytoductants. We suggest keeping two independent positive cytoductants to make a freezer stock. For this, grow an overnight YPD culture of cytoductants and prepare a 25% glycerol stock for the −70 °C freezer.
15. Amplify the inserted gene by PCR and sequence to verify correct integration of the $ARG8^{m}$ reporter on the desired *locus*.

3.2 Growth of Yeast Cells on Media Lacking Arginine to Assess ARG8^{m} mRNA Translation

All procedures are carried out under sterile conditions. Flame sterilization with a Bunsen burner is enough to prepare and plaque serial dilutions. Al materials and reagents must be sterile.

1. Grow the yeast strains in 2 mL liquid YPD medium at 30 °C overnight, with constant agitation.
2. Next morning, weight on analytical balance the number of 1.5 mL microtubes equivalent to the number of yeast strains to analyze.
3. Add 1 mL of culture to each pre-weighted 1.5 mL microtubes. Centrifuge on a tabletop equipment at 13,200 × *g* for 1 min. Completely eliminate supernatant.
4. Wash once with sterile water, and centrifuge as in **step 3**. Completely eliminate supernatant.
5. Weight tubes again to obtain the wet weight of the cells pellet.
6. Add water to a concentration of 0.02 g of cells/mL.
7. On a 96-well plaque, add 135 μL of sterile water to all the wells to analyze, starting from the second column and to the sixth one, and including rows A, B, C... depending on the number of strains to study (*see* Fig. 4). In other words, fill spots A2, B2, C2...A6, B6, C6...
8. On the first column, which is empty, add 150 μL of each yeast suspension to fill wells A1, B1, C1...
9. Take 15 μL of cell suspensions on column 1 (A1, B1, C1...) and transfer to column 2 (A2, B2, C2...). Mix with the 135 μL of water present in these wells. This column, thus, contains the first 10^{-1} dilution.
10. Repeat **step 9** by taking 15 μL of suspension from column 2 (A2, B2, C2...) into column 3 (A3 B3, C3...). Repeat until column 6.

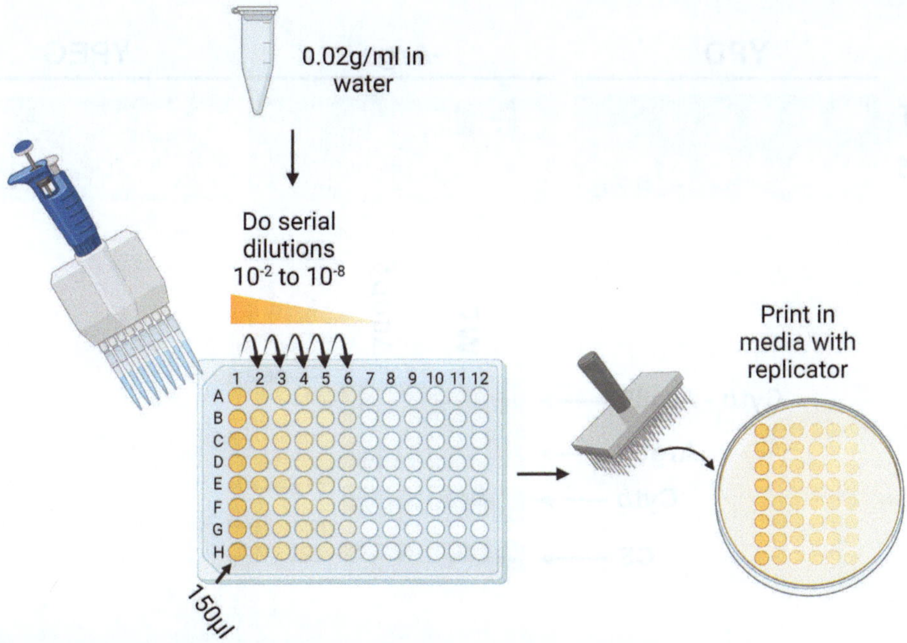

Fig. 4 Diagram of the process to make yeast serial dilutions. (Created with BioRender.com)

11. Immerse the poles from the replica plater on 100% ethanol (the use of a glass petri dish to pour ethanol is a good idea) and flame to sterilize it. Allow replicator to cool down to room temperature (1–2 min) before soaking it on the 96-well plate.
12. Take out replica plater and place it in contact with the surface of the indicated solid medium on petri dish. For each replica plate, it is necessary to soak the replica plater on the 96-well plate (*see* **Note** 7).
13. Incubate plates at 30 °C (or the desired temperatures) and follow cell growth for 5 days. It is recommended to take pictures every day starting on day 2.

3.3 Mitochondrial Translation in Whole Cells or in Isolated Organelles

This method is useful to analyze mitochondrial translation products, including translation of the *ARG8^m^* mRNA, and has been previously described [18]. One additional information to note is that, in our hands, Arg8 labels poorly, and the band migrates close to the Var1 band on SDS-PAGE. If Arg8 is fused to a mitochondrial coding region, then a higher size is expected. Detection of the Arg8 fusion protein depends on the *locus* in mtDNA where it is inserted, and might require different labeling times, from 10 to 40 min. For example, the fusion of Cox1-Arg8, which includes the Arg8 mitochondrial targeting sequence is detectable after 10–15 min of labeling, and some of the processed Arg8 is also detected [9]. In contrast, Cyt*b*-Arg8 requires 40 min labeling for optimal detection (*see* Fig. 5).

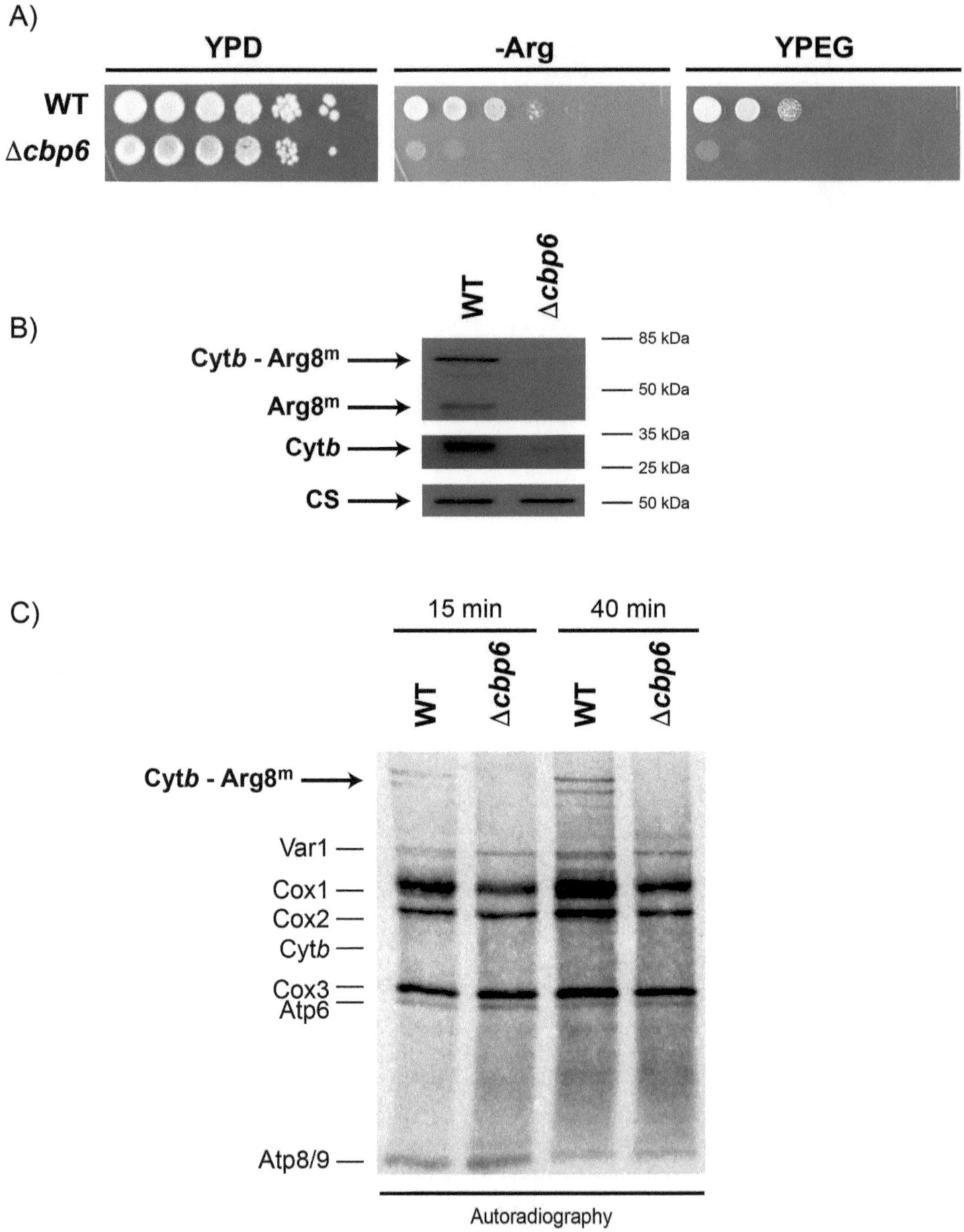

Fig. 5 Example of how the *ARG8^m* reporter can be used to study mitochondrial translation. The Arg8 reporter is fused to the C-terminal end of Cytochrome *b*. (**a**) Serial dilutions were incubated 3–4 days at 30 °C. Cbp6 is a chaperone involved in translational activation of the *COB* mRNA [24, 25]. Deletion of Cbp6 in cells containing the mtDNA construct abolished Cyt*b* synthesis. (**b**) Western blot of the indicated proteins. Citrate synthase (CS) was used as loading control. (**c**) Wild-type or *Δcbp6* cells were labeled for 10 or 40 min with (^{35}S)-Methionine in the presence of cycloheximide and analyzed by 16% acrylamide SDS-PAGE and autoradiography. Mitochondrial products were Cyt*b*, subunit from *bc*$_1$ complex; Cox1, Cox2 and Cox3, subunits 1, 2, and 3, respectively, from cytochrome *c* oxidase; Atp6, subunit 6 from ATP synthase; Var1 (small ribosomal subunit protein Var1)

3.4 Rapid Mitochondrial Preparations and Western Blot Assays to Detect Arg8 Production

One hour before starting the preparation, place the glass beads on the freezer. Use tabletop microcentrifuge throughout the procedure.

1. Grow 5 mL of liquid cultures on YPGal or YPRaf media at 30 ° C with constant agitation. Grow non-respiring strains for two nights instead.
2. Next day, make a 1:10 dilution (100 μL of culture and 900 μL of fresh medium) and read the Optical Density (OD) at 600 nm. Harvest the equivalent amount to ten ODs from each culture. Transfer to a 1.5 mL microtube.
3. Centrifuge at 13,200 × *g* for 5 min and eliminate supernatant.
4. Wash with 1 mL of water and repeat centrifugation.
5. Add 2 mL of MPZ buffer and incubate 1 h at 30 °C under constant agitation on a thermal mixer.
6. From this point, all the procedure is carried out at 4 °C. Centrifuge at 2000 × *g* for 5 min at 4 °C, take out supernatant.
7. Resuspend pellet in 300 μL of SHP buffer.
8. Add 1 volume of glass beads (Sigma-Aldrich, Z250465).
9. Mix on the vortex at maximum power for 30 s, incubate on ice for 30 s. Repeat the cycle vortex/ice two more times.
10. Centrifuge at 600 × *g* for 5 min. Save the supernatant in a fresh, ice-cold 1.5 mL microtube.
11. Add 300 μL of SHP buffer to the remaining glass beads, resuspend with a micropipette tip, and repeat three cycles of vortex 30 s/ice 30 s.
12. Centrifuge at 600 × *g* for 5 min. Save supernatant and combine it with the supernatant from **step 10**.
13. Centrifuge at 13,200 × *g* for 12 min and eliminate supernatant.
14. Resuspend mitochondrial pellet in 100 μL of SHP buffer. Quantify proteins using the modified Lowry method [19] or any standard method used in your laboratory. At this point, samples can be kept at −70 °C or can continue with SDS-PAGE.
15. For electrophoresis, take 25–100 μg of mitochondrial protein, centrifuge at 13,200 × *g* for 5 min, eliminate supernatant, and resuspend pellet in 15 μL of Laemmli buffer with 5% 2-mercaptoethanol. Heat at 65 °C for 5 min before loading onto gel. Do not boil samples.
16. Proteins are separated on a 12% acrylamide gel for SDS-PAGE electrophoresis [20].
17. Gel is transferred to a PVDF or nitrocellulose membrane using a semidry or wet transfer chamber.

18. The membrane is incubated with an antibody against Arg8 and washed 3 times for 10 min at room temperature, then the membrane is incubated with the secondary antibody and washed 3 times for 10 min. Incubations are carried out for 1–2 h on a tube rocker, at room temperature or at 4 °C (*see* **Note 8**). We use a secondary antibody coupled to horseradish peroxidase activity, since many commercial kits for detection of this enzymatic activity are available. The signal is revealed with a Western blot imager or autoradiography.
19. If required, strip the membrane with stripping buffer incubating for 30 min at 50 °C. Wash 3 times with washing buffer for 10 min. The membrane is ready to be incubated with another antibody.

Figure 5 shows an example of the information that can be obtained when using the *ARG8^m* reporter. In this case, the reporter was inserted at the *COB locus*, and a mutant lacking the translational activator Cbp6 was included. Action of Cbp6 on growth on a medium lacking arginine, in (^{35}S)-Methionine labeling of mitochondrial products and on Arg8 steady-state levels were analyzed. Fig. 6 shows an example of how the *ARG8^m* reporter can be used to differentiate between a translational defect (e.g., lack of Pet309 on *COX1* mRNA translation) and assembly defect (e.g., lack of Cox14 or Shy1 on Cox1 assembly).

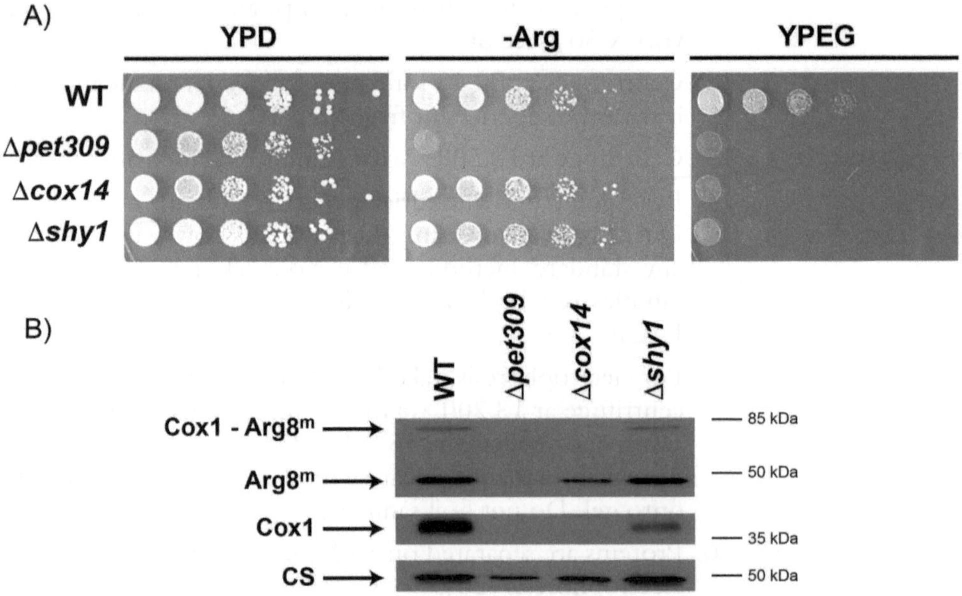

Fig. 6 The mitochondrial reporter gene *ARG8^m* can be used to distinguish between mutations affecting translation and mutations affecting assembly of mitochondrial encoded proteins. Pet309 is a protein necessary to activate translation of the *COX1* mRNA [23]. In contrast, Cox14 [26] and Shy1 [27] affect Cox1 assembly. (**a**) Serial dilutions of the indicated strains were incubated for 3–4 days at 30 °C. (**b**) Western blot of the indicated mutants. Citrate synthase (CS) was used as loading control

3.5 Search for Mitochondrial Translation Defect Suppressor Genes

Suppressors provide a large amount of information regarding your pathway of interest. The secondary mutation(s) leading to the suppressor phenotype can locate on the same *locus* where $ARG8^m$ is inserted in mtDNA, or can locate on a different gene, either nuclear or mitochondrial. In addition, these mutations can be dominant or recessive. Here we describe a method to identify spontaneous suppressors using the $ARG8^m$ reporter.

A mutant with no expression of the $ARG8^m$ reporter, leading to an arginine auxotrophy is better recommended, although mutants with leaky growth on media lacking arginine can also be used. All procedures are carried out under sterile conditions. Flame sterilization with a Bunsen burner is enough to prepare plates for suppressor mutants' growth. Al materials and reagents must be sterile.

3.5.1 Method to Find and Isolate Suppressors

1. Grow six independent colonies of the mutant of interest on six tubes with 3 mL of YPD medium. Incubate overnight at 30 °C under constant shaking.
2. Take 1 mL of culture, centrifuge 1 min at 13,200 × *g* and wash with sterile water. Resuspend on 150 μL of sterile water and evenly spread on media plates lacking arginine. Two or more plates per culture should be spread, increasing the possibility of finding spontaneous suppressors.
3. Grow plates 5–7 days at 30 °C. Some suppressor colonies should pop out at this time.
4. Streak putative suppressors on plates with medium lacking arginine (streak to obtain single colonies). These will be the master plates.
5. If colonies grow again, restreak them on YPD plates to check if the suppressor phenotype endures in the absence of selective media. In this case, after colonies grow on YPD, replica plate them with velveteen on YPD and on medium lacking arginine. If colonies retained the capacity to grow on media lacking arginine, it means that the suppressor mutation is stable (*see* **Note 9**).
6. From master plates, grow on liquid YPD overnight at 30 °C (or in medium lacking arginine if the suppressor phenotype is not stable on YPD) and prepare 25% glycerol stocks to maintain on the freezer (−70 °C).

From this point, several tests can be made to identify the nature of the suppressor and solve questions like: *Is the suppressor phenotype due to a single mutation or multiple mutations? Is the same mutated gene(s) in all the obtained suppressor colonies? If suppressor mutation (s) have a nuclear origin, are they dominant or recessive? Who is mutated and what is the mechanism of suppression?* To answer these questions, refer to [21, 22]. However, we provide a general view on how to address the next question.

3.5.2 Method to Determine Whether the Suppressor Is Nuclear or Mitochondrial

1. Cross on YPD your suppressors with a *rho*^{0} strain (*see* **Note 10**) that is wild type for the original mutation.
2. Purify diploids by streaking them on selective medium, so only diploids can grow (e.g., if one haploid has a *leu2Δ* mutation and the other has a *ura3Δ* mutation, then only the diploid will grow on media lacking both leucine and uracil).
3. Induce mitosis by streaking them on solid sporulation media. Incubate for ~3–7 days at room temperature.
4. Look under the microscope to detect presence of asci. They look like a tetrahedral sack with visible three to four spores inside.
5. As soon as asci show up, take few cells from the sporulation plate with a sterile toothpick and gently resuspend them in 50 μL of digest sporulation buffer.
6. Incubate at 37 °C for ~10 min. Gently add 450 μL of cold sterile water, do not vortex or mix. Keep at 4 °C.
7. Spot onto a line on a YPD plate with an inoculating loop. Gently put small drops along the line, in contact with each other, and allow them to dry before going to the dissecting microscope.
8. Incubate the resulting plates for 2 or 3 days at 30 °C until the dissected spores form visible colonies.
9. Replica plate on YPD and on medium lacking arginine. Incubate 24 h at 30 °C.
10. The suppressor mutation has a nuclear origin if the arg$^+$ phenotype segregates 2:2 for each tetrad. This is also a good opportunity to select and keep spores for further analysis containing the suppressor mutation together with the original mutant gene, but from the opposite mating type.

3.5.3 A Second Method to Analyze if the Suppressor has a Mitochondrial or Nuclear Origin

1. Deplete suppressors from their mtDNA (*rho*^{0}) (*see* **Note 10**).
2. By cytoduction (discussed on Subheading 3.1.4) add fresh mtDNA with the same *ARG8*^{m} construct originally used. If the suppressor phenotype (arg$^+$) is lost, it means that the suppressor mutation resides on the mitochondrial genome.

4 Notes

1. It is highly recommended that the plasmid containing *ARG8m* has minimum 100 bp of 5′ and 3′ flanking sequence with identical sequence to the target region on the mtDNA where the reporter is going to be inserted. Homologous recombination is more efficient if 500 bp or more are included on each flanking region.

2. Take macrocarriers with a pair of ethanol-sterilized-flamed tweezers. It is very easy to select two macrocarriers instead of one, so be particularly careful. If two macrocarriers are loaded on the Biolistic Particle Delivery System, then the experiment will fail.
3. During the 5–7 days incubation after bombardment, make everyday a visual inspection of the plates to ensure that no contamination is growing. If visible contamination pops up, flame and ethanol-sterilize spatula and cut the contaminated region.
4. Bombardment plates will be saturated with yeast colonies, and it can be difficult to locate the positive colonies that showed up on the replica plating screening. We recommend making labels on two different parts of the master plate, and aligning them with labels on the velveteen replicating disc tool. After every replica, make the same labels for each plate.
5. Make up to three rounds of restreaking and replica plating of synthetic *rho*$^-$ before preparing glycerol stocks for the −80 °C freezer. Additional rounds of purification may end in *rho*$^-$ mtDNA lost. Some synthetic *rho*$^-$ are not stable and tend to lose the transformed DNA. Indeed, if possible, after identification of correct synthetic *rho*$^-$, proceed to cytoduction as on Subheading 3.2 to integrate the reporter on the strain of interest. For this, carry out cytoduction from the same liquid YPD culture used for making freezer stocks.
6. At this point, it is essential to streak single colonies during cytoductant purification. Some colonies tend to contain a mix of correct cytoductants and original *rho*^{0} acceptors.
7. The use of multichannel micropipettes is highly recommended to avoid mistakes and to save time. To prevent contamination of the experiment, we recommend soaking the replicator on ethanol and flame it after every three to five replicas. Let it cool down to room temperature before using it. When replica platter is in contact with the replicating surface medium on a petri dish, do not apply pressure to the replicator, just let it rest for 10 s on the surface before retiring it. Check that all the dots are present as wet, small circles. If replicator stops printing round drops, wash it with water/lab detergent and completely dry before use.
8. Usually, incubation of 1 h at room temperature or overnight at 4 °C works well for antibodies against Arg8, Cox1, citrate synthase, and Cyt*b*.
9. Cells that have no stable suppressor phenotype on YPD might be worth to keep. Behavior of these suppressors could be due to the presence of heteroplasmic mtDNA. For an example of such suppressors, see [23]. In these kind of mutants,

appearance of an additional *rho*$^-$ genome, in combination with the original mtDNA provides the suppressor's phenotype. However, the novel *rho*$^-$ genome might not be stable on nonselective media like YPD.

10. To make a *rho*^{0} strain, incubate cells on 2 mL of liquid complete minimal media with 10 mg/mL of ethidium bromide overnight at 30 °C with constant shaking. The next day inoculate 50 μL of this culture into 2 mL of fresh minimal complete media with 10 mg/mL ethidium bromide. Incubate 2 days at 30 °C with constant shaking. Take a drop from this culture and streak it on YPD plates. Incubate 2 days at 30 °C. Then take one colony and restreak it on a YPD plate and grow for 2 days at 30 °C.

Acknowledgements

This work was supported by research grants from Consejo Nacional de Ciencia y Tecnología (CONACyT) [47514 to XP-M] and fellowship [777444 to D. F-M]; and Programa de Apoyo a Proyectos de Investigación e Innovación Tecnológica, (PAPIIT), UNAM [IN202720 and IN223623 to XP-M].

References

1. Bonnefoy N, Fox TD (2007) Directed alteration of Saccharomyces cerevisiae mitochondrial DNA by biolistic transformation and homologous recombination. Methods Mol Biol 372:153–166
2. García-Guerrero AE et al (2016) Chapter 6. In: Hernández G, Jagus R (eds) Evolution of the protein synthesis machinery and its regulation. Springer, Cham, Switzerland, pp 109–142
3. Steele DF, Butler CA, Fox TD (1996) Expression of a recoded nuclear gene inserted into yeast mitochondrial DNA is limited by mRNA-specific translational activation. Proc Natl Acad Sci U S A 93:5253–5257
4. Cohen JS, Fox TD (2001) Expression of green fluorescent protein from a recoded gene inserted into Saccharomyces cerevisiae mitochondrial DNA. Mitochondrion 1:181–189
5. Suhm T et al (2018) A novel system to monitor mitochondrial translation in yeast. Microb Cell 5:158–164
6. Mireau H, Arnal N, Fox TD (2003) Expression of Barstar as a selectable marker in yeast mitochondria. Mol Gen Genomics 270:1–8
7. Mookerjee SA et al (2013) In: Figurski D (ed) Genetic manipulation of DNA and protein -examples from current research, vol 19. Intech Open, London, pp 393–416
8. Bonnefoy N, Fox TD (2000) In vivo analysis of mutated initiation codons in the mitochondrial COX2 gene of Saccharomyces cerevisiae fused to the reporter gene ARG8m reveals lack of downstream reinitiation. Mol Gen Genet 262: 1036–1046
9. Perez-Martinez X, Broadley SA, Fox TD (2003) Mss51p promotes mitochondrial Cox1p synthesis and interacts with newly synthesized Cox1p. EMBO J 22:5951–5961
10. Sanchirico ME, Fox TD, Mason TL (1998) Accumulation of mitochondrially synthesized Saccharomyces cerevisiae Cox2p and Cox3p depends on targeting information in untranslated portions of their mRNAs. EMBO J 17: 5796–5804
11. Rak M, Tzagoloff A (2009) F1-dependent translation of mitochondrially encoded Atp6p and Atp8p subunits of yeast ATP synthase. Proc Natl Acad Sci U S A 106:18509–18514
12. Gruschke S et al (2011) Cbp3-Cbp6 interacts with the yeast mitochondrial ribosomal tunnel exit and promotes cytochrome b synthesis and assembly. J Cell Biol 193:1101–1114

13. Heimberg H et al (1990) Escherichia coli and Saccharomyces cerevisiae acetylornithine aminotransferase: evolutionary relationship with ornithine aminotransferase. Gene 90:69–78
14. Duvezin-Caubet S et al (2006) A "petite obligate" mutant of Saccharomyces cerevisiae: functional mtDNA is lethal in cells lacking the delta subunit of mitochondrial F1-ATPase. J Biol Chem 281:16305–16313
15. Zakharov IA, Yarovoy BP (1977) Cytoduction as a new tool in studying the cytoplasmic heredity in yeast. Mol Cell Biochem 14:15–18
16. Dorweiler JE, Manogaran AL (2021) Cytoduction and plasmiduction in yeast. Bio Protoc 11:e4146
17. Conde J, Fink GR (1976) A mutant of Saccharomyces cerevisiae defective for nuclear fusion. Proc Natl Acad Sci U S A 73:3651–3655
18. Carlström A, Rzepka M, Ott M (2021) The analysis of yeast mitochondrial translation. Methods Mol Biol 2192:227–242
19. Markwell MA et al (1978) A modification of the Lowry procedure to simplify protein determination in membrane and lipoprotein samples. Anal Biochem 87:206–210
20. Laemmli UK (1970) Cleavage of structural proteins during the assembly of the head of bacteriophage T4. Nature 227:680–685
21. Burke D, Dawson D, Stearns T, Cold Spring Harbor Laboratory (2000) Methods in yeast genetics: a Cold Spring Harbor Laboratory course manual, 2000th edn. Cold Spring Harbor Laboratory Press, Plainview, N.Y, p xvii, 205 p
22. Dunham MJ, Gartenberg MR, Brown GW (2015) Methods in yeast genetics and genomics: a Cold Spring Harbor Laboratory course manual/Maitreya J. Dunham, University of Washington, Marc R. Gartenberg, Robert Wood Johnson Medical School, Rutgers, The State University of New Jersey, Grant W. Brown, University of Toronto, 2015th edn. Cold Spring Harbor Laboratory Press, Cold Spring Harbor, New York, p xvii, 233 pages
23. Manthey GM, McEwen JE (1995) The product of the nuclear gene PET309 is required for translation of mature mRNA and stability or production of intron-containing RNAs derived from the mitochondrial COX1 locus of Saccharomyces cerevisiae. EMBO J 14:4031–4043
24. Dieckmann CL, Tzagoloff A (1985) Assembly of the mitochondrial membrane system. CBP6, a yeast nuclear gene necessary for synthesis of cytochrome b. J Biol Chem 260:1513–1520
25. Gruschke S et al (2012) The Cbp3-Cbp6 complex coordinates cytochrome b synthesis with bc(1) complex assembly in yeast mitochondria. J Cell Biol 199:137–150
26. Barrientos A, Zambrano A, Tzagoloff A (2004) Mss51p and Cox14p jointly regulate mitochondrial Cox1p expression in Saccharomyces cerevisiae. EMBO J 23:3472–3482
27. Mashkevich G et al (1997) SHY1, the yeast homolog of the mammalian SURF-1 gene, encodes a mitochondrial protein required for respiration. J Biol Chem 272:14356–14364
28. García-Guerrero AE et al (2018) Cbp3 and Cbp6 are dispensable for synthesis regulation of cytochrome. J Biol Chem 293:5585–5599
29. Ding MG et al (2009) Chapter 27: An improved method for introducing point mutations into the mitochondrial cytochrome B gene to facilitate studying the role of cytochrome B in the formation of reactive oxygen species. Methods Enzymol 456:491–506
30. Rak M et al (2007) Yeast cells lacking the mitochondrial gene encoding the ATP synthase subunit 6 exhibit a selective loss of complex IV and unusual mitochondrial morphology. J Biol Chem 282:10853–10864
31. Bietenhader M et al (2012) Experimental relocation of the mitochondrial ATP9 gene to the nucleus reveals forces underlying mitochondrial genome evolution. PLoS Genet 8: e1002876

Chapter 17

Four-Color STED Super-Resolution RNA Fluorescent In Situ Hybridization and Immunocytochemistry to Visualize Mitochondrial mRNAs in Context with Mitochondrial Ribosomes

Christin A. Albus, Rolando Berlinguer-Palmini, Zofia M. Chrzanowska-Lightowlers, and Robert N. Lightowlers

Abstract

High-resolution imaging has enabled scientists to explore the mitochondrial network at remarkable resolution. This has been exploited to help increase our knowledge of how mitochondrial gene expression is compartmentalized in cultured cells. Here, we provide detailed methodology to simultaneously visualize up to four components including mtDNA-encoded transcripts, submitochondrial marker proteins, mitoribosomal subunits, or core members of the translational apparatus using STED super-resolution nanoscopy.

Key words Super-resolution nanoscopy, Mitochondrial mRNA, STED super-resolution, RNA in situ hybridization

1 Introduction

1.1 Background

The human mitochondrial genome (mtDNA) is essential for aerobic life. Transcription generates 11 mt-mRNAs that are matured and translated within the mitochondrial matrix. All 13 resultant gene products are essential for coupling the production of ATP to respiration, a process termed oxidative phosphorylation. Over the past 40 years or so, our understanding of the molecular mechanisms underlying mtDNA maintenance and expression has increased exponentially, with many of the key players and processes now well described. Further, by exploiting the impressive technologies of confocal, super-resolution, cryo- and 3D electron microscopy, it is now possible to begin to conceptualize in three dimensions how mitochondrial gene expression is integrated within the organellar network [1]. We are aware how mtDNA is packaged and maintained in membrane-associated nucleoids, how transcripts are

Antoni Barrientos and Flavia Fontanesi (eds.), *The Mitoribosome: Methods and Protocols*,
Methods in Molecular Biology, vol. 2661, https://doi.org/10.1007/978-1-0716-3171-3_17,

processed and matured in highly dynamic mitochondrial RNA condensates or granules that are often juxtaposed to the nucleoid and may also be attached to the membrane [2]. The complex process of mitoribosomal assembly is a very active research field and detailed molecular mechanisms that describe the distinct stages of mitochondrial protein synthesis are mostly accepted [3–5].

Many questions, however, still remain. Recently, it has become possible to mark where mitochondrial protein synthesis is occurring in the mitochondrial network of intact cells [6, 7]. High-resolution imaging has led to the suggestion that functioning mitoribosomes are found in association with cristae membranes [7], although studies in yeast using antibodies to visualize translational activators have inferred that translation may also occur at the inner boundary membrane [8]. Mitoribosomal assembly at least in large part appears to occur in the RNA granule [9]. How does the mitoribosome move from the granule and localize to the mitochondrial inner membrane prior to translation? Although excellent imaging techniques are available to help address these questions, high-resolution imaging is often limited by the number of probes that can be used at one time. For example, it may be ideal to visualize markers of the mitochondrial ribosome or mitoribosomal RNA, the RNA granule, protein synthesis, and respiratory complexes at the same time. To address this issue, we have now produced images for four components of the mitochondrion that can be visualized together in fixed cultured cells. Below is a detailed protocol that we are currently using for this purpose.

2 Materials

When working with RNA, wear gloves at all times. Avoid touching materials (e.g., tip boxes, bottles, pipettes) with ungloved hands. Avoid touching contaminated surfaces with gloved hands. Decontaminate pipettes and bench surfaces with RNaseZAP™ (Sigma, R2020) or using UV. If possible, keep reagents (e.g., Triton-X, PBS-tablets) used for working with RNA separate from reagents used for other routine protocols in the same lab. RNase-free microcentrifuge tubes and filter tips should be used throughout.

2.1 General Supplies, Solutions, and Equipment

1. Cultured U2OS or HeLa cells were grown in Dulbecco's modified Eagle's medium (Sigma D6429, Burlington, MA, USA), supplemented with 10% fetal calf serum, 1 × nonessential amino acids, and 50 μg/mL uridine, at 37 °C in humidified 5% CO_2.
2. High-precision 170 ± 5 μm coverslips (Carl Zeiss, 474030-9020-000). 22 mm × 22 mm coverslips are suitable for growing cells in 6-well plates or 35 mm cell culture dishes. Sterilize by autoclaving in round glass dish.

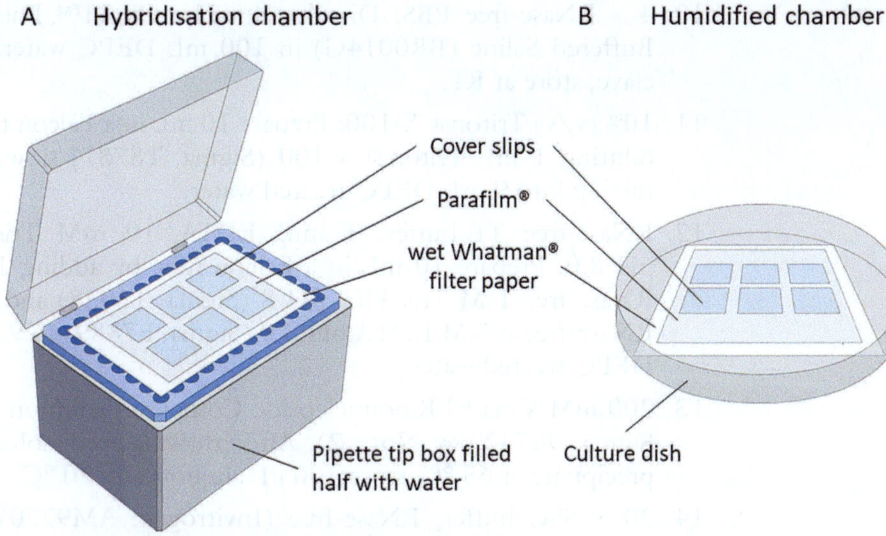

Fig. 1 (**a**) For overnight incubation of coverslips at 37 °C, half fill a pipette tip box with DEPC water and place a wet Whatman® filter paper on the tip grid. Place a new sheet of Parafilm® onto the filter paper so that the "clean" side that was protected by the backing paper faces up. This side will come in contact with the FISH probes and cells and must therefore be kept RNase-free. (**b**) For antibody incubation, it is sufficient to use a simple humidified chamber. Place a wet Whatman® filter paper in a cell culture dish that is big enough to hold your samples. Place a clean sheet of Parafilm® on top. During incubation, the dish is covered with the lid of the dish wrapped in aluminum foil. We perform antibody incubations with the coverslip's *cell side up*. If antibody incubations are carried out with the coverslip's *cell side down*, a new sheet of Parafilm® must be used and the side which was protected by the backing paper facing up. To improve protection of the samples from RNases, the assembled chambers can be pretreated with UV

3. Microscopic glass slides, 76 mm × 26 mm, 1–1.2 mm thick (Fisherbrand™, 12383118).
4. Fine-point forceps.
5. Hybridization chamber (*see* Fig. 1a).
6. Humidified chamber (optional, *see* Fig. 1b).
7. RNase-free glassware for buffer preparation: To decontaminate glass bottles from RNases, prepare DEPC water in the glass bottle making sure to shake vigorously so that DEPC also reaches the lid.
8. DEPC water: Prepare in a glass bottle by adding 0.1% DEPC to ultrapure water. Shake the bottle vigorously and let sit for at least 1 h at room temperature (RT), autoclave and let cool to RT, store at RT (*see* **Note 1**).
9. 1 × RNase-free PBS: 8.0 g/L NaCl, 0.2 g/L KCl, 1.15 g/L Na_2HPO_4, 0.2 g/L KH_2PO_4, pH 7.3. Prepare in RNase-free glass bottle. Dissolve 1 tablet Oxoid™ Phosphate Buffered Saline (BR0014G) in 100 mL DEPC water, autoclave, store at RT.

10. 4 × RNase-free PBS: Dissolve 4 tablet Oxoid™ Phosphate Buffered Saline (BR0014G) in 100 mL DEPC water, autoclave, store at RT.
11. 10% (v/v) Triton® X-100: Prepare 10 mL in a Falcon tube by diluting 1 mL Triton® X-100 (Sigma, T8787) slowly with mixing into 9 mL DEPC-treated water.
12. RNase-free TE-buffer: 1 mM EDTA, 10 mM Tris–HCl, pH 8.0. Prepare 10 mL in a Falcon tube by adding 100 μL RNase-free 1 M Tris–HCl pH 8 (Sigma, T3038) and 20 μL RNase-free 0.5 M EDTA pH 8.0 (Sigma, E7889) to 9.88 mL DEPC-treated water.
13. 200 mM Vanadyl Ribonucleoside Complexes solution (VDR, Sigma, 94742, *see* **Note 2**). After thawing, redissolve VDR precipitate at 65 °C, store 100 μL aliquots at −20 °C.
14. 20 × SSC buffer, RNase-free (Invitrogen, AM9770), 3 M NaCl, 300 mM Na citrate.
15. Formamide, extra pure (Thermo Scientific Acros Organics, 181090010).
16. Dextran sulfate, Mw > 500,000 (Sigma, D8906).
17. ProLongTM Glass Antifade Mountant (Invitrogen, P36984).

2.2 Buffers

All buffers are prepared with DEPC-treated water in RNase-free plastic or glassware. All VDR-containing buffers are thawed at 65 ° C to redissolve VDR precipitate and cooled to room temperature before using.

1. *Fixation buffer*: 8% (w/v) PFA, 2.5% (v/v) methanol, 0.05% (v/v) glutaraldehyde in 1 × PBS. Prepare 20 mL in 50 mL Falcon tube by mixing 10 mL 16% (w/v) methanol-free formaldehyde solution (Thermo Scientific, 28908), 0.5 mL methanol, 5 mL 4 × RNase-free PBS, 40 μL 25% glutaraldehyde solution (Sigma, G5882), 4.46 mL DEPC-treated water, store at 4 °C.
2. *Permeabilization buffer*: 0.3% (v/v) Triton® X-100, 2 mM VDR in 1× PBS. Prepare 10 mL in Falcon tube: 300 μL 10% (v/v) Triton® X-100, 100 μL 200 mM VDR, 2.5 mL 4× PBS, 7.1 mL DEPC-treated water, aliquot and store at −20 °C. After thawing, redissolve VDR precipitate at 65 °C.
3. *IF buffer*: 2 mM VDR in 1 × PBS. Prepare 10 mL in Falcon-tube: 100 μL 200 mM VDR, 2.5 mL 4 × PBS, 7.4 mL DEPC-treated water, aliquot and store at −20 °C. After thawing, redissolve VDR precipitate at 65 °C.
4. *Pre-hybridization buffer*: 2 × SSC, 10% (v/v) formamide. Prepare fresh every time as you need it, do not store overnight. For 10 mL: 1 mL 20 × SSC, 1 mL formamide, 8 mL DEPC-treated water.

5. *Hybridization buffer*: 2 × SSC, 10% (v/v) formamide, 10% (w/v) dextran sulfate (Mw > 500,000), 2 mM VDR. Prepare 10 mL in a Falcon tube by dissolving 1 g dextran sulphate in 7 mL DEPC-treated water, mix by rotation. Once dissolved, add 1 mL formamide, 1 mL 20 × SSC and 100 μL VDR, fill up to 10 mL by adding DEPC-treated water. Aliquot and store at −20 °C.
6. *Wash buffer*: 2 × SSC. Prepare 50 mL in Falcon tube by diluting 5 mL 20 × SSC with 45 mL DEPC-treated water.

2.3 RNA FISH Probes and Antibodies

2.3.1 RNA FISH Probes

1. Stellaris™ RNA FISH probes were purchased from Biosearch™ Technologies and designed using the Stellaris™ Probe designer tool provided by the company (https://www.biosearchtech.com/stellaris-designer) with the following settings: masking level 4, oligomer length 20 nt, and 2 nt minimum spacing. Probe sets with a minimum of 25 oligonucleotides were selected. We successfully used probes targeting human mitochondrial mRNAs *MT-ND2* (ENST00000361453.3) and *MT-CO1* (ENST00000361624.2) (*see* Table 1).
2. Probes were purchased either coupled with Quasar®570 or CAL Fluor Red 610 fluorophores. These dyes have previously been used successfully for STED microscopy [10].
3. Lyophilized probes (5 nmol) were redissolved in 50 μL RNase-free TE buffer to yield a 100 μM stock solution, store at −20 ° C.
4. Dilute 12.5 μL of the probe stock solution with 87.5 μL RNase-free TE-buffer to create a 12.5 μM probe solution, which is then aliquoted and stored at − 20 °C.

2.3.2 Mitochondrial Antibodies

1. For the visualization of mitochondrial ribosomes, we recommend using rabbit polyclonal antibodies against mitoribosomal proteins mL45 (Thermo Fisher, PA5-54778, 1:200), mS27 (Proteintech, 17280-1-AP, 1:200), uS17m (Proteintech, 18881-1-AP, 1:200) or uS15m (Proteintech, 17006-1-AP, 1: 200).
2. The mitochondrial inner membrane protein OXA1L which is required for the insertion of newly translated proteins was visualized using OXA1L rabbit polyclonal antibody (Proteintech, 21055-1-AP, 1:200).
3. Mitochondrial import receptor subunit TOM20 was visualized as a mitochondrial outer membrane marker using mouse monoclonal antibody (Abcam, ab56783, 1:400–1:200).

Table 1
Stellaris probe sequences

Gene name	*MT-ND2*	*MT-CO1*
Transcript	ENST00000361453.3	ENST00000361624.2
Probe sequences 5′ → 3′		
#1	atgagtgtgcctgcaaagat	aatagtcaacggtcggcgaa
#2	aaatcagtgcgagcttagcg	ctcatgcgccgaataatagg
#3	tttatttctaggcctactca	taaggaggcttagagctgtg
#4	gcagcttctgtggaacgagg	gtcgttacctagaaggttgc
#5	gcttgcgtgaggaaatactt	acaaatgcatgggctgtgac
#6	tagaaggattatggatgcgg	aaagcctccgattatgatgg
#7	gtccggagagtatattgttg	gcaccgattattaggggaac
#8	ggtagtattggttatggttc	atgcggggaaacgccatatc
#9	gggctattcctagttttatt	gagtcagaagcttatgttgt
#10	taacctctgggactcagaag	actatagcagatgcgagcag
#11	catgtgagaagaagcaggcc	gtagactgttcaacctgttc
#12	atgattgagatgggggctag	agatggttaggtctacggag
#13	tagtgagggagagatttggt	atagaggagacacctgctag
#14	gagagagtgaggagaaggct	tgtgatgaaattgatggccc
#15	gcctgctatgatggataaga	tatggcaggggggttttatat
#16	tttggtttaatccacctcaa	aagaggggcgtttggtattg
#17	gctaagattttgcgtagctg	aggactgctgtgattaggac
#18	cctatgtgggtaattgagga	ggactgggagagataggaga
#19	cggtagaactgctattattc	tagtatagtgatgccagcag
#20	gaatggttatgttagggttg	agaaggtggtgttgaggttg
#21	ggaatgcggtagtagttagg	ggtgttggtatagaatgggg
#22	ggtgctggagtttaagttga	ataaacttcagggtgaccga
#23	ttcaggtgcgagatagtagt	cgaagcctggtaggataaga
#24	ggtgttagtcatgttagctt	ctcagaccatacctatgtat
#25	agaggagggtggatggaatt	aaccctaggaagccaattga
#26	ttgggcaaaaagccggttag	tatatggtgtgctcacacga
#27	gaattcttcgataatggccc	ctacgtctattcctactgta
#28	tggggatgatgaggctattg	cggaggtgaaatatgctcgt
#29	agggtgatggtggctatgat	ggggatagcgatgattatgg

(continued)

Table 1
(continued)

Gene name	*MT-ND2*	*MT-CO1*
#30	gcgtaggtagaagtagaggt	gtggcgagtcagctaaatac
#31	tgtgattgaggtggagtaga	catttcatattgcttccgtg
#32	cgttgttagatatggggagt	atgaatcctagggctcagag
#33	ggtgggttttgtatgttcaa	gccacctacggtgaaaagaa
#34	cgatgagtgtggggaggaat	tgtctagtgatgagtttgct
#35	aaggggagataggtaggagt	gtagtacgtgtcgtgtagta
#36		acatagtggaagtgggctac
#37		ggcaaatacagctcctattg
#38		agtgaatgaagcctcctatg
#39		tagggtgtagcctgagaata
#40		atggattttggcgtaggttt
#41		agttagatttacgccgatga
#42		taggccgagaaagtgttgtg
#43		tttcatgtggtgtatgcatc
#44		tgagcctacagatgatagga
#45		acttttcgcttcgaagcgaa
#46		ggagggttcttctactatta
#47		catatagtcactccaggttt
#48		ttcgaatgtgtggtagggtg

4. Secondary antibodies used here were goat anti-mouse AF488 (Invitrogen, A-21121, 1:200) and goat anti-rabbit abberior STAR RED (abberior, STRED-1002, 1:200).

2.3.3 Microscopy

Images were acquired on a Leica TCS SP8 gSTED 3X point scanning confocal microscope equipped with a while light super continuum laser (WLL), three STED depletion lasers (592, 660 and 775 nm), a 405 nm solid-state laser, two hybrid detectors, and a STED WHITE HC PL APO CS2 100×/1.40 OIL lens.

3 Methods

3.1 Cell Culture and Fixation

1. Place coverslips in wells of a 6-well plate or in individual 35 mm dishes using sterile forceps.

2. Adherent cells (i.e., U2OS or HeLa) are grown at a low density to avoid compromising cell morphology by crowding. We found that seeding cells at 10% confluency resulted in a suitable cell density the next day. To this end, dilute 1/10 of a 90–100% confluent T75 flask (75 cm^2 growth area) in 15 mL growth medium and plate 2 mL of the cell suspension per well (9.6 cm^2 growth area). Place the plate into the incubator and leave to grow overnight.
3. Remove cells from incubator and wash twice with warm 1 × PBS to avoid cell shock.
4. Remove 1 × PBS and fix cells with 1 mL room-temperature fixation buffer for 10 min.
5. Remove fixation buffer and wash three times with 1 × PBS (*see* **Note 3**).

3.2 Immunofluorescence (IF) Staining

When attempting the IF and RNA-FISH protocols together, it is important that the IF staining is carried out without the use of BSA as a blocking agent. Therefore, all primary antibodies must be tested for background staining by comparing the performance using the BSA-free protocol versus a protocol where 5% BSA is added to the permeabilization buffer and IF buffer. Background staining of secondary antibodies must be tested by IF staining without primary antibody.

For IF staining without BSA blocking, we found that Alexa-Fluor 488 and abberior STAR RED (also named KI14 in the literature) fluorescence dyes resulted in low background staining while being suitable for STED-imaging. Abberior STAR 635P can be tested as a substitute for abberior STAR RED.

The following steps can be carried out using the hybridization chamber (*see* Fig. 1a) or in a humidified chamber (*see* Fig. 1b) with slides placed onto the parafilm *cell side up* and buffers being applied on top of the slides (volume 130–140 μL per 22 × 22 mm slide) and incubated for the indicated time. If coverslips are incubated cell side down for incubation with antibody, use a fresh sheet of Parafilm® each time and the side which was protected by the cover paper facing up. Avoid handling too many samples at a time to avoid drying of the fixed cells on the coverslips. We found that placing two to three coverslips onto the Parafilm® before applying the buffer to each allows for efficient workflow without risking drying of the samples. Thaw VDR-containing buffers at 65 °C to redissolve VDR precipitate and cool to room temperature before using.

1. Using fine point forceps, place the coverslips onto the Parafilm® with the fixed cells facing up. Add 150 μL permeabilization buffer on top of each coverslip. Incubate for 10 min and place back into the 6-well plate containing 1× PBS with the *cell side facing up* (*see* **Note 4**).

2. Remove the PBS and wash with 1 mL fresh PBS. Repeat this wash step and let sit in 1× PBS while preparing the primary antibody.
3. Dilute the primary antibody in IF buffer, 150 μL per 22 × 22 mm slide.
4. Place the coverslips back onto the Parafilm® with the *cell side facing up* and apply 140–150 μL of the diluted primary antibody. Cover the chamber with the lid and incubate for 60 min.
5. The secondary antibody dilution can be prepared during the incubation with the primary antibody: Dilute the secondary antibody in IF buffer, 150 μL per 22 × 22 mm slide. Keep on ice and in the dark until needed.
6. Place the coverslips back into the 6-well plate containing 1× PBS with the *cell* side *facing up*. Remove the PBS and replace with 1 mL 1× PBS, incubate for 1 min. Repeat the wash step two times.
7. Place the coverslips back onto the Parafilm® with the *cell side facing up* and apply 140–150 μL of the diluted secondary antibody. Cover the chamber with the lid and incubate in the dark for 40 min.
8. Place the coverslips back into the 6-well plate containing 1× PBS with the *cell side facing up*. Remove the PBS and replace with 1 mL 1× PBS, incubate for 1 min. Repeat the wash step two times.
9. Remove the 1× PBS and replace with 1 mL pre-hybridization buffer, incubate for 5 min. Remove the pre-hybridization buffer and replace with 1 mL pre-hybridization buffer and incubate for 20–30 min in the dark. During this incubation, prepare the RNA-FISH probes.

3.3 RNA Fluorescence In Situ Hybridization

The *in situ* hybridization step is carried out at 37 °C overnight in a hybridization oven or incubator which must not be used for growing bacteria at the same time. The coverslips are placed onto the Parafilm® of the hybridization chamber (*see* Fig. 1a) with the *cell side facing down*. Thaw hybridization buffer at 65 °C to redissolve VDR precipitate and cool to room temperature before using.

1. For each coverslip, prepare 80–100 μL Stellaris™ RNA FISH probe by diluting the 12.5 μM probe solution 1:100 in hybridization buffer (final probe concentration 125 nM).
2. Place the diluted probe as a droplet onto the parafilm and gently lower the coverslip onto the probe with the *cell side facing down* towards the probe solution, repeat for all samples (*see* **Note 5**).

3. Close the lid of the hybridization chamber and place the chamber onto a sheet of Saran wrap and carefully wrap the box to seal the chamber; use two sheets if needed. Place the sealed chamber onto a sheet of aluminum foil and wrap for incubation in the dark.
4. Place the hybridization chamber into the hybridization oven and incubate at 37 °C overnight.
5. Remove the leftover pre-hybridization buffer from the 6-well plate and keep the plate for wash steps the following day.
6. The next day, place 1 mL freshly prepared pre-hybridization buffer into the 6-well plate. Carefully remove the hybridization chamber from the hybridization oven and gently unwrap without disturbing the coverslips. Using the fine-pointed forceps, carefully lift the coverslips off the parafilm and place them *cell side facing up* into the pre-hybridization buffer in the 6- well plate (*see* **Note 6**).
7. For all subsequent wash steps, the plate can be gently agitated on a platform rocker. Remove pre-hybridization buffer and replace with 1 mL hybridization buffer, incubate in the dark for 15 min, repeat once.
8. Remove pre-hybridization buffer and replace with 1 mL 2× SSC, incubate for 5 min.
9. Remove 2× SSC and wash twice with 1× PBS for 5 min in the dark.
10. During the wash steps, prepare glass slides. Clean slides with 70% ethanol, do not use tissue containing dyes to avoid depositing color onto the slides.
11. Place a drop of mounting medium in the middle of the slide and mount coverslips with the *cell side facing down* (*see* **Note 7**).
12. Place the slides horizontally into a slide holder boy and let cure at least 24 h, better 48 h at room temperature in the dark (*see* **Note 8**).
13. Longer term storage at 4 °C in the dark.

3.4 Imaging

We used 4 color super-resolution STED microscopy for the visualization of mitochondrial mRNAs in context with mitochondrial ribosomes. STED overcomes the diffraction limit of conventional confocal microscopy [11] (*see* Fig. 2).

This yields resolutions of ≥200 nm for visible light in the lateral dimensions (in *x–y*) and ≥500 nm, in the axial direction (in *Z*). STED resolution is typically approximately 50 nm in *XY* and 150 nm in *Z*. Images were acquired on a Leica TCS SP8 STED 3X point scanning confocal nanoscopy with while light

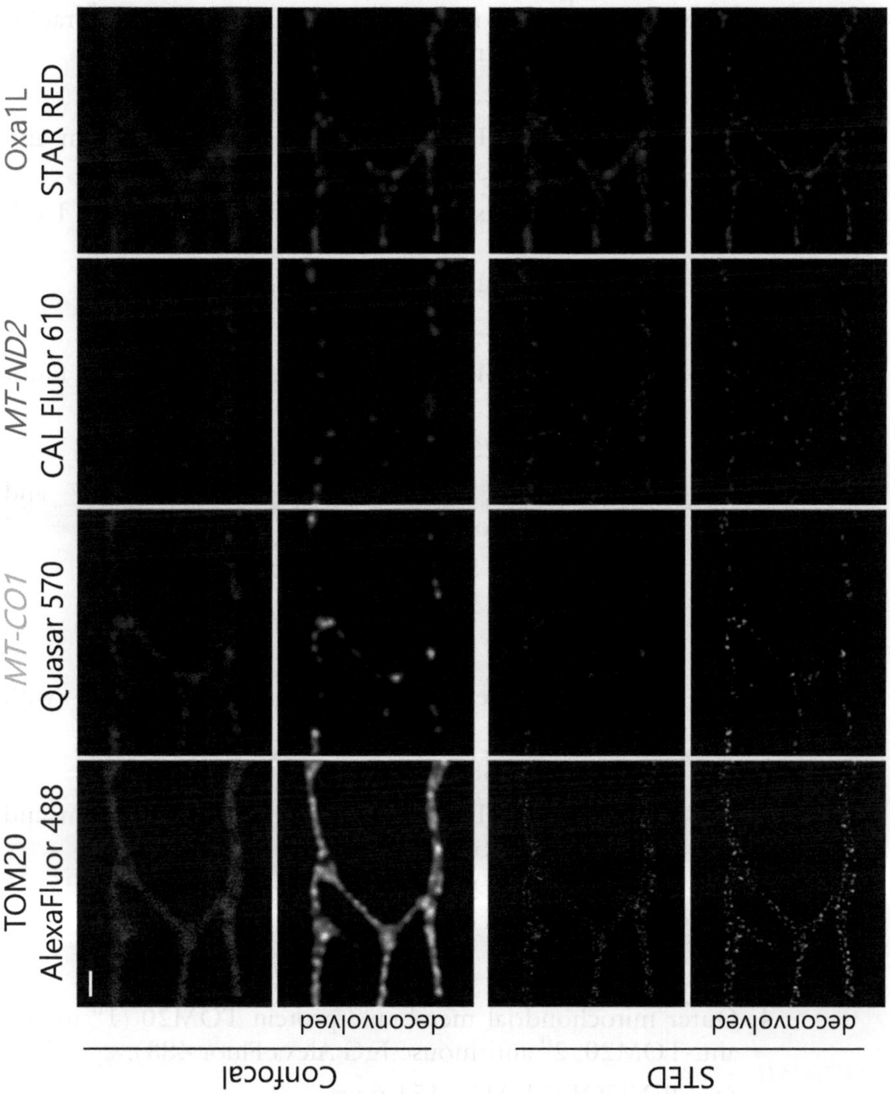

Fig. 2 *4 Color STED imaging of mitochondrial translation components.* U2OS cells were fixed and stained to reveal the distribution of the outer membrane protein TOM20 (AlexaFluor 488), mt-mRNAs *MT-CO1* (Quasar ® 570), and *MT-ND2* (CAL Fluor 610) as well as the protein insertase OXA1L (abberior STAR RED). Images were deconvolved using Huygens 21.02 software. Size bar = 1 µm

supercontinuum lasers and three STED depletion lasers (592, 660, and 775 nm) using STED WHITE HC PL APO CS2 100×/1.40 OIL lens. Colocalization analysis could also be performed following the protocol [12]. Deconvolution and particle analysis was performed with Huygens 21.02 from SVI (www.svi.nl). Details of acquisitions setting for visualizing the following proteins or mRNA are:

1. Inner mitochondrial membrane insertase OXA1L (1^0 rabbit anti-OXA1L; 2^0 anti-rabbit IgG abberior STAR RED):
   (a) PINHOLE 0.82309 AU – 124.8 μm
   (b) Acquisition on HYD detector 3 between 645 nm and 750 nm. Detector gain = 146%
   (c) White Light Laser (WWL) exciting at 637 nm. Laser power 10%
   (d) STED laser 775 – 65% (main power 80%)
   (e) HYD gating 0.5–6 ns
2. *MT-ND2* (antisense *MT-ND2* probe labelled with CAL Fluor 610):
   (a) PINHOLE 0.82309 AU – 124.8 μm
   (b) Acquisition on HYD detector 3 between 597 and 630 nm. Detector gain = 137%
   (c) WLL exciting at 590 nm. Laser power 7%
   (d) STED laser 775 – 55% (main power 80%)
   (e) HYD gating 0.5–6 ns
3. *MT-CO1* (antisense *MT-CO1* probe labelled with Quasar®570)
   (a) PINHOLE 0.935 AU – 141.88 μm
   (b) Acquisition on HYD detector 1 between 554 nm and 580 nm. Detector gain = 152%
   (c) WLL exciting at 548 nm. Laser power 10%
   (d) STED laser 660 – 50% (main power 95%)
   (e) HYD gating 0.9–6 ns
4. Outer mitochondrial membrane protein TOM20 (1^0 mouse anti-TOM20; 2^0 anti-mouse IgG Alexa Fluor 488):
   (a) PINHOLE 1 AU – 151.6 μm
   (b) HYD 1 502–540, G150
   (c) Acquisition on HYD detector 1 between 502 nm and 540 nm. Detector gain = 150%
   (d) WLL exciting at 496 nm. Laser power 10%

(e) STED laser 592 – 70% (main power 95%)

(f) HYD gating 1–6 ns

All analyses are in standard confocal mode using the same acquisition windows, no STED lasers, no gating, and the minimum possible laser power to allow targeting and monitoring the image and setting the Z-stack.

4 Notes

1. We have found that tabletop autoclaves sometimes do not properly inactivate DEPC.
2. Vanadyl ribonucleoside complexes solution (VDR) is an RNase inhibitor. Vortex vigorously, store 100 μL aliquots at −20 °C. After thawing, redissolve VDR precipitate at 65 °C.
3. To lower the risk of RNase contamination caused microbial growth we strongly recommend to continue immediately with Subheading 3.2 and avoid storing fixed cells at 4 °C.
4. We had success permeabilizing without VDR. In those cases, permeabilization was carried out in the culture dish by removing the 1× PBS and adding 0.3% (v/v) Triton® X-100 in 1× PBS. Incubate for 10 min and proceed with 1× PBS washes.
5. Several droplets of probe solution can be placed onto the parafilm before lowering the coverslips onto it. However, take care not to disturb the hybridization chamber to avoid the possibility that the droplets roll into each other and combine.
6. If the coverslips attach too firmly onto the parafilm, place a pipette tip with pre-hybridization buffer next to the coverslip, gently press it into the surface, and apply the buffer under the coverslip to lift it.
7. It is not advisable to dab off excess buffer on tissue paper since the high-precision coverslips easily slip off the forceps. Instead, briefly hover the coverslip over the well and let the excess buffer collect in a drop at one corner, dab the drop onto the buffer surface in the well to let the surface tension pull the drop off the coverslip.
8. According to the manufacturer, the curing time for ProLong™ Antifade Mountant for final refractive index under coverslip is approximately 48–60 h.

Acknowledgments

This work was supported by The Wellcome Trust [203105/Z/16/Z] RNL and ZCL.

References

1. Jakobs S, Stephan T, Ilgen P et al (2020) Light microscopy of mitochondria at the nanoscale. Annu Rev Biophys 49:289–308
2. Pearce SF, Rebelo-Guiomar P, D'Souza AR et al (2017) Regulation of mammalian mitochondrial gene expression: recent advances. Trends Biochem Sci 42(8):625–639
3. Lopez Sanchez MIG, Kruger A, Shiriaev DI et al (2021) Human mitoribosome biogenesis and its emerging links to disease. Int J Mol Sci 22(8):3827
4. Maiti P, Lavdovskaia E, Barrientos A et al (2021) Role of GTPases in driving mitoribosome assembly. Trends Cell Biol 31(4):284–297
5. Rackham O, Filipovska A (2022) Organization and expression of the mammalian mitochondrial genome. Nat Rev Genet 23:606
6. Yousefi R, Fornasiero EF, Cyganek L et al (2021) Monitoring mitochondrial translation in living cells. EMBO Rep 22(4):e51635
7. Zorkau M, Albus CA, Berlinguer-Palmini R et al (2021) High-resolution imaging reveals compartmentalization of mitochondrial protein synthesis in cultured human cells. Proc Natl Acad Sci U S A 118(6):e2008778118
8. Stoldt S, Wenzel D, Kehrein K et al (2018) Spatial orchestration of mitochondrial translation and OXPHOS complex assembly. Nat Cell Biol 20(5):528–534
9. Tu YT, Barrientos A (2015) The human mitochondrial DEAD-box protein DDX28 resides in RNA granules and functions in mitoribosome assembly. Cell Rep 10:854
10. Zorkau M, Proctor-Kent Y, Berlinguer-Palmini R et al (2021) Visualizing mitochondrial ribosomal RNA and mitochondrial protein synthesis in human cell lines. Methods Mol Biol 2192:159–181
11. Abbe E (ed) "Beiträge zur Theorie des Mikroskops und der mikroskopischen Wahrnehmung" [Contributions to the theory of the microscope and of microscopic perception]. Archiv für Mikroskopische Anatomie, vol 9 (1). von Max Cohen & Sohn, Bonn, pp 413–468
12. Dunn KW, Kamocka MM, McDonald JH (2011) A practical guide to evaluating colocalization in biological microscopy. Am J Physiol Cell Physiol 300(4):C723–C742

Chapter 18

Digital RNase Footprinting of RNA-Protein Complexes and Ribosomes in Mitochondria

Danielle L. Rudler, Stefan J. Siira, Oliver Rackham, and Aleksandra Filipovska

Abstract

RNA-binding proteins and mitochondrial ribosomes have been found to be linchpins of mitochondrial gene expression in health and disease. The expanding repertoire of proteins that bind and regulate the mitochondrial transcriptome has necessitated the development of new tools and methods to examine their molecular functions. Next-generation sequencing technologies have advanced the RNA biology field through application of high-throughput methods to study RNA-protein interactions. Here we describe a digital RNase footprinting method to analyze protein and ribosome interactions with mitochondrially encoded transcripts that provides insight into their mechanisms and minimal binding sites. We provide details on RNase digestion and next-generation sequencing, along with computational analyses and visualization of the binding targets within the mitochondrial transcriptome.

Key words RNA-Seq, Footprinting, RNA-binding proteins, Mitoribosome, mtDNA, Bioinformatics

1 Introduction

Mitochondrial gene expression is a key driver of energy conversion, as the 13 mitochondrially encoded polypeptides form essential components of the oxidative phosphorylation system (OXPHOS). The mitochondrial genome is transcribed along its entire length producing polycistronic transcripts that are processed by tRNA-cleaving enzymes to release individual mRNA, tRNA, and rRNA transcripts [1–4]. Consequently, posttranscriptional processing of mitochondrial RNAs (mt-RNAs) is essential for maintaining RNA metabolism in mitochondria. Nuclear-encoded and posttranslationally imported RNA-binding proteins (RBPs) regulate the lifecycles of mt-RNAs and are essential components of the mitochondrial gene expression and translation machinery. Mitochondrial RBPs regulate transcription, processing, modification, polyadenylation, translation, and degradation of mt-RNAs [1]. A

Antoni Barrientos and Flavia Fontanesi (eds.), *The Mitoribosome: Methods and Protocols*,
Methods in Molecular Biology, vol. 2661, https://doi.org/10.1007/978-1-0716-3171-3_18,

range of different next-generation sequencing techniques have provided insights into detailed and mitochondria-specific mechanisms of RNA regulation and translation in transgenic animal or cell models of mitochondrial dysfunction and disease [5–12]. Nevertheless, the interactions between mt-RBPs and their RNA targets are not well-defined and new, high-throughput approaches would provide greater insights into their binding modes. Recent advances in genome editing have made it possible to modulate mtDNA expression [13–15]; however, the nature of these modifications has not been explored in detail and would benefit from nucleic acid-protein analyses.

We developed a digital RNase footprinting method that can provide either a general profile of RNA-protein interactions in mitochondria or can be applied to investigate the binding targets of a specific RBP. We have used this method to profile diverse RNA-protein interactions in human mitochondria and those of ribosomes when they are stalled during translation [7]. Furthermore, we have applied the method to the LRPPRC-SLIRP protein complex to reveal its role as a global mt-RNA chaperone required to relax secondary structures and, through its association with mitoribosomes, to facilitate mRNA translation [16].

Here we provide a detailed protocol for digital RNase footprinting that can be applied to mitochondria isolated from cells or tissues. We outline how to isolate and RNase-treat mitochondria from both cells and tissues, prepare RNA for next-generation sequencing, and how to analyze the binding sites of RBPs and ribosomes. This protocol can be used to identify transcriptome-wide footprints of RBPs of interest, specific motifs they may bind, to investigate the consequences of RBP loss on mt-RNAs, or the effects of mt-RNA editing. Our approach uses RNases to target unbound regions of mt-RNA and can reveal insights into many aspects of mtRNA regulation or degradation.

2 Materials

RNA work requires sterile equipment to prevent contamination and degradation of RNA samples. Plasticware should be RNase-free and additional equipment should be treated with RNaseZap (Ambion). All reagents are prepared in DEPC-treated ultrapure water.

2.1 Isolation of Mitochondria from Tissues

1. Dounce tissue grinder set, 7 mL.
2. STE buffer (for liver tissues): 250 mM sucrose, 5 mM Tris–HCl, 1 mM EGTA, pH 7.4 at room temperature and store at 4 °C.

3. MST buffer (for heart tissues): 210 mM mannitol, 70 mM sucrose, 0.1 mM Tris–HCl, 0.5 M EDTA, pH 7.4 at room temperature and store at 4 °C.
4. Lysis buffer: 100 mM Tris–HCl, 100 mM NaCl, 40 mM $MnCl_2$, 2 mM dithiothreitol pH 7.5, 0.1% TritonX-100.

2.2 Isolation of Mitochondria from Cells

1. Trypsin.
2. Swelling buffer: 10 mM NaCl, 1.5 mM $MgCl_2$ and 10 mM Tris–HCl, pH 7.5.
3. Dounce tissue grinder set, 7 mL.
4. $T_{10}E_{20}$ buffer: 10 mM Tris–HCl and 1 mM EDTA, pH 7.6, with 2 M sucrose added to 250 mM final concentration by adding 1 ml of 2 M sucrose in 10 mL after adjusting the pH.

2.3 RNase Treatment and Library Construction

1. RNase A, RNase T1 and RNase If.
2. miRNeasy Mini Kit.
3. Illumina TruSeq Small RNA Sample Prep Kit.
4. Discontinuous sucrose gradient (sucrose adjusted to 1.0 and 1.7 M final concentration in $T_{10}E_{20}$ buffer), for cell mitochondria.
5. Standard Illumina TruSeq primers:

   Read 1 AGATCGGAAGAGCACACGTCTGAACTC CAGTCA

   Read 2 AGATCGGAAGAGCGTCGTGTAGGGAAA GAGTGT

2.4 Software Packages Required for Analyses

BEDtools <= v2.26.0

Custom scripts available from https://github.com/ORAFLAB/RNaseFootprinting

3 Methods

3.1 Isolation of Mitochondria from Mouse Tissues

Different tissues may require different buffers for optimal mitochondrial extraction. The subsequent protocol describes extraction of mitochondria from heart using MST buffer or liver using STE buffer. Additional tissues may require buffer modifications. All steps for mitochondrial isolation are performed on ice or centrifugation at 4 °C (*see* Fig. 1).

1. Rinse dounce and plungers with STE/MST buffer and chill on ice.
2. Collect fresh tissues and immediately place in STE/MST buffer and keep on ice.

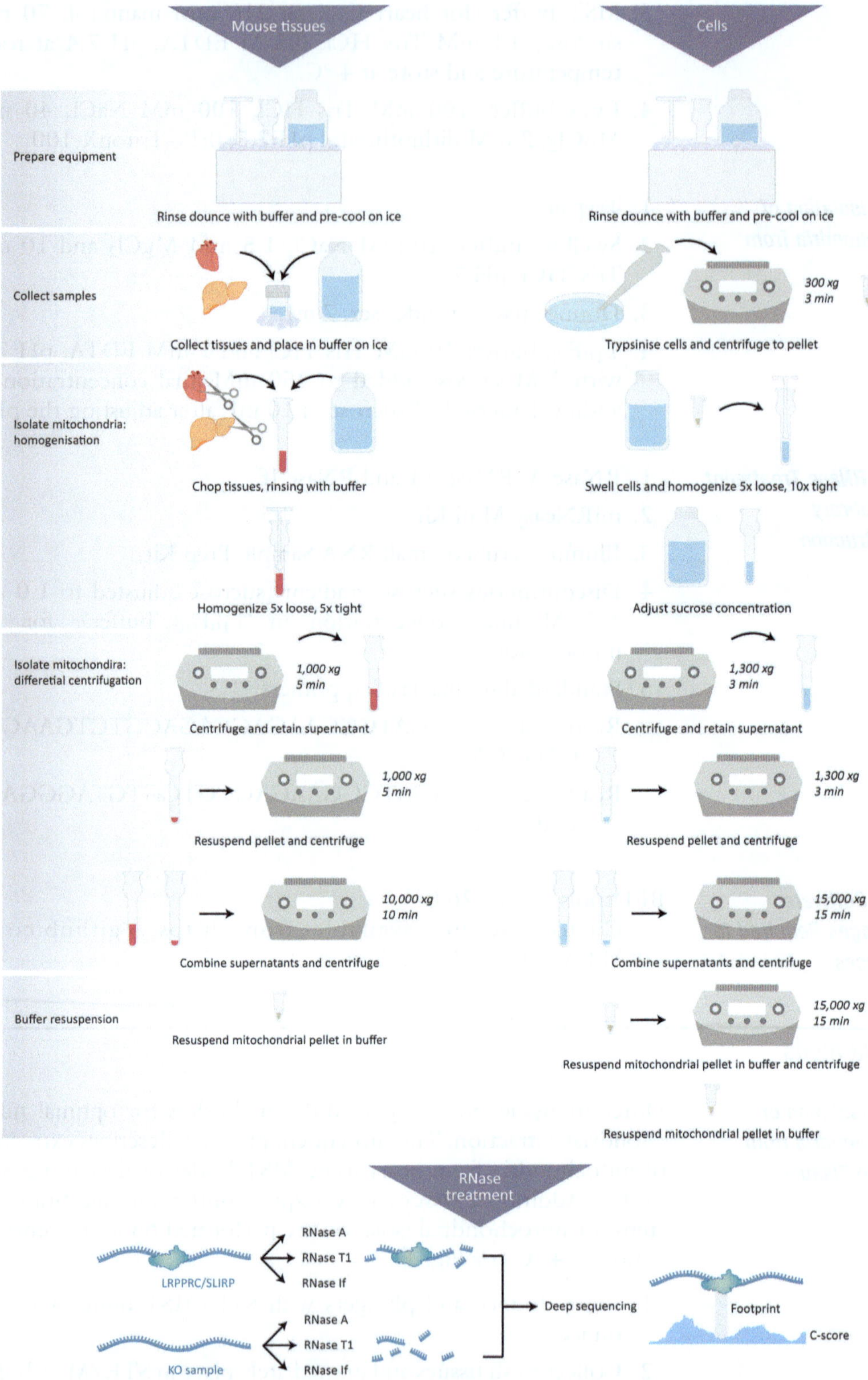

Fig. 1 A schematic showing the isolation and preparation of mitochondria from cells or tissues, for subsequent RNase treatment

3. Chop the tissue into fine pieces with a scalpel, rinsing with STE/MST buffer to remove excess blood and connective tissue prior to homogenization.
4. Transfer tissue and ~3 mL of STE/MST to glass homogenizer and homogenize five times with a loose glass homogenizer and five times with a tight glass homogenizer.
5. Centrifuge the suspension at 1000 × *g* for 5 min, retaining the supernatant and resuspending the remaining pellet in STE/MST, and centrifuge again at 1000 × *g* for 5 min. Pool the two supernatants and use in **step 6**.
6. Centrifuge the pooled suspension at 10,000 × *g* for 10 min to sediment the mitochondria.
7. Add 90 μL of MST to heart mitochondrial pellet and 300 μL of STE buffer to the liver mitochondrial pellet for resuspension (*see* **Note 1**).

3.2 Isolation of Mitochondria from Cells

1. Rinse dounce and plungers with swelling buffer and precool on ice.
2. Collect cells from confluent 15 cm dishes by adding trypsin and pellet by low-speed centrifugation at 300 × *g* for 3 min.
3. Resuspend the cell pellet in 7 mL of swelling buffer, allowing cells to swell for 30 min on ice.
4. Transfer cell suspension to glass homogenizer and homogenize five times with a loose glass homogenizer and then ten times with a tight glass homogenizer.
5. Add 1 mL of 2 M sucrose solution (diluted in $T_{10}E_{20}$ buffer) to the cells, resulting in a final sucrose concentration of 250 mM.
6. Centrifuge the suspension at 1300 × *g* for 3 min, retaining the supernatant and resuspending the remaining pellet in 250 mM sucrose solution (diluted in $T_{10}E_{20}$ buffer), and centrifuge again at 1300 × *g* for 3 min. Pool the two supernatants and use in **step** 7.
7. Centrifuge the supernatant at 15,000 × *g* for 15 min to sediment the mitochondria, discarding the supernatant and resuspending the mitochondrial pellet in 1 mL of 250 mM sucrose solution (diluted in $T_{10}E_{20}$ buffer).
8. Centrifuge the mitochondrial solution at 15,000 × *g* for 15 min. Add 100 μL of 250 mM sucrose solution (diluted in $T_{10}E_{20}$ buffer) to the pellet.

3.3 RNase Treatment of Mitochondria from Mouse Tissues

1. Add 200 μl of lysis buffer to mitochondria (2 mg/mL).
2. Add 10 U/μL of RNase A, 0.1 U/μL of RNase T1 or 0.01 U/μL of RNase If to each mitochondrial lysate in addition to purified RNA as a control (*see* **Note 2**) and incubate at 37 °C for up to 30 min.

3. Isolate RNA using the miRNeasy Mini Kit (Qiagen), incorporating on-column DNase digest, according to the manufacturer's instructions.

3.4 RNase Treatment of Mitochondria from Cells

1. Add 10 U/μL of RNase A, 0.1 U/μL of RNase T1 or 0.01 U/μL of RNase If to each mitochondrial lysate in addition to purified RNA as a control (*see* **Note 3**) and incubate on ice for 30 min (*see* **Note 4**).
2. Prepare a discontinuous sucrose gradient of 1 M and 1.7 M sucrose solution (diluted in $T_{10}E_{20}$ buffer) and layer the mitochondrial suspension on to this. Centrifuge at 70,000 × *g* for 40 min.
3. After centrifugation, the mitochondrial fraction will be present at the interface between the two sucrose cushions. Using a Pasteur pipette recover the mitochondria and transfer to a new tube (300–400 μL is typically recovered).
4. Wash the recovered mitochondria in 1–2 mL of 250 mM sucrose solution (diluted in $T_{10}E_{20}$ buffer) and centrifuge at 15,000 × *g* for 10 min.
5. Isolate RNA using the miRNeasy Mini Kit (Qiagen), incorporating on-column DNase digest, according to the manufacturer's instructions.

3.5 Library Construction

1. Confirm RNA concentration, purity, and fragmentation using a BioAnalyser (Agilent).
2. Construct libraries with the Illumina TruSeq Small RNA Sample Prep Kit following manufacturer's instructions. To facilitate capture of small RNAs between 15 and 55 nt, sample isolation should be performed from a PAGE gel or using a Pippin Prep (Sage Sciences) after adaptor ligation.
3. Sequence the products to a depth of at least five million reads using an appropriate Illumina sequencing platform (e.g., a NovaSeq 6000 or NextSeq 550) according to manufacturer's instructions.

3.6 Sequence Alignment

Prior to footprinting analysis, ensure sequenced reads are quality-checked and adapter-trimmed. Align trimmed reads against a genome masked for nuclear mitochondrial sequences and generate BED files containing strand-specific mitochondrial 5′ end read counts (or fragment counts, if paired-end) normalized to total reads mapped to mitochondria (*see* **Note 5**).

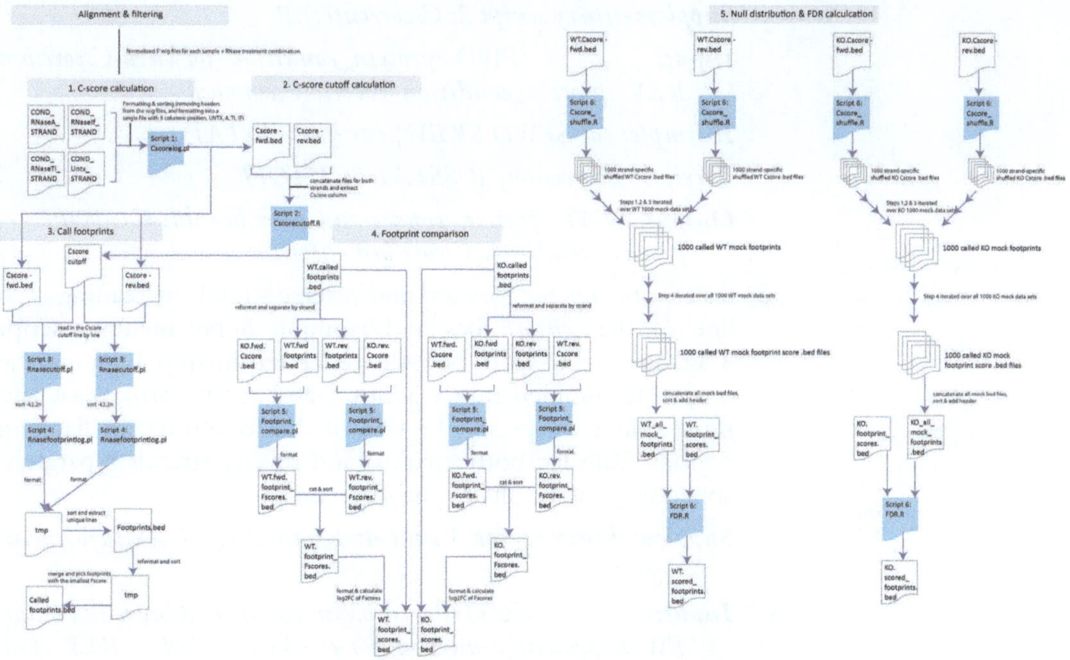

Fig. 2 A diagram outlining the bioinformatic workflow for digital RNase footprinting

3.7 Footprint Analysis

The following protocol describes the use of RNase footprinting in mouse mitochondria and as such the genome lengths provided in examples and supplementary files may need to be adjusted for applications in other genomes (*see* Fig. 2).

1. Calculate the C-score (cleavage score) for each treatment, experimental condition, and strand by passing the normalized 5′ end count from each RNase treatment, condition, and strand to Supplementary script 1 for C-score calculation.

   Supplementary script 1: *Cscorelog.pl*

   Input: *UNTX=protein_condition_RnaseUNTX_strand.bed A=protein_condition RnaseA_strand.bed TI–protein_condition_RnaseTI_strand.bed IF=protein_condition_RnaseIF_strand.bed*

   Example: *paste $UNTX $A $TI $IF | cut -f1,2,4,6,8 | Cscorelog.pl - | awk 'BEGIN {FS="\t";OFS="\t"} {if (($1-2) == -1) {print "chrM",16298,16299,$6} else{print "chrM,$1-2,$1-1,$6}' | sort -k2,2n > $CSCORE*

   Output: *CSCORE=protein_condition_strand.Cscore.bed*

2. Combine strand files by concatenating forward and reverse strand C-score files and pass to Supplementary script 2 to calculate C-score cutoffs.

Supplementary script 2: *Cscorecutoff.R*

Input: *FWD=protein_condition_forward.Cscore.bed REV=protein_condition_reverse.Cscore.bed*

Example: *cat $FWD $REV | cut -f4 > $STAT*

Rscripts Cscorecutoff.R $STAT $CUTOFF

Output: *STAT=protein_condition.Cscore.bed.stat CUTOFF=-protein_condition.Cscore.bed.cutoff*

3. Call footprints for forward and reverse strands by reading each line of the *.cutoff* files and running Supplementary script 3 against the C-score bed file generated in **step 1** piping the output to Supplementary Script 4 and setting *min, max,* and *flank* values to specify the size of the footprint and flanking regions. Run for both forward and reverse strands separately, appending the results to the same output file.

Supplementary script 3 and 4: *Rnasecutoff.pl Rnasefootprintlog.pl*

Input: *CUTOFF=protein_condition.Cscore.bed.cutoff FWD=protein_condition_forward.Cscore.bed REV=protein_condition_reverse.Cscore.bed MIN=8 MAX=40 FLANK=3*

Example: *cat $CUTOFF | unique | cat - | while read myline; do Rnasecutoff.pl $FWD $myline | sort -k2,2n | Rnasefootprintinglog.pl - $FWD $MIN $MAX $FLANK | awk ‘BEGIN {FS=”\t”;OFS=”\t”} {print $1,$2,$3,”footprints_” $4 “_” $5 “_” $6,$6,”’+’”,$5,$7,$8}’ >> $TMP; done*

cat $CUTOFF | uniq | cat - | while read myline; do RNasecutoff.pl $REV $myline | sort -k2,2n | Rnasefootprintlog.pl - $REV $MIN $MAX $FLANK | awk ‘BEGIN {FS=”\t”; OFS=”\t”} {print $1,$2,$3,”footprints_” $4 “_” $5 “_” $6, $6,”’-’”,$5,$7,$8}’ >> $TMP; done

sort -k2,2n $TMP | uniq > $FOOTPRINTS && rm $TMP

Output: *TMP=protein_condition.footprints.tmp FOOTPRINTS=protein_condition.footprints.bed*

4. Merge any overlapping footprints into a single footprint with BEDtools.

Input: *FOOTPRINTS=protein_condition.footprints.bed*

Example: *awk ‘BEGIN{FS=”\t”;OFS=”\t”} {if($3>$2){print $0}}’ $FOOTPRINTS | sort -k2,2n > $TMP*

bedtools merge -s -c 1 -o count -i $TMP | awk ‘BEGIN{FS=”\t”; OFS=”\t”} {print $1,$2,$3, “I”,$5,$4}’ | bedtools intersect -s -wa -wb -a - -b $TMP > $MERGED

Output: *TMP=protein_condition.footprints.merged MERGED=protein_condition.footprints.merged.bed*

5. For overlapping footprints, select the footprint with the smallest F-score.

   Input: *MERGED=protein_condition.footprints.merged.bed*

   Example: *awk 'BEGIN{FS="\t"; OFS="\t"} NR==FNR { ID=$1 $2 $3 $3; if(array[ID] == ""){ array[ID]=$11} else {if (array[ID] > $11){array[ID]=$11}} } NR > FNR {ID=$1 $2 $3 $4; if (array[ID] == $11) print $7,$8,$9, $10,$11,$12,$13,$14,$15}}' $MERGED $MERGED $CALLED*

   Output: *CALLED=protein_condition.called_footprints.bed*

6. Edit the files for circular mitochondrial chromosomes at the end of the files for forward ("+") and reverse ("−") strands and append to a single output file.

   Input: *FOOTPRINTS=protein_condition.footprints.bed*

   Example: *awk 'BEGIN{FS="\t";OFS="\t"}{if($3<$2 && $6 == "+"){print $0}}' $FOOTPRINTS | sort -k5,5n | head -1 >> $CALLED*

   awk 'BEGIN{FS="\t";OFS="\t"}{if($3<$2 && $6 == "-") {print $0}}' $FOOTPRINTS | sort -k5,5n | head -1 >> $CALLED

   sort -k2,2n -o $CALLED $CALLED

   Output: *CALLED=protein_condition.called_footprints.bed*

7. Compare footprints by first splitting footprints into strand files and convert to BED format.

   Input: *CALLED=protein_condition.called_footprints.bed*

   Example: *awk 'BEGIN{FS="\t";OFS="\t"} {if ($6=="+" && $2>$3) {print $1,$2,$3,(16299-$2)+$3} else if ($6=="+" && $3>$2) {print $1,$2,$3,$3-$2} else {}}' $CALLED > $FWD*

   awk 'BEGIN{FS="\t";OFS="\t"} {if ($6=="-" && $2>$3) {print $1,$2,$3,(16299-$2)+$3} else if ($6=="-" && $3>$2) {print $1,$2,$3,$3-$2} else {}}' $CALLED > $REV

   Output: *FWD=protein_condition_forward.called_footprints.bed REV=protein_condition_reverse.called_footprints.bed*

8. Using footprints in the control data files, call F-scores from experimental data C-score files with Supplementary script 5, setting *min, max,* and *flank* parameters. Do the same for control C-score data using experimental footprints.

   Supplementary script 5: *footprint_compare.pl*

   Input: *FWDwt=protein_WT_forward.called_footprints.bed FWDkoscore=protein_KO_forward.Cscore.bed REVwt=protein_WT_reverse.called_footprints.bed*

REVkoscore=protein_KO_reverse.Cscore.bed MIN=8 MAX=40 FLANK=3

Example: *footprint_compare.pl $FWDwt $FWDkoscore $MIN $MAX $FLANK | awk 'BEGIN {FS="\t";OFS="\t"} {print $1,$2,$3,"footprints_" $4 "_" $5 "_" $6,$6, "+", $5,$7,$8}' > protein_WT_forward.footprint_Fscores.tmp*

footprint_compare.pl $REVwt $REVkoscore $MIN $MAX $FLANK | awk 'BEGIN {FS="\t";OFS="\t"} {print $1, $2,$3,"footprints_" $4 "_" $5 "_" $6,$6, "-",$5,$7,$8}' > protein_WT_reverse.footprint_Fscores.tmp

cat protein_WT_forward.footprint_Fscores.tmp protein_WT_reverse.footprint_Fscores.tmp | sort -k2,2n -o $FSCORES

Output: *FSCORES=protein_WT.footprint_Fscores.bed or protein_KO.footprint_Fscores.bed*

9. Create footprint data arrays for each condition, calculating the log2 fold change of F-scores between the *called_footprints.bed* files and the *footprint_Fscore.bed* files.

Input: *WTfp=protein_WT.called_footprints.bed WTscore=protein_WT.footprint_Fscores.bed KOfp=protein_KO.called_footprints.bed KOscore=protein_KO.footprint_Fscores.bed*

Example: *paste $WTfp $WTscore | awk 'BEGIN{FS="\t"; OFS="\t"} {print $1,$2,$3,$4,$5,$6,$7,$14,$16,log($5/$14)/log(2)}' - > $WTfpscored*

sed -i '1iChr\tStart\tEnd\tName\tFscore\tStrand\tCentreCscore\tWTFscore\tWTCentreCscore\tlog2FC_Fscore' $WTfpscored

paste $KOfp $KOscore | awk 'BEGIN{FS="\t";OFS="\t"} {print $1,$2,$3,$4,$5,$6,$7,$14,$16,log($14/$5)/log(2)}' - > $KOfpscored

sed -i '1iChr\tStart\tEnd\tName\tFscore\tStrand\tCentreCscore\tKOFscore\tKOCentreCscore\tlog2FC_Fscore' $KOfpscored

Output: *WTfpscored=protein_WT.called_footprint_scores.bed KOfpscored= protein_KO.called_footprint_scores.bed*

10. Construct an empirical null model for false discovery rate (FDR) calculation of footprints by randomly shuffling the C-score bed files 1000 times using Supplementary script 6.

Supplementary script 6: Cscore_shuffle.R

Input: *CSCORE=protein_condition_strand.Cscore.bed*

Example: *for ((i=1;i<=1000;i++)); do Rscript Cscore_shuffle.R $CSCORE protein_condition_strand.Cscore.shuf_$i.bed; done*

Output: *protein_condition_strand.Cscore_shuf_[1-1000].bed*

11. Repeat **steps 3–9** for all shuffled mock C-score files and collate all iterations into a single sorted file formatted with headers: *protein_condition.called_null_footprints.shuf_all.bed.*

12. Calculate the FDR for the F-score log2 fold changes by comparing real and mock footprint scores with Supplementary script 7 for footprints called in each condition.

    Supplementary script 7: *FDR.R*

    Input: *WTfscore=protein_WT.footprint_Fscores.bed WTnull=-protein_WT.called_null_footprints.shuf_all.bed KOfscore=protein_KO.footprint_Fscores.bed KOnull=protein_KO.called_null_footprints.shuf_all.bed*

    Example: *FDR.R $WTfscore $WTnull $WTscored*

    FDR.R $KOfscore $KOnull $KOscored

    Output: *WTscored=protein_WT.scored_footprints.txt KOscored=protein_KO.scored_footprints.txt*

    Output: *protein_WT.scored_footprints.txt protein_KO.scored_-footprints.txt*

4 Notes

1. The use of at least 10 mg of isolated mitochondria from either cells or tissues is required for the protocol. This depends on the cells used and typically 15 dishes that are 10 cm in diameter are sufficient, one mouse liver or two to three mouse hearts homogenized in 20 mL or 10 mL of buffer, respectively.

2. The amount of final buffer volume for mitochondrial resuspension in can be adjusted based on the size of the pellet and the required final concentration of mitochondria. The size of the pellet will depend on the starting material and type of cells or tissues used. Generally, one mouse liver produces 300 μL of 40 mg/mL mitochondria and cells from 15 dishes will generate 100 μL of 10 mg/mL mitochondria.

3. RNase treatment of purified control RNA is required to ensure activity of RNases prior to sequencing.

4. Ideally, all three RNases should be used to gain a complete profile of mitochondrial RNA-binding sites, alternatively for ribosome profiles the use of RNase T1 is sufficient. The concentrations of RNases have been optimized for ideal fragmentation based on the protein amount input. If further optimization is required, it is best to analyze the fragments using a Tapestation to identify the fragment sizes of the RNAs following RNA isolation.

5. BEDtools <= v2.26.0 is required for this pipeline due to later changes to the software that alter the outputs.

Acknowledgments

This project was supported by fellowships and project grants from the NHMRC, ARC, and CCWA (to OR and AF). DR was supported by a UWA Postgraduate Scholarship.

References

1. Rackham O, Filipovska A (2022) Organization and expression of the mammalian mitochondrial genome. Nat Rev Genet 23:606
2. Rackham O, Busch JD, Matic S et al (2016) Hierarchical RNA processing is required for mitochondrial ribosome assembly. Cell Rep 16:1874–1890
3. Siira SJ, Rossetti G, Richman TR et al (2018) Concerted regulation of mitochondrial and nuclear non-coding RNAs by a dual-targeted RNase Z. EMBO Rep 19:e46198
4. Holzmann J, Frank P, Lofler E et al (2008) RNase P without RNA: identification and functional reconstitution of the human mitochondrial tRNA processing enzyme. Cell 135: 462–474
5. Rackham O, Shearwood AJ, Mercer TR et al (2011) Long noncoding RNAs are generated from the mitochondrial genome and regulated by nuclear-encoded proteins. RNA 17:2085–2093
6. Mercer TR, Neph S, Dinger ME et al (2011) The human mitochondrial transcriptome. Cell 146:645–658
7. Liu G, Mercer TR, Shearwood AJ et al (2013) Mapping of mitochondrial RNA-protein interactions by digital RNase footprinting. Cell Rep 5:839–848
8. Rudler DL, Highes LA, Perks KL et al (2019) Fidelity of translation initiation is required for coordinated respiratory complex assembly. Sci Adv 5:eaay2118
9. Perks KL, Rossetti G, Kuznetsova I et al (2018) PTCD1 is required for 16S rRNA maturation complex stability and mitochondrial ribosome assembly. Cell Rep 23:127–142
10. Kuznetsova I, Sira SJ, Shaerwood AJ et al (2017) Simultaneous processing and degradation of mitochondrial RNAs revealed by circularized RNA sequencing. Nucleic Acids Res 45: 5487–5500
11. Suzuki T, Yashiro Y, Kikuchi I et al (2020) Complete chemical structures of human mitochondrial tRNAs. Nat Commun 11:4269
12. Schöller E, Marks J, Marchand V et al (2021) Balancing of mitochondrial translation through METTL8-mediated m(3)C modification of mitochondrial tRNAs. Mol Cell 81: 4810–4825.e12
13. Mok BY, de Moraes MH, Zeng J et al (2020) A bacterial cytidine deaminase toxin enables CRISPR-free mitochondrial base editing. Nature 583:631–637
14. Bacman SR, Kauppila JHK, Pereira CV et al (2018) MitoTALEN reduces mutant mtDNA load and restores tRNA(Ala) levels in a mouse model of heteroplasmic mtDNA mutation. Nat Med 24:1696–1700
15. Wallis CP, Scott LH, Filipovska A et al (2020) Manipulating and elucidating mitochondrial gene expression with engineered proteins. Philos Trans R Soc Lond B Biol Sci 375: 20190185
16. Siira SJ, Spahr H, Shearwood MJ et al (2017) LRPPRC-mediated folding of the mitochondrial transcriptome. Nat Commun 8:1532

Chapter 19

Dead-Seq: Discovering Synthetic Lethal Interactions from Dead Cells Genomics

Joan Blanco-Fernandez and Alexis A. Jourdain

Abstract

Pooled genetic screens have revolutionized the field of functional genomics, yet perturbations that decrease fitness, such as those leading to synthetic lethality, have remained difficult to quantify at the genomic level. We and colleagues previously developed "death screening," a protocol based on the purification of dead cells in genetic screens, and used it to identify a set of genes necessary for mitochondrial gene expression, translation, and oxidative phosphorylation (OXPHOS), thus offering new possibilities for the diagnosis of mitochondrial disorders. Here, we describe *Dead-Seq*, a refined protocol for death screening that is compatible with most pooled screening protocols, including genome-wide CRISPR/Cas9 screening. *Dead-Seq* converts negative-selection screens into positive-selection screens and generates high-quality data directly from dead cells, at limited sequencing costs.

Key words Genome-wide screening, Synthetic lethality, Drop-out screen, Mitochondria, sgRNA, shRNA, RNAi, ORFeome, Annexin V, MACS, Auxotrophy, Systems genetics, Metabolism, Galactose, Apoptosis, Necroptosis, Mitochondrial translation

1 Introduction

The complete sequencing of the human genome and the development of high-throughput gene perturbation technologies, such as RNA interference and CRISPR/Cas9, have transformed the field of functional genomics. These new tools have significantly advanced the discovery of genes and helped assign new gene functions in response to environmental and drug challenges. Many screening methodologies have taken advantage of the synthetic lethal interactions that occur between gene pairs [1, 2], between genes and the environment [3, 4], or between genes and drugs [5–8] (Table 1, *see* **Note 1**). Yet, and somewhat paradoxically, the most common screening modalities used for studying gene essentiality have been based on cell growth [4, 9–11]. In this experimental setting, genes whose depletion confers a growth advantage are enriched in the population after several days of selection (positive-

Antoni Barrientos and Flavia Fontanesi (eds.), *The Mitoribosome: Methods and Protocols*,
Methods in Molecular Biology, vol. 2661, https://doi.org/10.1007/978-1-0716-3171-3_19,

Table 1
Examples of synthetic lethal interactions resulting in phosphatidylserine exposure

	1st hit	2nd hit	Reference
Gene vs. gene	*BCL2L1, BLC2*	*MCL1*	[2, 23]
	EP300	*CREBBP*	[24]
Gene vs. environment	OXPHOS inhibition	Glucose deprivation	[3, 19, 25]
	AKT2	*Salmonella* infection	[26]
Gene vs. drug	*IDH1* mutation	Triptolide (NRF2 inhibitor)	[27]
	MYC overexpression	Tigecycline (mitochondrial translation inhibitor)	[28]

selection), whereas genes whose depletion decreases fitness are lost (negative-selection, or drop-out). Identifying drop-out genes however can be challenging given their low abundance in the cell population, or when the fitness difference is modest. Consequently, high-sequencing coverage is required for drop-out screens, with attendant high costs and often low efficiency [12].

With this in mind, we and colleagues previously developed "death screening," a pooled screening methodology based on the selection of dead cells [3]. Death screening relies on the Annexin V-mediated purification of phosphatidylserine-positive cells, characteristic of certain forms of cell death such as apoptosis and necroptosis [13–15]. Death screening has enabled the identification of a set of genes necessary for mitochondrial energy production, with among them 68 bona fide OXPHOS complexes subunits and 108 genes involved in the transcription and translation of the mitochondrial genome. This genetic screen also led to the discovery of a protein module involved in mitochondrial 16S rRNA maturation and translation [3], as well as novel nuclear pre-mRNA splicing mechanisms involved in energy metabolism [16].

We now present *Dead-Seq*, a refined protocol of death screening to study synthetic lethality at the genomic level (*see* Fig. 1). *Dead-Seq* is a versatile tool for converting negative-selection screens into positive-selection screens, thus improving the identification of drop-out genes with lower sequencing costs. It can be applied to a wide variety of genetic, environmental, or drug-induced genetic interactions which result in exposure of phosphatidylserine on the cell surface, and can be used in combination with all types of genetic libraries (e.g., CRISPR-based, shRNA, ORFeome). In this protocol, we provide a detailed procedure for *Dead-Seq*, using as an example a genome-wide CRISPR/Cas9 screen applied to the synthetic lethality between OXPHOS inhibition and reduced glycolysis.

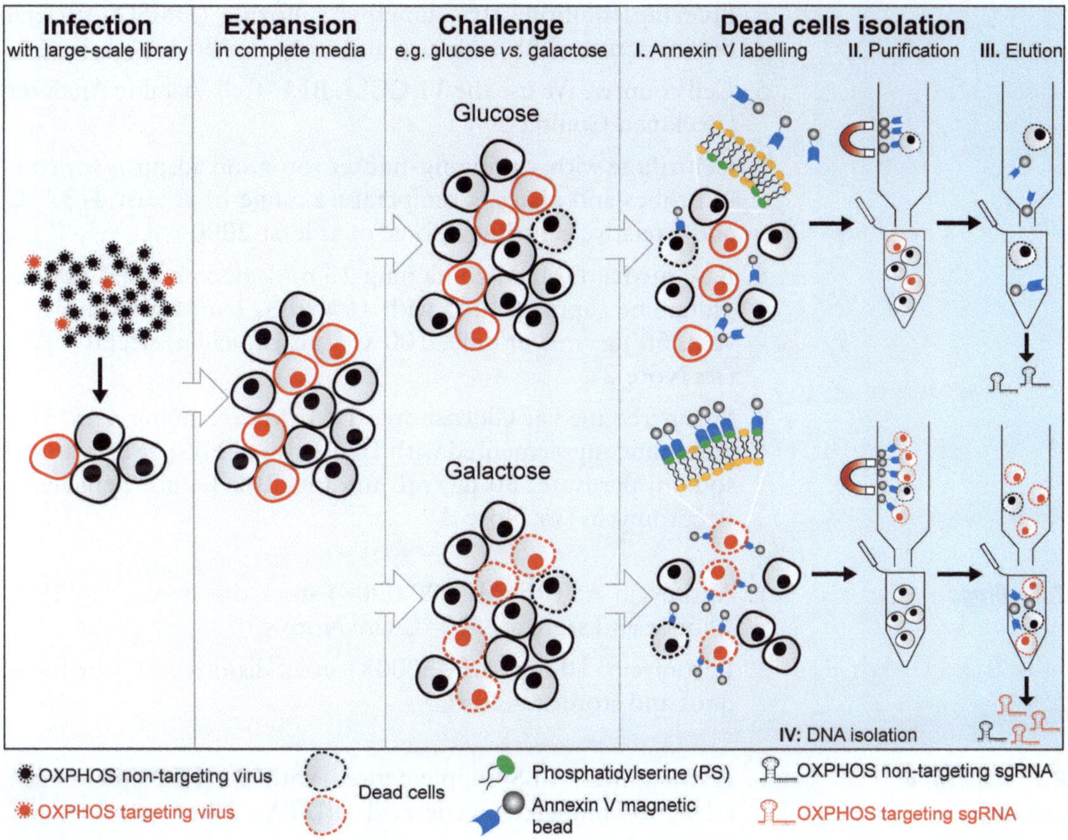

Fig. 1 Schematic overview of *Dead-Seq*. A large number of cells are infected with a genome-scale library carrying sgRNAs, shRNAs, or any barcoded constructs. After several days of expansion, the cells are divided into control and treatment groups (in our example, in glucose- and galactose-containing media). Shortly after treatment, dead cells from the population are purified using Annexin V-conjugated magnetic microbeads in a magnetic field. Total genomic DNA is then isolated from each condition and the sgRNAs/shRNAs/barcodes from the libraries are identified by next-generation sequencing. PS: phosphatidylserine, OXPHOS: oxidative phosphorylation

2 Materials

All solutions are prepared in a laminar flow cabinet using sterile milli-Q water and tissue culture-grade reagents.

2.1 Cell Culture

1. A cell line of interest. For genome-wide screening, we routinely use K562 cells (ATCC CCL-243).
2. Tissue culture (TC) flasks (25–300 cm^2) and multi-well plates (6- and 12-well format).
3. Disposable conical tubes (15–500 mL).
4. Phosphate-buffered saline (PBS).

5. Freezing solution: 10% dimethyl sulfoxide (DMSO) in fetal bovine serum (FBS). Aliquot and store at −20 °C.
6. Cell counter. We use the Vi-CELL BLU Cell Viability Analyzer (Beckman Coulter).
7. Centrifuge with a swinging-bucket rotor and adaptors for conical tubes and plates, a temperature range of at least 4–37 °C and a relative centrifugal force of at least 2000 × *g*.
8. Base media: DMEM containing 25 mM glucose and 4 mM L-glutamine supplemented with 10% FBS, 1 mM sodium pyruvate, 50 μg/mL uridine, 100 U/mL penicillin/streptomycin (*see* **Note 2**).
9. Sugar-free media: Glucose-free DMEM containing 4 mM L-glutamine supplemented with 10% dialyzed FBS (dFBS), 1 mM sodium pyruvate, 50 μg/mL uridine, 100 U/mL penicillin/streptomycin (*see* **Note 3**).

2.2 Drugs

1. Antimycin A: 10 mM (100,000×) stock dissolved in DMSO. Aliquot and store at −20 °C (*see* **Note 4**).
2. Puromycin: 10 mg/mL (5000×) stock dissolved in water. Aliquot and store at −20 °C.

2.3 Annexin V Labeling

1. FACS buffer: PBS supplemented with 2% FBS and 5 mM ethylenediaminetetraacetic acid (EDTA). Filter, sterilize, and store at 4 °C.
2. Annexin V-FITC.
3. Annexin V binding buffer: PBS containing 10 mM HEPES, 150 mM NaCl, 2.5 mM $CaCl_2$ (pH 7.4).

2.4 Genetic Perturbation Library

1. Genetic perturbation library of interest. We have obtained good results using the human Avana [17], Brunello [17] and ORFeome [18] libraries. Several libraries are available at Addgene as ready-to-use lentiviral preparations carrying the puromycin resistance gene.
2. Polybrene (1000×) stock: Commercially available; or alternatively dissolve polybrene in water to 10 mg/mL. Aliquot and store at −20 °C.

2.5 Annexin V-Based MACS Purification

1. LS columns (Miltenyi Biotec).
2. Annexin V-coated microbeads (Miltenyi Biotec).
3. 20× binding buffer stock solution (Miltenyi Biotec).
4. QuadroMACS Separator (Miltenyi Biotec).

3 Methods

Perform all procedures under sterile conditions at room temperature. The protocol described is based on the use of K562 cells. It can be readily adapted to other cell lines, including adherent cells, using appropriate modifications to media and culture conditions. Follow local institute biosafety guidelines when handling viruses.

3.1 Cell Line Maintenance

1. Thaw cells in base media.
2. Maintain in flasks in an incubator at 37 °C and 5% CO_2.
3. Passage the cells every 3–4 days by dilution to 0.1×10^6 cells/mL.
4. Periodically check for the presence of mycoplasma.

3.2 Prescreen: Validating a Suitable Synthetic Lethal Interaction

Before performing a screen, it is essential to confirm phosphatidylserine exposure in the model of choice (*see* examples in Table 1). In the example below, we investigate the synthetic lethality conferred by OXPHOS inhibition during glucose deprivation [19].

1. Dilute antimycin A stock 1:100 in DMSO (*see* **Note 4**).
2. Prepare 4× 15 mL tubes each containing 4 mL of sugar-free media.
3. Add 180 μL of 1.11 M glucose or galactose, and 8 μL of DMSO or diluted antimycin A from **step 1** to obtain the following four conditions:
   (a) Glucose + DMSO
   (b) Glucose + Antimycin A
   (c) Galactose + DMSO
   (d) Galactose + Antimycin A
4. Centrifuge 40×10^6 K562 cells for 3 min at $300 \times g$.
5. Wash cells twice in PBS (*see* **Note 5**).
6. Resuspend cells in 20 mL of sugar-free media.
7. Add 4 mL of cells to each of the four media prepared in **step 3**. At this point, the final concentration of cells is 10^6 cells/mL, glucose and galactose are at 25 mM, and antimycin A, when present, is at 100 nM.
8. From each condition, transfer 3 × 2 mL cells into 3 wells of a 6-well plate (technical triplicates) and transfer to the incubator.
9. After 24 h, harvest the cells from each well into a 15 mL tube.
10. Centrifuge the cells for 3 min at 2000 × g (*see* **Note 6**).
11. Wash the pellets in FACS buffer.
12. Repeat **steps 10** and **11**.

13. Resuspend each pellet in 1 mL Annexin V binding buffer.
14. For each condition, transfer 100 μL of cells into a FACS-compatible tube and add 5 μL of Annexin V-FITC (*see* **Note 7**).
15. Protect the cells from light and incubate for 15 min at room temperature.
16. After 15 min of incubation, add a further 400 μL of annexin V binding buffer to each tube.
17. Measure FITC fluorescence by flow cytometry (*see* **Note 8**).

3.3 Screen

Perform all procedures under sterile conditions at room temperature. In this example, we use K562 cells and the genome-wide Brunello CRISPR/Cas9 library with 76,441 sgRNAs targeting 19,114 coding genes and 1000 noncoding sites [17]. The Brunello library carries Cas9 and the puromycin resistance gene for selection in mammalian cells and is available at Addgene as ready-to-use lentiviral supernatants (73179-LV). The protocol below is fully compatible with other libraries with appropriate modification of the resistance marker selection conditions.

3.3.1 Library Titration for the Cell Line Used

1. Centrifuge 21×10^6 cells for 3 min at $300 \times g$.
2. Discard the supernatant and resuspend the pellet in 7 mL of base media supplemented with 14 μL of polybrene.
3. Plate 1 mL of cells in 6 wells of a 12-well plate. Each well contains 3×10^6 cells.
4. Quickly thaw an aliquot of the viral library in a 37 °C water bath (*see* **Note 9**).
5. Label and prepare and 6 × 1.5 mL Eppendorf tubes as follows:
   (a) V0: 1 mL of base media.
   (b) V25: 975 μL of base media + 25 μL of lentiviral supernatant.
   (c) V50: 950 μL of base media + 50 μL of lentiviral supernatant.
   (d) V100: 900 μL of base media + 100 μL of lentiviral supernatant.
   (e) V200: 800 μL of base media + 200 μL of lentiviral supernatant.
   (f) V400: 600 μL of base media + 400 μL of lentiviral supernatant.
6. Add the contents of each tube prepared in **step 5** to one well of cells from **step 3**.
7. Centrifuge the plate at $1000 \times g$, 30 °C for 2 h using a swing-out rotor (spinfection).

8. Carefully aspirate all the media from the wells.
9. Add 2 mL of base media to each well and transfer the cells to the incubator.
10. 24 h postinfection, collect and count the cells from each well.
11. Centrifuge 1.5×10^6 cells from each of the six conditions for 3 min at $300 \times g$.
12. Resuspend the cell pellets in 3 mL of base media.
13. For each condition, add 1 mL of cells to two wells of a 12-well plate.
14. Prepare the selection buffer in a 15 mL tube by combining 7 mL of base media with 2.8 μL of puromycin.
15. To one well of each replicate pair, add 1 mL of base media alone; to the other add 1 mL of selection buffer. The final concentration of cells in each well is 0.25×10^6 cells/mL, and the final concentration of puromycin is either 0 or 2 μg/mL.
16. Transfer cells to the incubator.
17. After 72 h, determine the concentration of viable cells in each well. From the viral concentrations from **step 5**, determine the amount of virus required to obtain ~25% infection (*see* **Note 10**). This corresponds to a multiplicity of infection (MOI) of ~0.3 (*see* **Notes 11** and **12**).

3.3.2 Infection

Before starting the infection, it is important to determine the number of cells that need to be infected based on the library size, the desired infection rate per sgRNA, and the MOI. In this example, we use the Brunello library (76,441 sgRNAs) to target 500 cells/sgRNA at an MOI of 0.3 (*see* **Note 13**). In this case, we will infect 127×10^6 cells (*see* **Note 14** for the calculation). All steps are described for a single sample, but the screen should be performed in at least two infection replicates.

1. *Day 0*: Centrifuge 135×10^6 K562 cells for 5 min at $300 \times g$ (*see* **Note 15**).
2. Resuspend the cell pellet in 45 mL of base media (3×10^6 cells/mL).
3. Add 90 μL of polybrene to the cells.
4. Plate 44×1 mL of cells in 12-well plates. 42 wells will be used for the screen and 2 wells will serve as controls.
5. Quickly thaw the viral library in a 37 °C water bath (*see* **Note 9**).
6. In a 50 mL conical tube, add the volume of virus calculated to obtain the desired infection rate (MOI ~0.3) and add base media to a final volume of 45 mL.

7. Add 1 mL of the diluted virus to the 42 screening wells and to one of the control wells (+virus control).
8. Add 1 mL of base media alone to the remaining control well (−virus control).
9. Centrifuge the plates for 2 h at 1000 × *g*, 30 °C (spinfection).
10. Carefully remove all the supernatant and add 2 mL of base media to each well.
11. Transfer cells to the incubator.

3.3.3 Selection with Puromycin and Confirmation of the MOI

1. *Day 1*: 24 h postinfection, collect and pool the cells from the 42 screening wells (screen cells).
2. Collect the cells from each of the two control wells (+virus and −virus controls).
3. Count cells in all three conditions.
4. For the +virus and −virus controls: spin down 1.5×10^6 cells and resuspend in 3 mL of base media. The final cell concentration is 0.5×10^6 cells/mL. For each condition, add 1 ml of cells to 2 wells of a 12-well plate (there will be 4 wells in total).
5. For the screen cells: dilute to 0.5×10^6 cells/mL.
6. For selection, prepare a sufficient amount of base media containing puromycin at a 2× concentration (4 μg/mL).
7. For +virus and −virus controls: add 1 mL of base media to one replicate from each condition, and 1 mL of puromycin selection buffer to the other. The final concentration of cells is 0.25×10^6 cells/mL, and the final concentration of puromycin is 0 or 2 μg/mL.
8. For screen cells: Add a volume of puromycin-containing media equal to the volume of screen cells resuspended at 0.5×10^6 cells/mL from **step 5** and transfer to large TC flasks (e.g., 300 cm^2) to ensure adequate oxygenation. The final concentration of cells is 0.25×10^6 cells/mL, and the final concentration of puromycin is 2 μg/mL.
9. Transfer cells to the incubator.
10. *Day 4*: After 72 h, count the cells from the screen and the two controls with and without puromycin selection.
    - The concentration of the screen cells should be equal to the concentration of cells in the +virus control cells treated with puromycin, and ~25% of the +virus controls cells not treated with puromycin (for MOI ~0.3, *see* **Notes 10** and **16**).
    - The −virus control cells treated with puromycin should be mostly dead.

11. Centrifuge the screen cells for 5 min at 300 × *g* (*see* **Note 17**). Resuspend the cells at 0.1×10^6 cells/mL in base media. Use several large TC flasks to ensure adequate oxygenation.
12. Transfer cells to the incubator.

3.3.4 Cell Expansion, Early Time Point, and Preparation of Cell Stocks

1. *Day 7*: Count cells in each preparation.
2. "Day 7" sample: Centrifuge 40×10^6 viable cells for 5 min at 300 × *g*. Label the tube "day 7." Remove the supernatant and freeze the cell pellet at −80 °C for DNA isolation. This step is optional (*see* **Note 18**).
3. Screen: Centrifuge another 40×10^6 viable cells for 5 min at 300 × *g* and resuspend the pellet in 400 mL of base media. The concentration of cells is 0.1×10^6 cells/mL. Transfer to large 300 cm^2 flasks to ensure adequate oxygenation and return to the incubator.
4. Freezing cells for future use: At this point, it is also possible to freeze cells for future screens. We recommend freezing as many aliquots as possible, with a minimum of 1000–1500 cells/guide (80–120 × 10^6 cells) (*see* **Note 19**) in 10 mL of cell freezing solution (*see* **Note 20**).

3.3.5 Glucose/Galactose Treatment

1. *Day 10*: For each replicate, count the cells.
2. "Day 10" sample: Centrifuge 40×10^6 viable cells for 5 min at 300 × *g*. Label the tube "day 10." Remove the supernatant and freeze the cell pellet at −80 °C for DNA isolation. This step is optional (*see* **Note 18**).
3. Screen: For each replicate, centrifuge 360×10^6 viable cells for 5 min at 300 × *g*.
4. Wash the pellets twice in PBS by centrifugation for 5 min at 300 × *g* (*see* **Note 5**).
5. Resuspend cells in a total of 720 mL sugar-free media.
6. Prepare two tubes each with 320 mL of cells and add either 7.2 mL of 1.11 M glucose or 7.2 mL of 1.11 M galactose stock solutions. The final concentration of glucose or galactose is 25 mM, the cells are at a concentration of 0.5×10^6 cells/mL and there are 2000 cells/sgRNA.
7. Transfer cells to several large TC flasks to ensure oxygenation and place the flasks in the incubator.
8. Incubate for 24 h before harvesting the cells (*see* **Note 21**).

3.3.6 Select the Annexin V Positive (Dead) Cells by MACS for DNA Isolation

1. Prepare 120 mL of 1× Annexin V-microbead binding buffer by diluting 6 mL of the 20× stock solution into 114 mL of water.
2. Count the glucose and galactose cells and centrifuge both cell preparations for 5 min at 2000 × *g*.

3. Resuspend the cell pellets for each condition in 80 μL of 1× Annexin binding buffer per 10^7 cells. Transfer cells to a 50 mL tube.
4. Save a 10 μL aliquot from each condition labelled as "input."
5. Add 20 μL of Annexin V microbeads per 10^7 total cells from each condition.
6. Mix well and incubate 15 min at 4 °C.
7. Add 1 mL of 1× Annexin V-microbeads binding buffer per 10^7 cells to each condition.
8. Centrifuge cells for 5 min at 2000 × *g*.
9. Resuspend the cell pellet in Annexin V-microbead binding buffer at 2×10^8 cells/mL, or in a minimum volume of 0.5 mL.
10. Keep cells on ice.
11. Assemble a separate LS column for each condition in the magnetic field of a suitable MACS Separator (e.g., QuadroMACS).
12. Equilibrate the columns using 3 mL of 1× Annexin V-microbeads binding buffer.
13. Apply the cell suspension to the column, making sure that the maximal column capacity (2×10^9 cells/column) is not exceeded. In case the maximal column capacity is reached, split the sample in multiple columns.
14. Collect the flow-through containing the unlabeled cells.
15. Wash the columns 4 times with 3 mL 1× Annexin V-microbeads binding buffer. Collect the unlabeled cells from all the washes and combine with the flow-through from **step 14**. Label the tube as "viable cells." Keep on ice.
16. Remove the columns from the magnetic separator and fit them to a suitable collection tube.
17. Add 5 mL of 1× binding buffer to each column and recover the retained cells by applying firm pressure to a plunger inserted into the column.
18. If the number of cells reached the maximal column capacity in 13, pool the cells from similar columns.
19. Label the tubes as "dead cells" and keep on ice.
20. Transfer 10 μL from the "input", "viable cells" and "dead cells" tubes to 1.5 mL tubes and proceed to FACS-based analysis as in Subheading 3.2, **steps 10–17**. The viability (i.e. Annexin V-negative cells) is expected to be >90% in the "input" and "viable cells," and <10% in "dead cells" (*see* **Note 22**).
21. Spin down the "dead cells" for 5 min at 2000 × *g*.
22. Remove the supernatant and freeze the pellets at −80 °C for DNA isolation.

3.4 DNA Isolation and Data Analysis

We recommend contacting the local institute genomics platform regarding the preferred methods for gDNA isolation, PCR amplification, and next-generation sequencing. In our hands, good results have been obtained using the NucleoSpin Blood Kit from Macherey Nagel, and the Herculase polymerase for PCR amplification. We recommend sequencing all the samples, including "day 7" and "day 10" for comparative analysis. For data analysis, we have obtained highly consistent results using both MAGeCK [20] and Z-score-based methods [16, 21]. Comparisons can be made between both arms of the screen (e.g., glucose vs. galactose), or between each arm and corresponding earlier time points.

4 Notes

1. Synthetic lethality is defined as "a type of genetic interaction where the co-occurrence of two genetic events results in organismal or cellular death" [22]. *See* Table 1 for examples.
2. OXPHOS-deficient cells are auxotrophic for glucose, pyruvate, and uridine [19].
3. Normal FBS contains glucose and other nutrients which need to be excluded in glucose-free media. Using dFBS provides greater control over media composition.
4. Antimycin A is an inhibitor of the mitochondrial respiratory chain complex III. Mitochondrial protein synthesis inhibitors, such as chloramphenicol, may also be used. It is also possible to use genetic models of OXPHOS deficiency, such as those resulting from the genetic ablation of respiratory chain or mitoribosomal subunits.
5. PBS washes are important to remove any cell debris and leftover media from the previous passage.
6. Apoptotic cells are not completely recovered at 300 × *g*. Centrifugation at 2000 × *g* increases the collection of all cells without significantly reducing viability.
7. It may be necessary to titrate the amount of Annexin V-FITC used. Refer to the instructions from the supplier.
8. K562 cells in glucose should show low Annexin V staining (<10% Annexin V positive cells). Before analyzing Annexin V levels, cells can be counted in trypan blue to control for their viability.
9. Thaw at 37 °C and for no longer than necessary as lentiviruses are heat-sensitive. Viral supernatants from Addgene are usually supplied with a small 2 mL aliquot for titration.
10. Use the following formula to calculate the infection rate:

$$\text{Infection rate} = \frac{\text{Infected cells in selection media}}{\text{Infected cells in normal media}} \times 100$$

11. We use a MOI of ~0.3 to minimize the risk of a cell being infected with two different gRNAs.
12. The MOI varies between cell lines. Libraries need to be titrated for each cell line.
13. A high cells/sgRNA ratio will lead to a better signal-to-noise ratio.
14. In this example, about 127×10^6 cells need to be used with the Brunello library (76,441 sgRNA), targeting 500 cells/sgRNA and at an MOI of ~0.3:

$$\text{Guides in the library} \times \text{desired} \frac{\text{cells}}{\text{sgRNA}} \times \frac{1}{\text{MOI}}$$

$$= \text{Number of cells for infection.}$$

$$76{,}441 \text{ guides} \times 500 \frac{\text{cells}}{\text{sgRNA}} \times 3.33 = 127 \times 10^6 \text{cells}$$

15. Centrifugation time needs to be slightly longer for large volumes.
16. An infection rate of up to 50% is acceptable. Make sure that at least 500 cells/sgRNA are infected.
17. It is important to always centrifuge cells to avoid the presence of dead cells.
18. Cells can be used for DNA isolation to provide a baseline for guide representation after selection.
19. To obtain the number of cells required for a specific cells/sgRNA value, use the following formula:

$$\text{Desired} \frac{\text{cells}}{\text{sgRNA}} \times \text{guides in the library} = \text{Number of cells needed}$$

20. Freezing a batch of cells after selection provides a stock population of cells infected with the chosen library.
21. The duration of the synthetic lethality treatment depends on the experiment and the biological question addressed. In K562 cells, 24 h is sufficient for comparing viability in glucose and galactose.
22. For our synthetic lethal interaction, we expect more dead cells in the input of the galactose condition than in the glucose.

Acknowledgments

The authors would like to thank J. Arroyo and V. Mootha for their role in developing death screening, and M. Lisci, A. Strefeler and K. Maundrell for their critical comments on the manuscript. This work was supported by grants from the Fondation Pierre-Mercier pour la Science and the Swiss National Science Foundation (310030_200796).

References

1. Thompson NA, Ranzani M, van der Weyden L et al (2021) Combinatorial CRISPR screen identifies fitness effects of gene paralogues. Nat Commun 12:1302
2. Najm FJ, Strand C, Donovan KF et al (2018) Orthologous CRISPR-Cas9 enzymes for combinatorial genetic screens. Nat Biotechnol 36: 179–189
3. Arroyo JD, Jourdain AA, Calvo SE et al (2016) A genome-wide CRISPR death screen identifies genes essential for oxidative phosphorylation. Cell Metab 24:875–885
4. Rossiter NJ, Huggler KS, Adelmann CH et al (2021) CRISPR screens in physiologic medium reveal conditionally essential genes in human cells. Cell Metab 33:1248–63 e9
5. Ji Z, Mei FC, Lory PL et al (2009) Chemical genetic screening of KRAS-based synthetic lethal inhibitors for pancreatic cancer. Front Biosci (Landmark Ed) 14:2904–2910
6. Wang T, Wei JJ, Sabatini DM et al (2014) Genetic screens in human cells using the CRISPR-Cas9 system. Science 343:80–84
7. Shalem O, Sanjana NE, Hartenian E et al (2014) Genome-scale CRISPR-Cas9 knockout screening in human cells. Science 343:84–87
8. Guieze R, Liu VM, Rosebrock D et al (2019) Mitochondrial reprogramming underlies resistance to BCL-2 inhibition in lymphoid malignancies. Cancer Cell 36:369–84 e13
9. Wang T, Birsoy K, Hughes NW et al (2015) Identification and characterization of essential genes in the human genome. Science 350: 1096–1101
10. Tsherniak A, Vazquez F, Montgomery PG et al (2017) Defining a cancer dependency map. Cell 170:564–76 e16
11. McDonald ER 3rd, de Weck A, Schlabach MR et al (2017) Project DRIVE: a compendium of cancer dependencies and synthetic lethal relationships uncovered by large-scale, deep RNAi screening. Cell 170:577–92 e10
12. Hartenian E, Doench JG (2015) Genetic screens and functional genomics using CRISPR/Cas9 technology. FEBS J 282: 1383–1393
13. Martin SJ, Reutelingsperger CP, McGahon AJ et al (1995) Early redistribution of plasma membrane phosphatidylserine is a general feature of apoptosis regardless of the initiating stimulus: inhibition by overexpression of Bcl-2 and Abl. J Exp Med 182:1545–1556
14. Vermes I, Haanen C, Steffens-Nakken H et al (1995) A novel assay for apoptosis flow cytometric detection of phosphatidylserine expression on early apoptotic cells using fluorescein labelled Annexin V. J Immunol Methods 184: 39–51
15. Zargarian S, Shlomovitz I, Erlich Z et al (2017) Phosphatidylserine externalization, "necroptotic bodies" release, and phagocytosis during necroptosis. PLoS Biol 15:e2002711
16. Jourdain AA, Begg BE, Mick E et al (2021) Loss of LUC7L2 and U1 snRNP subunits shifts energy metabolism from glycolysis to OXPHOS. Mol Cell 81:1905–19 e12
17. Doench JG, Fusi N, Sullender M et al (2016) Optimized sgRNA design to maximize activity and minimize off-target effects of CRISPR-Cas9. Nat Biotechnol 34:184–191
18. Yang X, Boehm JS, Yang X et al (2011) A public genome-scale lentiviral expression library of human ORFs. Nat Methods 8:659–661
19. Robinson BH, Petrova-Benedict R, Buncic JR et al (1992) Nonviability of cells with oxidative defects in galactose medium: a screening test for affected patient fibroblasts. Biochem Med Metab Biol 48:122–126
20. Li W, Xu H, Xiao T et al (2014) MAGeCK enables robust identification of essential genes from genome-scale CRISPR/Cas9 knockout screens. Genome Biol 15:554
21. To TL, Cuadros AM, Shah H et al (2019) A compendium of genetic modifiers of

mitochondrial dysfunction reveals intra-organelle buffering. Cell 179:1222–38 e17
22. Nijman SM (2011) Synthetic lethality: general principles, utility and detection using genetic screens in human cells. FEBS Lett 585:1–6
23. Senft D, Weber A, Saathoff F et al (2015) In non-transformed cells Bak activates upon loss of anti-apoptotic Bcl-X_L and Mcl-1 but in the absence of active BH3-only proteins. Cell Death Dis 6:e1996
24. Ogiwara H, Sasaki M, Mitachi T et al (2016) Targeting p300 addiction in CBP-deficient cancers causes synthetic lethality by apoptotic cell death due to abrogation of MYC expression. Cancer Discov 6:430–445
25. Bayona-Bafaluy MP, Sanchez-Cabo F, Fernandez-Silva P et al (2011) A genome-wide shRNA screen for new OxPhos related genes. Mitochondrion 11:467–475
26. Kum WW, Lo BC, Yu HB et al (2011) Protective role of Akt2 in Salmonella enterica serovar typhimurium-induced gastroenterocolitis. Infect Immun 79:2554–2566
27. Yu D, Liu Y, Zhou Y et al (2020) Triptolide suppresses IDH1-mutated malignancy via Nrf2-driven glutathione metabolism. Proc Natl Acad Sci U S A 117:9964–9972
28. Oran AR, Adams CM, Zhang XY et al (2016) Multi-focal control of mitochondrial gene expression by oncogenic MYC provides potential therapeutic targets in cancer. Oncotarget 7: 72395–72414

Index

Antoni Barrientos and Flavia Fontanesi (eds.), *The Mitoribosome: Methods and Protocols*,
Methods in Molecular Biology, vol. 2661, https://doi.org/10.1007/978-1-0716-3171-3,

G

H

I

K

L

M

MIX
Papier aus verantwortungsvollen Quellen
Paper from responsible sources
FSC® C105338

If you have any concerns about our products,
you can contact us on
ProductSafety@springernature.com

In case Publisher is established outside the EU,
the EU authorized representative is:
Springer Nature Customer Service Center GmbH
Europaplatz 3, 69115 Heidelberg, Germany

Printed by Libri Plureos GmbH
in Hamburg, Germany